Religion, Spirituality and Health: A Social Scientific Approach

Volume 3

The relationship between religious/spiritual belief or behaviour and health behaviour has been explored over several decades and across various disciplines. Religious variables have consistently been found to have a direct relationship to physical and mental health. At the same time - research has also indicated potential societal tensions that can exist between religion and health – we have seen this in relation to family planning, HIV/AIDS, and reproduction. This book series aims to uncover the impact of religion on individual health behaviours and outcomes but also the influence of religion on health practices at the community level. This book series uncovers the impact of religion on individual health behaviors and outcomes, as well as the influence of religion on health practices at the community level. It consists of volumes that are based on multi-methodological approaches, provide quantitative and qualitative forms of analysis, and advance the understanding of the intersection between religion and health beyond the correlation of religious belief and health outcomes. Building on earlier research, the series explores the direct relationship between religious variables and physical and mental health, as well as the potential societal tensions that have been shown to exist between religion and health – for example in relation to family planning, HIV/AIDS, and reproduction. Spoken values are often shared within religious communities; however, religious influence can at times be extended outside of the community in instances of service provisions such as hospital ownership, various research active think tanks, political action, and the development of community mores.

More information about this series at http://www.springer.com/series/13437

Marcin Moskalewicz
Editor-in-Chief
Ute Caumanns • Fritz Dross
Editors

Jewish Medicine and Healthcare in Central Eastern Europe

Shared Identities, Entangled Histories

Springer

Editor-in-Chief
Marcin Moskalewicz
Department of Social Sciences
Poznan University of Medical Sciences
Poznan, Poland

The Oxford Research Centre in the
Humanities (TORCH) University of Oxford
Oxford, UK

Res Publica Foundation
Warsaw, Poland

Editors
Ute Caumanns
Department of Historical Sciences
Heinrich-Heine-Universität Düsseldorf
Düsseldorf, Germany

Fritz Dross
Institute for the History of Medicine
and Medical Ethics
Friedrich-Alexander-Universität
Erlangen-Nürnberg
Erlangen, Germany

Religion, Spirituality and Health: A Social Scientific Approach
ISBN 978-3-030-06441-9 ISBN 978-3-319-92480-9 (eBook)
https://doi.org/10.1007/978-3-319-92480-9

This Springer imprint is published by the registered company Springer International Publishing AG part of Springer Nature.
The registered company address is: Gewerbestrasse 11, 6330 Cham, Switzerland

Foreword

Scholars have thoroughly examined Jewish history within the framework of the secularization and emancipation movement of Western European and German Jewry, especially in regard to the history of Jewish medicine and health care in Judaism. However, as Dan Diner mentioned on several occasions, such one-sided approach proves insufficient, because only studies that focus primarily on the history of spaces will manage to encompass not only an integrated European history but also a new perspective on such Jewish history, one that exceeds the usual paradigm of the nation state. It is because the transnational and transterritorial Jewish life-worlds and migrations reveal "a comprehensive view of history as it were" (Diner 2003, p. 12). It is important that, when thinking of Jews in European space and history, we account for their situation beginning from Europe's easternmost parts. Jewish history is particularly suitable for this approach, because Jews lived for many years both in the East and the West, split into different Jewries. Within the frame of this view, a habitus emerges from the peripheries into the center – "a habitus woven out of textuality, urbanity, mobility, and transterritoriality, slowly carved through the prolonged validity of premodern traditions; a habitus influenced by the homogenization shifts of the 19th and 20th century yet differing from the ethnically and territorially structured and etatistically constructed populations" (Diner 2003, p. 246). In addition, the nineteenth-century confessionalization of Judaism and its transformation into a private faith in the West enabled the formation of a modern, individualistic, and tolerant constitutional state. This process contrasted with the secularization in the East, where ethnic-national understanding of Judaism combined with humanitarian interventions in support of oppressed minorities and overseas migratory movements.

The book *Jewish Medicine and Healthcare in Central Eastern Europe. Shared Identities, Entangled Histories* offers exactly such a new interpretation of Jewish history in general and of Jewish health care in particular. The editors of this volume – Marcin Moskalewicz, Ute Caumanns, and Fritz Dross – have already thoroughly examined the contributions gathered. I may only draw attention to the scope of the articles, which range from analyses of the relations between Halakhic authority and the Haskalah-introduced early modernity to considerations of the modern

Jewish health care in the German-Jewish-Polish relations in the nineteenth and twentieth centuries, reflections on the problems of German-Jewish-Polish identity, and, finally, to descriptions of the dilemmas of Jewish doctors in the ghettos and death camps.

All in all, this volume makes an important contribution in the direction of study Dan Diner called for: history and memory, society and memory, as well as the methodical transfer of hybrid cultural experience in the process of secularization. This book represents an important building block in the formation of a history of memory of the nineteenth and twentieth centuries in the area of Jewish-German-Polish network of relationships and Jewish-German-Polish culture.

Institute for the History of Medicine and Ethics in Medicine
Charité – Universitätsmedizin Berlin
Berlin, Germany

Gerhard Baader

Reference

Diner, D. (2003). Gedächtniszeiten. *Über jüdische und andere Geschichten*. München: Beck.

Acknowledgements

Since 2005 the board and several members of the German-Polish Society for the History of Medicine have been considering discussing the historical-medical relationships of Germans and Poles by focusing on Jewish physicians and Jewish medicine in a broader comparative perspective. The German-Polish Society for the History of Medicine is a group of academics aiming at sharing their research with each other and with a wider audience. Originally, the cooperation between historians of medicine in Poland and East Germany (GDR) began in 1977 under the difficult conditions of the Cold War. In the 1990s, the German unification broadened the society's academic base, including social historians of medicine both in Poland and Germany. The society started to organize biannual conferences taking place alternately at German and Polish universities. On the occasion of the 70th anniversary of the outbreak of the Second World War by German troops invading Poland, an international conference on "Medicine and War" took place in 2009 in Düsseldorf. Other past conferences included "Medicine and Technology" (Poznań 2011) and "Medicine and Language" (Magdeburg 2013). Alongside our meetings, we came to a conclusion that Jewish culture is one of the most pervasive links in the history of Poland and Germany. The history of Jewish healthcare and medicine appeared as a perfect topic to demonstrate the complex historical framework of the two countries.

Nevertheless, ideas grow slowly. During the meeting in Magdeburg in 2013 the concept crystallized. The society decided to devote the upcoming conference to the theme of Jewish medicine and healthcare in Germany and Poland. In September 2015, a conference entitled "Jews–Germans–Poles: Histories and Traditions in Medical Cultures" took place at the German Historical Institute and the Emanuel Ringelblum Jewish Historical Institute in Warsaw. It gathered over 40 participants, experts on many aspects of Jewish studies from 7 countries. Selected papers from the conference became a point of departure for this book, which took three more years to complete. Almost 20 authors of different cultural backgrounds worked toward the final result.

We wish to express our gratitude to all those who helped and supported us during the course of this project. A grant received from the German-Polish Science

Foundation made the conference and the book project financially feasible. We wish to thank Ruth Leiserowitz, the deputy director of the German Historical Institute in Warsaw, and her team for hosting the first day of the conference; Paweł Śpiewak, the director of the Jewish Historical Institute in Warsaw, and his team for hosting the next two days of the conference and organizing the cultural programme; Zuzanna Lewandowska and Joanna Lusek for their work on the organizing committee; and Mikołaj Golubiewski and Anna Topolska for the translation and language editing of selected papers.

Projekt wspierany przez Polsko-Niemiecką Fundację na rzecz Nauki
Gefördert aus Mitteln der Deutsch-Polnischen Wissenschaftsstiftung

DEUTSCH POLNISCHE WISSENSCHAFTS STIFTUNG | POLSKO NIEMIECKA FUNDACJA NA RZECZ NAUKI

Poznan, Poland — Marcin Moskalewicz
Düsseldorf, Germany — Ute Caumanns
Erlangen, Germany — Fritz Dross

Contents

Chapter 1
Jewish – German – Polish: Histories and Traditions in Medical Culture

Marcin Moskalewicz, Ute Caumanns, and Fritz Dross

Abstract This chapter introduces the content and the methodological premises of the book *Jewish Medicine and Healthcare in Central Eastern Europe*. Integrating academic disciplines from medical history to philology and Jewish studies, the book geographically concentrates on what is today Poland and Germany (and the former Russian, Prussian and Austro-Hungarian Empires). In this significant zone of ethnic, religious and cultural interaction Jewish, Polish, and German traditions and communities were more entangled, and identities were shared to an extent greater than anywhere else. Starting with early modern times and the Enlightenment, through the nineteenth century, up until the horrors of medicine in the ghettos and concentration camps, the book aims at collecting a variety of perspectives on the question of how Judaism and Jewish culture was dynamically related to medicine and healthcare.

Keywords Jewish medicine · Healthcare · Central-Eastern Europe · Halakha

M. Moskalewicz (✉)
Department of Social Sciences, Poznan University of Medical Sciences, Poznan, Poland

The Oxford Research Centre in the Humanities (TORCH), University of Oxford, Oxford, UK

Res Publica Foundation, Warsaw, Poland
e-mail: moskal@ump.edu.pl

U. Caumanns
Department of Historical Sciences, Heinrich-Heine-Universität Düsseldorf, Düsseldorf, Germany
e-mail: ute.caumanns@uni-duesseldorf.de

F. Dross
Faculty of Medicine, Institute for the History of Medicine and Medical Ethics, Friedrich-Alexander-Universität Erlangen-Nürnberg, Erlangen, Germany
e-mail: fritz.dross@fau.de

M. Moskalewicz et al. (eds.), *Jewish Medicine and Healthcare in Central Eastern Europe*, Religion, Spirituality and Health: A Social Scientific Approach 3,
https://doi.org/10.1007/978-3-319-92480-9_1

1.1 A Jewish Medicine?

Is 'Jewish medicine' a valid historical category? How could medicine and healthcare be Jewish (Nevins 2006)? This volume aims at answering this question historically. Does the concept of 'Jewish medicine' denote a historically consistent phenomenon constituted by the continuous interplay of medical, ethnic and religious cultures? Or rather, was Jewish medicine subject to epochal and cultural changes to an extent precluding the very possibility of speaking of it in any sort of essentialized way? Obviously, there were Jewish doctors and religious laws concerning physical care. Certainly, segregated Jewish communities had to develop special kinds of welfare. And last but not least, anti-Semitic racial medicine was a major factor in legitimizing the Shoah. But does it all make a coherent story? In the introduction to the first part of *The Origins of Totalitarianism,* which was devoted to anti-Semitism, Hannah Arendt warned against the retrospective illusion of continuity that combines different, incommensurable forms of anti-Semitism into a unity of eternal anti-Semitism (Arendt 1967). Such an ahistorical view, Arendt argued, could have been comforting to the victims but at the same time it could have been helping the perpetrators, and it simply was not true.

Analogical jeopardy concerns the idea of Jewish medicine. Undoubtedly, there is a connection between Jewish religion, Jewish tradition, Jewish communities and the care for endangered and damaged bodies and souls. However, we might easily succumb to the illusion that this connection is essential and stable. It is not, and we hope the present volume will prove it.

Integrating academic disciplines from medical history to philology and Jewish studies, the book geographically concentrates on Central Eastern Europe, mostly on what is today Poland and Germany (and the former Russian, Prussian and Austro-Hungarian Empires). In this significant zone of ethnic, religious and cultural interaction Jewish, Polish, and German traditions and communities were more entangled, and identities were shared to an extent greater than anywhere else (Płonka-Syroka and Dross 2010). Starting with the early modern times and the Enlightenment, through the nineteenth century, up until the horrors of medicine in the ghettos and concentration camps, the book aims at collecting a variety of perspectives on the question how Judaism and Jewish culture was dynamically related to medicine and healthcare. The volume discusses the *Halakhic* traditions, hygiene-related stereotypes, the organization of healthcare within specified communities, academic carriers, hybrid identities, and diversified medical practices.

Jewish culture has been an integral part of European culture as present in the history of both Eastern and Western Europe. Learned Jews played an eminent role within the vast process of transformation from the medicine of the ancient Mediterranean to the medieval medicine of transalpine Europe (Freudenthal 2011). Starting in the Middle Ages the disintegration of Jews led to the formation of special Jewish communities that would develop their own institutional frameworks of healthcare services, complying to their religious beliefs and practices. Therefore, we can notice strong relations between religion, medicine and healthcare in those times. However, welfare was very much contingent on the social structure of the communities practising it as a common challenge. From the nineteenth century, academic training promising a professional career as a physician or medical scientist was a major factor of Jewish acculturation (Wolff 2014). But Jewish academics had still been underprivileged regarding appointments to professorships in Prussian universities in the late nineteenth century, and even in the Weimar Republic (as Catholic academics, as well). For this reason, about 70% of Prussian professors of Jewish descent had been baptized (Ebert 2008, p. 626).

At the same time, dating back to the medieval discourse on the *limpieza de sangre* and up to the twentieth-century discourses on racial hygiene and eugenics, medicine was engaged in defining and deprecating the 'Jewish body' as a foreign and dangerous object, thus fuelling anti-Judaic and racist social practices. The latter had a multifaceted impact on medical history: legally, through bioethical considerations influencing science, and socially, through emigration of Jewish medical scientists from the German Reich, among others.

Jewish culture constitutes a uniting element of the history of Poland and Germany. When in the medieval times Jews migrated from Germany (*Ashkenaz*) to Poland, their German dialect evolved into Yiddish as a fusion language of German, Hebrew, and Slavic components. Jewish communities developed in different regional neighbourhoods of Austrian, Belarusian, Prussian, Lithuanian, Polish, Ukrainian, etc. origin. Jews were thus cross-border mediators right since the medieval times. On the other hand, however, Jewish culture and communities differed in German and Polish historical contexts. While in German territories the Jewish part of the population was comparatively small, in Warsaw, until the German invasion of 1939, it constituted about one-third of the population. The East European *Shtetl,* a Jewish community of its own, is one unparalleled by a comparable phenomenon in German history (Basch et al. 1994).

The slogan 'We didn't cross the border, the border crossed us' (Basch et al. 1994) accompanying global migration and transnational practices nowadays, would already have been clearly true in the sixteenth century and later if stated by the inhabitants of Central Europe. Some pre-national cultural practices had obviously existed, which preceded the emergence of the nation-state and should not be considered in its categories. Cultural encounters and even cultural melting pots are not that "modern": these processes have been present throughout all of history (Burke 2016). Furthermore, we should distinguish the concept of Jewish/German/Polish as an "other" from the self-consciousness of being Jewish/German/Polish. Following Sander Gilman's (Gilman 1991) and John M. Efron's statement that 'being Jewish meant having not just a spiritual identity, but a physical one as well' (Efron 2001: 4), we are not prone to accept the idea of 'German Jews'. The concept of 'German nation' does not work on the unstable territorial organisation of Central Eastern Europe. An integrating German-speaking academic culture (Baader 1997), for instance in Breslau (today Wrocław) and Vienna, was functioning without subjecting its participants to the idea of 'Germany'. Jewish history as a part of Polish and

German history and vice versa is thus to be considered as a matter of *longue durée* – including the problem of essential change regarding intra-cultural coherence, as well as the inter-cultural relatedness of German, Polish, Jewish, and medical histories.

1.2 Jews in Central Eastern Europe

On the one hand, today's sociological categories (Hirschauer 2014: 170–191) might not adequately serve the comprehension of divergent historical circumstances that we are considering here. On the other hand, according to the idea of hybridization (for a historical approach see: Burke 2010; see also Stockhammer 2012), historical subjects could have effectively belonged to several different and coherent social groups, updating and refreshing their "identity" situation by situation (Wolff 2014: 244–258). A physician and a medical scientist descending from a Polish-speaking family with a Jewish religious background, who had his academic medical training in Vienna, and belonged to a German-speaking academic network, could serve as an example.

From the Jewish studies perspective, Central Eastern Europe (see: Halecki 1952, Kłoczowski and Łaszkiewicz 2009) is a both conflicting and productive historical region. This large cross-border zone, relatively open to neighbouring regions, is extraordinary regarding its duration, scope and diversity. The phenomenon of shared identities and entangled histories has become evident here on an unprecedented scale. While Jews were being expelled from England, Spain and German territories, the medieval Piast Poland, and later the entire Polish-Lithuanian Commonwealth, including Western Ukraine, became a place of hope and opportunities for them. The comparatively ample ritual and religious freedom granted by a relatively liberal constitution as well as the small-scale diversity of the region gave the Jewish population a degree of self-government which had not been previously seen in Europe. This was taking place both on a supra-regional level in the Jewish *Va'ad* (Council of Four Lands) and in the local *Kehillot*. Within this framework, it could have been (or was) possible to safely cultivate traditional rules of Jewish medicine. At the same time, the (multi-ethnic) non-Jewish environment provided opportunities for spatial mobility and language transfer unknown anywhere else. These conditions made Jewish scholars cross-border mediators. This phenomenon can be observed in the impact of *Halakhic* discourse, shifting at the end of the seventeenth century from Poland to Central Europe, but also in the transfer of language, and thus knowledge. An example of the latter is the bilingual Polish-Yiddish translation of the *Regimen sanitatis Salernitanum*.

With the partitions of the Nobles' Republic in the late eighteenth century, this Polish-Lithuanian-Ruthenian-Jewish exchange area fell under the rule of three hegemonic powers and became a part of the Russian, Habsburg and Prussian-German empires. The East-Jewish population lost both the scope of self-determination and political participation, and, in the Russian Pale of Settlement, much of its mobility. The already entangled history of the commonwealth expanded into new spatial dimensions and included new (religious, ethnic, national and governmental) actors. This process affected both traditional healthcare and academic medicine (Sinkoff 2004). In the West, the medical customs of *Ashkenaz* folk medicine have to be seen as a complex heterogeneous system interacting with official medicine and local German folklore. As a result, German medical culture was influencing the Jewish population living among the Slavic peoples.

At that time, the difficulties of the nation building process, running parallel to the process of medicalization, were shaping the general political framework (Hagen 1980). The concept of 'nation' as commonly used in academic scholarship concerning Western does not apply here. Jewish medicine and history, with its trans- and supra-national contexts, is distinctive for Central Eastern Europe (Guesnet 1998). It is precisely this entangled situation that invites questions that transcend those concerning the nation-state.

In the twentieth century, the issue of religious, social, and cultural orientation was still relevant for Jewish medicine and its actors. In the nineteenth century, medicine changed its traditional academic base and opened up to the evolving sciences. The subsequent process of scientification had a huge impact on medical professionalization. Modernised medical occupations brought new, multiple loyalties, especially among Jewish physicians. As a result, their activities within Jewish healthcare organisations in the interwar and the early post-war period were sometimes conflictive.

The demise of the three empires dominating Central Eastern Europe at the end of the First World War and the emergence of the "small states" transformed the existing political combination. The Jewish population became a minority living next to a new dominant nation. In the Second Polish Republic, this was the case in regard to the state as a whole, but not necessarily regionally or locally. The *Shtetl* remained a traditional locus of medical culture until the Second World War (Kassow and Katz 2007). Tradition and modern medicine co-existed, albeit to a different extent, in large cities, competing with, but also supplementing each other. The increase in discrimination, leading as far as to open anti-Semitic boycotts, intensified the process of politicization of the Jewish population. In these circumstances, even assimilated physicians found themselves under political pressure.

It was mostly in Central Eastern Europe that the National Socialists built and operated the great ghettos. Later on, when concentration and extermination camps became a reality, a new chapter of this entangled history opened up, a chapter hitherto unimagined in its dark dimension. It has been argued that far less attention has been devoted to Jewish medical resistance than to racial German medicine (Grodin 2014: xvi). The same applies to the questions of identity. At the time of the Shoah, racial concepts forcibly imposed a redefinition of what it meant to be Jewish. Terror and extermination raised the question of multiple loyalties in a totally transformed context and led to the processes of transformation of self-awareness.

Regarding the latter, it was possible that a Polish assimilated physician with a white and blue armband visited a Catholic church standing within the Warsaw Ghetto walls that reflected arbitrary boundaries imposed by the occupiers. Ludwik Hirszfeld, the Ghetto inmate, an internationally renowned epidemiologist trained in Germany, was now working to fight epidemic diseases in cooperation with an externally appointed *Judenrat* in a community that was socially, politically, ethnically, religiously and culturally diverse (Dross 2015: 49–54).

1.3 The Contents of the Book

The first part of the volume, entitled *Between Religious and Medical Authority: Early Modern Jewish Care for Body and Soul* describes the historical background of later developments as well as introduces the methodological perspective of entangled cultures between lay and learned, religious and medical, and Jewish, Polish, and German cultures in the emerging modernity.

The opening chapter by Ewa Geller presents an unknown early modern adaptation of the well-known European dietetic and medical *Regimen sanitatis Salernitanum*. The presented Polish-Yiddish *regimen sanitatis* is an extraordinary example of the transfer and adaptation of scientific medical knowledge. Passing the linguistic barrier between Latin and Yiddish, it crosses the boundaries between Jews and Gentiles. Its unknown author, a former student of the renowned medical school of Padua, proves that the medical practices of Jews in Eastern Europe were well-informed by up-to-date medical knowledge, and thus challenges many widely held assumptions. The next chapter by Eliezer Sariel on differing attitudes to medical diagnosis among *Halakhic* authorities discusses the credibility of medical diagnosis by Polish and German *Poskim* (Jewish law scholars) over a comparatively long period. The need to decide whether a woman leaving a red mark on her clothes outside her menstruation period was to be regarded as being "clean" or not was a tricky situation requiring a religious judgement on a medical opinion. Sariel presents this discourse beginning with the sixteenth and seventeenth centuries *Halakhic* tradition, through the Enlightenment, up to the early nineteenth century. The last chapter of the first part examines a typical Enlightenment discourse, a debate on the early burial practices in the Jewish community. Hans-Uwe Lammel focuses on a controversy that was termed "the first halakhic dispute in the era of emancipation".

The second part of the book brings together scholars in the social history of medicine from Poland and Germany and focuses on the nineteenth and twentieth centuries. Under investigation in *Modern Jewish Healthcare: Community and the State* are especially the interactions among the welfare state, academic medicine, and social and cultural groups.

Marek Tuszewicki analyses *Ashkenaz* folk medicine of the Eastern Askenaz as a complex heterogeneous system, one interacting with the 'official' academic medicine and local folklore(s) simultaneously. Beyond enumerating excerpts from the early modern German-Yiddish medical literature, published in Germany by the Polish Jews, Tuszewicki focuses on historical and ethnographic sources, mainly in Yiddish and Hebrew, and amplifies our understanding of Jewish folk medicine in the context of the Slavic-Germanic borderlands. Katharina Kreuder-Sonnen's chapter on the 1893 cholera epidemic in the *Fin de Siècle* Polish Kingdom gives a detailed analysis of how Polish medical discourse conceived of the Jewish body. The concept of 'national indifference' helps in gaining a differentiated picture of medical Jewish 'othering' in the imperial context of the Tsarist Empire and the nationalized medical discourse of the Polish Kingdom. In the next chapter, Beata Szczepańska studies early twentieth-century activities aimed at a 'healthy' population and including eugenic as well as pedagogical beliefs through the lens of school hygiene discourse. Jewish pupils were a group of a special interest of that discourse. The last chapter in this part, written by Ignacy and Jakub Einhorn, depicts the activities of the Society of Protection of Health of the Jewish People in Poland (TOZ). Founded in 1921 and re-activated in 1945, the society functioned under the difficult political conditions of post-war Poland. It combined the best Jewish traditions of self-help organisations with the need to support Polish-Jewish Holocaust survivors.

The third part of the book, entitled *Shared Identities*, aims at analysing a more personal perspective. It concerns persons of mixed heritage, for example of Jewish cultural and often religious background, but educated at German-speaking medical universities, speaking Polish and German, and belonging to a world-wide scientific community.

The part starts with an article by Philipp Teichfischer who, while studying German physicians who joined the Dutch colonial health service in East India in mid-nineteenth century, discovered a number of physicians of Jewish background. Comparing Jewish and Christian German physicians, Teichfischer asks about the similarities and differences regarding their motivation to leave Germany and to enter Dutch military service in order to work in the East Indies as military or civilian doctors. In the next chapter, Joanna Lusek and Horst Doležal investigate Jewish students from Silesia studying at the medical faculty of Vienna University in the context of supra-cultural assimilation. The chapter discusses the coexistence of Jews in the academic circles of Vienna as well as the development of their professional careers in the German-speaking lands. The subsequent paper written by Naomi Menuhin has a biographical character and focuses on the physician Izrael Milejkowski. As the head of the Jewish physicians' organization, Milejkowski participated in the First World Congress of Jewish Physicians in 1936 in Tel Aviv. Milejkowski struggled to retain his dual, Polish-Jewish identity, and the visit inspired him to do soul-searching concerning Zionism. The following chapter by Monika Rice is also biographical and compares four versions of Julian Aleksandrowicz's (1908–1988) memoirs from 1955, 1962, 1967, and 1983. These memoirs concentrate on the shifting Polish-Jewish identities of its author, who ultimately reached his "final identity". The paper emphasises the dynamic work of Aleksandrowicz's subjective memory in his creation of himself. Yet another chapter on the construction of personal identity closes this part of the book. Miłosława Borzyszkowska-Szewczyk interprets the autobiographical texts of Polish-Jewish-British immunologist Leslie Baruch Brent. Brent was born in 1925 in pre-war Köslin and was saved from the Holocaust through one of the "Kindertransports" from Berlin to Great Britain. Köslin – after 1945 named Koszalin (in Polish) – is located in Pomerania, a region that spans the historical and cultural Slavic-Germanic, and contemporary administrative Polish-German borderlands.

The final part of the book is dedicated to Jewish medicine during the Holocaust. Unlike many existing historical analyses, it does not concern biological and medical aspects of national socialist racism, but concentrates on a special aspect of the history of the victims of terror and extermination, namely Jewish doctors practising in the concentration camps and in the ghettos. *Jewish Doctors in the face of Terror and Extermination* comprises three chapters.

Ross Halpin opens the part by giving an overview of the function of Jewish physicians in labour, concentration and extermination camps. Once introduced into the camp's hospital system, the prisoner-physician had the dual role of a prisoner sentenced to death, and a carer whose responsibility was to treat prisoners in order to return them to work or select for execution. Subsequently, Maria Ciesielska provides us with a detailed picture of Jewish Doctors in the Warsaw Ghetto. They were working in the ghetto's health administration, in its hospitals, and managed to secretly teach medical students and carry out research on the hunger disease. The third and final chapter of this part analyzes the developmental roots of the Jewish medical system in the ghettos. Miriam Offer reveals how the Jewish community succeeded in establishing a professional medical system using modern perceptions based on public health principles, despite the inhuman and genocidal conditions to which their practice was subject during the war.

Jewish medical culture has never been isolated. For a vast number of contingent, historical reasons, it has had special connections to German and Polish history and culture. This is true with respect to tolerance, assimilation, and acculturation as well as regarding the history of segregation and persecution. In the twentieth century, the latter culminated in mass-murder and genocide, and continued after World War II, which led to new waves of Jewish migration which had great impact on medicine and science. The context of medicine and health provides a starting focal point to re-consider the impact of Jews and Jewish culture on Central European history as a whole, and vice versa. It is this central region of transfer and contact that proves to be highly suitable for analyzing the question of Jewishness as well as the issues of loyalty and identification. Considering the fact that identities here have been numerous, challenged and even threatened, Central Eastern Europe presents not just a local melting pot, but a complex cross-border region of entangled Jewish, German, and Polish medical histories.

References

Arendt, H. (1967). *The origins of totalitarianism* (3rd ed.). New York: Harcourt Brace and Co.

Baader, G. (1997). The contribution of central European doctors to medical science. In N. Berger (Ed.), *Jews and medicine: Religion, culture, science* (pp. 167–183). Philadelphia: Jewish Publication Society.

Basch, L., Glick Schiller, N., & Blanc-Szanton, C. (1994). *Nations unbound. Transnational projects, postcolonial predicaments, and deterritorialized nation-states*. Langhorne: Gordon and Breach.

Burke, P. (2010). *Cultural hybridity*. Cambridge: Polity.

Burke, P. (2016). *Hybrid renaissance. Culture, language, architecture. (The Natalie Zemon Davis annual lectures)*. Budapest: Central European University Press.

Dross, F. (2015). Blutforschung oder Blutwahn? Ludwik Hirszfelds Theorie und Praxis im historischen Kontext. *Konferenzbericht. Historie, 8*(9), 49–54.

Ebert, A. D. (2008). *Jüdische Hochschullehrer an preussischen Universitäten (1870–1924). Eine quantitative Untersuchung mit biografischen Skizzen*. Frankfurt am Main: Mabuse.

Efron, J. M. (2001). *Medicine and the German Jews: A history*. New Haven: Yale University Press.

Freudenthal, G. (Ed.). (2011). *Science in medieval Jewish cultures*. New York: Cambridge University Press.

Gilman, S. L. (1991). *The Jew's body*. New York: Routledge.

Grodin, M. A. (Ed.). (2014). *Jewish medical resistance in the holocaust*. New York/Oxford: Berghahn.

Guesnet, F. (1998). *Polnische Juden im 19. Jahrhundert: Lebensbedingungen, Rechtsnormen und Organisation im Wandel*. Köln/Weimar/Wien: Böhlau Verlag.

Hagen, W. (1980). *Germans, Poles, and Jews. The nationality conflict in the Prussian East, 1772–1914*. Chicago/London: University of Chicago Press.

Halecki, O. (1952). *Borderlands of western civilization. A history of East Central Europe*. New York: The Ronald Press.

Hirschauer, S. (2014). Un/doing differences. The contingency of social belonging. *Zeitschrift für Soziologie, 43*(3), 170–191.

Kassow, S., & Katz, S. (2007). *The shtetl: New evaluations*. New York/London: New York University Press.

Kłoczowski, J., & Łaszkiewicz, H. (2009). *East-Central Europe in European history. Themes & debates*. Lublin: Instytut Europy Środkowo-Wschodniej.

Nevins, M. (2006). *Jewish medicine: What it is and why it matters*. New York: iUniverse.

Płonka-Syroka, B., & Dross, F. (Eds.). (2010). *Sąsiedztwo i granica: medycyna, historia i kultura w dialogu polsko-niemieckim/Nachbarschaft und Grenze: Medizin, Geschichte und Kultur im Polnisch-Deutschen Dialog* (Biblioteka Polsko-Niemieckej Historii Medycyny; 2). Wrocław: Oficyna Wydawnicza Arboretum.

Sinkoff, N. (2004). *Out of the shtetl: Making Jews modern in the Polish borderlands*. Providence: Brown University.

Stockhammer, P.W. (2012). *Conceptualizing cultural hybridization. A transdisciplinary approach* (Transcultural research – Heidelberg studies on Asia and Europe in a global context). Berlin: Springer.

Wolff, E. (2014). *Medizin und Ärzte im deutschen Judentum der Reformära. Die Architektur einer modernen jüdischen Identität* (Jüdische Religion, Geschichte und Kultur; Bd. 15). Göttingen: Vandenhoeck & Ruprecht.

Part I
Between Religious and Medical Authority: Early Modern Jewish Care for Body and Soul

Chapter 2
Yiddish "Regimen sanitatis Salernitanum" From Early Modern Poland: A Humanistic Symbiosis of Latin Medicine and Jewish Thought

Ewa Geller

Abstract The hitherto unknown early modern Eastern Yiddish adaptation of a well-known European dietetic and medical *Regimen sanitatis Salernitanum* printed in Poland in 1613, is to be found in the Austrian National Library in Vienna under the Ashkenazi Hebrew title *Seyfer derekh eyts ha-khajim*. This anonymous treatise written by a Jewish medical doctor reveals a remarkable symbiosis of medical and religious ideas of the great Jewish philosopher and physician Maimonides and typical occidental dietetic and hygienic rules propagated by the Latin genre *regimen sanitatis*. The Yiddish *regimen sanitatis*, presented here, is an extraordinary example of the transfer and adaptation of scientific medical knowledge that crosses the boundaries of Jews and Gentiles. In this remarkable attempt, the unknown author, a former student of the renowned medical school of Padua was certainly inspired by the humanistic spirit of that time. It is especially noteworthy considering the fact that medical practices of Jews in Eastern Europe, from where the book originates, are considered to be particularly superstitious and obscure at that time. *Seyfer derekh eyts ha-khajim* proves the contrary and an analysis of the work challenges many of the widely held assumptions.

Keywords Jewish medicine · Regimen sanitatis salernitanum · Eastern Yiddish · Polish Jews · Padua medical school · Maimonides · Jews and gentiles · Early modern medicine · Medical knowledge transfer

---

E. Geller (✉)
Department of German Studies, University of Warsaw, Warsaw, Poland
e-mail: e.geller@uw.edu.pl

M. Moskalewicz et al. (eds.), *Jewish Medicine and Healthcare in Central Eastern Europe*, Religion, Spirituality and Health: A Social Scientific Approach 3,
https://doi.org/10.1007/978-3-319-92480-9_2

2.1 Introduction

The Yiddish book under consideration is a discovery of significant cultural and historical value in both the study of Yiddish as well in the history of medicine. Published in 1613 in Poland, the Jewish medical self-help book entitled in Ashkenazi Hebrew *Seyfer derekh eyts ha-khayim*, (*A Path to the Tree of Life*), is—in a multitude of ways—a unique piece of work. Firstly, its uniqueness lies in the fact that we are dealing with a *rarum* which has only been discovered relatively recently. A unique copy of the book was discovered some 18 years ago in the Austrian National Library in Vienna.[1] I have been working on this remarkable text for almost 8 years. Its first annotated translation appeared 2015 in Poland,[2] where–some 400 hundred years ago—it was first published anonymously. My preliminary findings were presented at a conference in Jerusalem and subsequently published in the journal 'European Judaism' in 2009 (Geller 2009). However, as my professional interests at the time lay in the analysis of the linguistic fabric of the text (c.f. Geller 2013a), I did not consider the significance of the fact that the book in question is in fact a unique early Yiddish adaptation of a medieval European codex of the preservation of health known as *Regimen sanitatis Salernitanum* (*The Salernitan Rule of Health*). Therefore, in this paper I would like to focus on the symbiosis of pre-modern Jewish rational, medical and religious ideas and the typically occidental dietetic and hygienic rules propagated in the genre of Latin *regimen sanitatis*, as it was presented in this old Yiddish printed book.

Secondly, although anonymous, the text – as can be concluded from its content – was certainly written by a Jewish physician who, at least for some period of time, had been living in the Polish-Lithuanian Commonwealth. Moreover, drawing from the author's own words, we can hypothesize that—similarly to other Jewish scholars, such as Delmedigo or Toviyah Kohn (Kats) from Early Modern period—he had been educated at the University of Padua, were the inspiration for publishing this kind of preventive-dietetic manual for Ashkenazi Jews might have come from.[3] Furthermore, despite its Hebrew title, giving the impression that the work is a

---

[1] Notwithstanding the annotations of a prominent Jewish bibliographer Moritz Steinschneider regarding the existence of another copy of this old print in the famous Oxford's Bodleian Library, the only confirmed copy of this work to date can be found in the Austrian National Library in Vienna (Steinschneider 1852–1869, position nr. 3452!). In addition, while Steinschneider has suggested the unnamed place of print might have been Cracow or Prague, it has in fact recently been identified and proven to be Lublin (Poland), for more details see Bendowska (2017).

[2] Geller, E. (ed.) (2015). *'Sejfer derech ejc ha-chajim'. Przewodnik po drzewie żywota [O sposobie zachowania dobrego zdrowia w języku jidysz przez anonimowego żydowskiego doktora w Polsce, w roku 1613 drukiem ogłoszony]*.[Translation, introduction and critical edition by Ewa Geller]. Warsaw: Muzeum Króla Jana III w Wilanowie. The original Yiddish book is now available also online under: http://digital.onb.ac.at/OnbViewer/viewer.faces?doc=ABO_%2BZ158306707 [25.08.2016].

[3] On the meaning and spiritual atmosphere of the medical school of Padua cf. especially Bylebyl (1979) and Ruderman (1995: 100–118). On Christian medicine and the Jews during the Renaissance see also Goody (2010).

Hebrew old print, the book in question is the oldest known secular text of practical nature which originated on the Polish territory and was written in early Eastern (Polish) Yiddish, which became the second, after Hebrew, most important Jewish language in this part of Europe.

2.2 The Genre of *Regimen sanitatis Salernitanum*

The recognition of this self-help medical manual as being a Yiddish version of a treatise known by its original Latin name *Regimen sanitatis Salernitanum,* is an issue of most relevance in the context of the history of Jewish medicine. *Regimen* was an independent genre of practical-preventative medical literature, a kind of 'catechism of healthy living', which emerged from Christian Europe.[4] Its Jewish adaptation, at the time scientific par excellence, illustrates the processes of the intercultural transmission of knowledge which were taking place in late Renaissance Europe. Regardless of religion or ethnicity, these processes were occurring according to the concept that all people are equal or equally helpless in the face of an illness or death. Thereby, the author of the text, an anonymous Jewish medic, shows himself to be one of the forerunners of the Enlightenment attitude. Advocated by *Haskalah*, the Jewish Enlightenment movement based on the ideas of Moses Mendelssohn, this attitude would become widespread among Polish Jews only some 200 years after the emergence of our anonymous old Yiddish edition of the self-help medical manual.

Collections of dietary and medical advice addressed to individual monarchs and popes started to appear as early as in the early Middle Ages. As it happens, the original Latin *Regimen sanitatis Salernitanum* contains references to an unidentified 'King of England'. Although becoming more individualized over time, such *regimina* would not only contain identical advice, but would also be of a similar structure, including the order of the afore-mentioned health risks as well as proposed remedies. In a straightforward and at times overly simplistic manner all works of this genre referred to the principles of Galen medicine which were prevailing at that time, and which were dominated by dietetics and hygiene. Their main objective was to prevent disease through promotion of the principles of healthy lifestyle, particularly through taking foods suitable to one's bodily make-up (the so-called *complexions*). One must remember that in the early Modern Era the understanding of health was based on the Humoral Theory attributed to Hippocrates, with the idea of bodily fluid balance at its core. Medical intervention, including dietary regimes, was aimed at achieving balance between the four basic humours (from Greek 'sap'), i.e. blood, phlegm, yellow bile and black bile, and their respective combinations of four qualities: warm, cold, dry and moist. The same qualities (to a different extent) were attributed to individual types of food and herbs, which—when added to or eliminated

[4] For comprehensive research studies regarding this genre, its meaning and spreading throughout Europe see, inter alia, Adamson (1995) and Hagenmeyer (1995).

from a patient's diet—were supposed to restore humoral balance and, in turn, restore health. It is important to note, however, that dietetics in this case was defined very broadly. Contrary to the present understanding of the term, which is limited exclusively to food and drink, it also encompassed other factors which are contained in the scheme *sex res non naturales*, i.e. the six non-natural things, which nowadays would constitute a part of a healthy lifestyle. Dietetics at the time not only referred to the quality and amount of food as well as to the time of eating, but also to mental states, internal cleanliness (secretions), external cleanliness (excretions), balance between activity (exercise) and rest (sleep), etc., in other words, the aspects of life dependent on the actions of a person and not imposed by nature.

The very name of the genre *Regimen sanitatis Salernitanum* is indicative of its connection with the first European medical school, which was established in the Italian city of Salerno and which became famous as early as in the twelfth century. However, neither the 'original' text nor its author, nor any specific addressee has been identified and the creation of the work itself is surrounded by many legends.[5] Over the course of five centuries it was published in print more than 250 times (and additionally in hundreds more manuscripts) in numerous languages. The authors of these editions, who would typically stay anonymous, were local compilers and translators. They each provided this popular self-help book with a unique flavour and shape. Its original form was a rather frivolous verse written most likely around mid thirteenth century in Medieval Latin. Originally, its 364 lines contained some basic advice relating to a healthy lifestyle, formulated on the basis of the aforementioned classic principles of Galen. The period marked by the influence of the *Salernitan school,* which lasted for many centuries, saw this work expand over a hundredfold, ultimately filling thick volumes and becoming available in verse and prose. Its original content dedicated to the practices of 'medical home remedies' was soon supplemented by theoretical and philosophical discourse. Thereby, the genre of a domestic self-help medical manual became widespread and although it was not always of the highest standard, it nevertheless provided the foundations for European medicine at that time. Even though the genre was based directly on ancient Greek and Arabic traditions, these manuals were, nonetheless, clearly adapted to the lifestyle and dietary customs of Christian Europe.

[5] More detailed information on the speculation regarding the origin and dissemination of this work are contained in EMG (III,1224f). Apart from the masters of the earliest medical schools in Salerno, Montpellier or Paris, the person most often mentioned as the author is a famous Catalan physician and alchemist Arnald de Villanova (ca. 1235–1331).

2.3 *Seyfer derekh eyts ha-khayim* – Eastern Yiddish Secular Codex of Healthy Life

At first glance, the arrangement of the content and the extent of the topics discussed in the Yiddish *Seyfer derekh eyts ha-khayim (A Path to the Tree of Life)* seem accidental and arbitrary. In reality, they closely replicate the original structure of the *Regimen sanitatis Salernitanum* retaining the order and the content of the original chapters. Broadly speaking, the content of the book is presented in 87 chapters of varying lengths, ranging from few lines to few pages. In line with the classical structure of the genre it is possible to identify its two main parts. The first part, consisting of chapters 1–49, discusses general principles of dietetics according to Galen's classical medicine. Presented here are remedies for particular conditions and diseases originating from person's non compliance with the rules of healthy lifestyle.[6] The title of the first and longest chapter is: *This chapter speaks of worries and anger and of how a man should manage an evening meal and also postprandial sleep. The path that will benefit a man.* Thus, three out of the six non-natural things (*sex res non naturales*): mental state (*afectus animi*) – anger and worry, eating and drinking (*cibus et potus*) – evening meal, sleep and wakefulness (*somnus et vigilia*) – postprandial sleep are introduced at the very beginning. Subsequent chapters discuss and elaborate on individual aspects of the 'non-natural things' (*res non naturales*); for instance, chapter 2 is dedicated to exercise (*motus et quies*) and reads: *Of what should a man do as he first rises from bed*, which includes a visit to a bath. Chapter 3 discusses sleep and wakefulness (*Of how a man should take postprandial sleep*), chapter 4 entitled *Of incarcerated flatus* examines secretions and excretions (*secreta et excreta*), chapter 5 *Of curtailing evening fare* deals with eating (*cibus et potus*) etc. A large part of this section of the book is dedicated to the remedial properties of individual food products which were popular at the time, as well as to the ways they should be prepared and order in which they should be consumed. On the one hand, in contrast to the Latin version, the Jewish author removes all references to foods forbidden to Jews according to *kashrut,* Jewish religious dietary laws. We will not find in the book any dishes prepared from pork, eel or certain types of birds, and, moreover, dietary recommendations and recipes provided do not defy the laws of

[6] It is worth noting, that a similar theme is discussed in one of the first known self-help guidebook printed in Poland, namely *Problemata Aristotelis. Gadki z pisma wielkiego filozofa [Problemata. Aristotelis. Discourse from the works of a great philosopher]...* written by Andrzej Gelber (1535). It is, in contrast, created in the form of questions and answers, for example: 'Why excess in fare begets ill-health?' 'What is the order of supping conducive to vitality when different fare presents?' (Rostafiński 1893: 61–76). This and similar dietary rules can also be found in many late medieval and Renaissance German medical self-help guidebooks, for example in: *Büchlein der Gesundheit*, where we can find the following titles of individual chapters: 'Of the volume of fare, Of the order of supping, What fare a man should eat in harmony with his temperament (*complexion*), Of the order supping and volume of fare, what is an accurate portions and how much should a man eat, Good lessons on how a man should ready for supping' etc. (Hagenmeyer 1995: 222–229).

kashrut.[7] On the other hand, interesting enough, the text lacks any specific cultural terms characteristic of Jewish diet, such as 'kosher' or 'treif'. Apart from general criticism of overeating, there are no reminders of, or even allusions to, traditional and restrictive Jewish dietary laws.

Through this lack of explicit reference to the most characteristic principles of Judaism, the dietary approach proposed by the author becomes, as it were, non-religious. Furthermore, the book is firmly based on rational premises and, as such, *Seyfer derekh eyts ha-khayim* references a universally humanistic science. In order to illustrate the point I am making, allow me to quote a paragraph taken from the chapter which discusses dietary principles regarding bread:

> Thusly, we should not forget that man needs to eat more bread with fish, vegetables and fruit then with meat or eggs. Similarly, less bread should be eaten with meat than with eggs and even more bread should be eaten with eggs that are cooked soft rather than with those cooked hard. For verily, fish, vegetables and fruit contain more moisture, thus, the bread draws inside itself and, thereby, these things do not create excesses in phlegm. Thus, more bread is needed for eggs cooked soft than for those cooked hard as the latter are hard to digest on their own, let alone with a lot of bread. (*Seyfer derekh eyts ha-khayim* [fol. 25])

The rules presented here and their rational justification could certainly appear on the pages of any contemporary dietary self-help manual, however, they would be difficult to find in any self-help book addressed to orthodox Jews, either past or present, where, as we could imagine, mainly religious matter would be stressed.

The second section of the book, consisting of chapters 40–83, is dedicated to herbalism. It contains descriptions—at times laconic—of the advantages and medicinal properties of certain herbs, fruits and vegetables as well as methods of their use. This section of the book is an example of a concise herbal manual. Such herbals were known in Europe since at least the times of Hildegard of Bingen (1098–1179), an abbess and healer from the Rhineland. The very concise nature of the descriptions which appear in this part of the book can be explained by the fact that more detailed information was available in specialised and usually more comprehensive herbals dedicated exclusively to this subject.[8] Quite unexpectedly, the two penultimate chapters, i.e. 84 and 85, are dedicated to the most severe diseases affecting children at the time, i.e. chickenpox and measles, as well as to parasites in children. The fact that the subject of childhood diseases is raised in *Seyfer derekh eyts ha-khayim* distinguishes this book from classical forms of the *Regimen sanitatis* genre which ordinarily would not deal with this issue.[9] The book concludes with

---

[7] This restriction, however, does not apply to the remedies prepared from, for example, lard cf. Geller (2015: 207 ft. 324).

[8] In Renaissance Poland, for example, numerous compendia of herbal medicine were written, the most renowned of them being the first herbal written in Polish *O ziołach i mocy ich* [Of Herbs and their Powers] (1534) by Falimirz, and, the most comprehensive of them, *Zielnik Herbarzem z języka łacińskiego zowią...* [Herbal, from the Latin Called Liber Herbalis] by Simon Syrenius, published in 1613, the same year as *Seyfer derekh eyts ha-khayim*.

[9] The subject of the author's attitude towards the health of children is discussed in a separate paper, see Geller (2013b).

two longer chapters that discuss blood-letting which follows the classical structure of a *Regimen sanitatis*.

2.4 The Jewish Author and the Latin Original

Since no specific text has been found so far, that could have served as a direct basis for translation, we have to assume—for the time being—that we are dealing here with author's original adaptation. It is important to note at this point, that the effort he put into creating this piece of work was indeed very significant. The book takes the form of a didactic lecture which is enriched and augmented by theoretical discussions with the creators of the source texts that our author references, including Aristotle, Hippocrates or Galen, to name but a few. In this respect, in comparison to light-hearted style and simple content of its Latin original *Seyfer derekh eyts ha-khayim* constitutes not only a practical medical self-help book, but also a piece of scientific work of its time and a platform of expressing author's own opinion in discussion with the medical giants.

Numerous aspects of the book can provide us with a better understanding of its anonymous author. These are: extensive nature and erudite character of writing, rationality of the arguments presented, work's overt didacticism and genuine concern of the author for the well-being of its Jewish readers. The impression of the author that emerges from the pages of our manual is that of a faithful son of Israel who is an expert on Judaism and its writings, but above all, the picture is of a person of extensive knowledge regarding the standards of academic medicine that prevailed in the Christian Europe of that time. The overriding objective of the author was to introduce modern secular knowledge regarding maintenance of good health into idiosyncratic community of Central and Eastern European Jewry. It remains beyond question that at the time the book was written this task was practically impossible to accomplish, since the influence of rabbinical authorities—concerned with restrictive safeguarding of Jewish laws—was extremely powerful in seventeenth century Poland.[10] *Halakha*, a collection of Jewish religious laws derived from the Torah and Tanakh, regulated—without exception—all aspects of daily life, in particular issues related to diet and hygiene, the very ones discussed in our book. At the time, these regulations remained the domain of Talmudic and Torah scholars, and any attempts to break their monopoly on the control of these issues could have resulted in the excommunication of any dissatisfied free-thinker. This situation could have been the

[10] In this regard the situation in Poland was somewhat different compared to that of other countries, especially the countries of southern Europe (i.e. among Sephardic Jews). This difference has been emphasised by scholars researching this subject. Compare, for example, Efron (2001: 42): "What allowed for this situation to develop in Poland is the fact that precisely because secular knowledge was so undervalued, Jewish physicians had only minimal intellectual impact on Polish Jewry. Supreme intellectual authority in early modern Poland resided with the talmudic scholar. He who knew halakha, not Hippocrates, was assumed pride of place in Jewish society".

motive behind our author's decision not to disclose his identity and the fact that his work bore all the hall-marks of an illegal publication.

Let us, therefore, attempt to construct a portrait of our mysterious author. In this exercise, we can both make use of writer's own words and also use our factual knowledge of the social and cultural background of the times in which he lived.

2.5 Polish Context of *Seyfer derekh eyts ha-khayim* and Its Anonymous Author

At this point, it is important to emphasise the fact that *Seyfer derekh eyts ha-khayim* was written at the height of the Polish-Lithuanian Commonwealth, a period which saw the greatest flourishing of the Jewish community, the time known in Jewish historiography as the Golden Age, and in Polish scholarly historical research as a 'Jewish paradise', *paradisus judaeorum*. The Jewish community of the Early Modern Age enjoyed considerable autonomy in self-governance and economy which was ensured by royal charters that were issued repeatedly by subsequent monarchs. Passionate and dogmatic religiosity of the then-contemporary Jews flourished as a direct result of the acceptance of religious diversity in Poland, virtually unknown outside the country's borders (Rosman 2002: 519n). This embrace of religious diversity resulted in widespread institutionalisation and popularity of Talmudic studies. The Polish-Lithuanian Commonwealth was a wealthy and well-respected European power where persecuted refugees, as well as victims of religious conflicts from both East and West, hoped to find a peaceful and prosperous haven. Consequently, the Commonwealth was perceived throughout Europe as economically attractive and politically stable. It appeared as such to many foreign newcomers, including Ashkenazi Jews arriving mainly from the Germanic countries and the Czech Kingdom. At the turn of the fifteenth century, following the expulsion of the Sephardic Jews from the Iberian Peninsula which – in the Middle Ages – was the centre of Jewish religious and philosophical thought. These Polish Ashkenazi Jews proudly assumed the role of the community's spiritual leaders. Talmudic academies that were created at that time in Krakow and Lublin enjoyed the fame throughout the whole Europe. Although their curriculum was nearly exclusively devoted to the study of Talmud and *Halakha* the schools attracted vast numbers of young Jewish men from all over the continent. Intellect, wisdom, as well as knowledge and the ability to assimilate and process it, when employed for the purpose of a deeper understanding of the intricate rules of the Jewish law and Judaism, were highly regarded among the Jewish population of the time. Any young man possessing these attributes was obliged to dedicate his life to Talmudic studies, whilst the local community was required to provide him with the necessary support and appropriate conditions for learning and spiritual growth. Individuals who, according to this social norm, misappropriated their intellectual talents, using them to attain secular knowledge—and such people could be found in every epoch—were at a risk of social ostracism, stigmatisation or even excommunication. In answer to a frequently posed question regarding the lack of secular studies in yeshivas, Jewish rabbinical scholars pointed to Tanach as the reservoir of complete and absolute knowledge about the world. Therefore, a meticulous study and thorough understanding of the Jewish holy scriptures was, in their opinion, an entirely sufficient source of wisdom and competence. In this context, it should be of no surprise that neither the position held by Jewish medics, nor their scientific knowledge was appreciated in rabbinical circles of pre-modern Poland. Yet, a somewhat tolerant attitude towards practising Jewish doctors was solely due to the absolute obligation in Jewish law to preserve life, the rule known as *pikuakh nefesh*.

Secular sciences, such as astronomy, mathematics or—significant to our discussion—medicine, which flourished during the period of Renaissance, were perceived by the Jewish religious elite as strictly Christian disciplines. After all, these areas of science were practised by Christian scholars, who frequently – as in the case of famous astronomer Nicolaus Copernicus – were members of the Christian clergy, and, more importantly, these subjects were taught at Christian universities. The majority of rabbis, therefore, perceived any involvement in these sciences as an endangerment to Talmudic studies and, consequently, a direct risk to the Jewish religiosity.[11] These concerns were not unreasonable, since young Jewish men who decided to study medicine at foreign universities not only resided in a multi-national and multi-denominational environment, often beyond the influence of rabbinical authorities, but were also forced to learn *lashon am zar,* ('the language of foreign people') i.e. Latin, which, when translated literally, meant 'strange' and therefore dangerous language. Latin was a window to the world, so to speak, and a gateway to the science accessible beyond the Jewish world, and for that reason this language was forbidden to Jewish people. Rabbis as well as Talmudic scholars were convinced that the study of the Torah and Talmud was entirely sufficient for the pursuit of medical profession and they strove to rigidly enforce this belief among their students. And, although Jewish holy scriptures contain many references to human body as well as rules regarding diet and hygiene, they do not—by any means—constitute a coherent system of medical knowledge.[12]

[11] Such attitudes, however, were at times challenged, cf. Ruderman (2005:126):" ... certain circles in central and eastern Europe pursued scientific learning, especially astronomy, as a desirable supplement to their primary curriculum of rabbinics. [...] Two rabbinic luminaries Mosses Isserles (1525–72) and Judah Loew ben Bezalel (the Maharal, c. 1525–1609) openly encouraged the acquisition of scientific knowledge."

[12] More on this topic see, inter alia, Preuss (1911) and Shatzmiller (1994).

2.6 Early Modern Padua Medical Canon and the Legacy of Maimonides

Nevertheless, with time, a distinctive Jewish model of a rabbi-physician emerged, a man who used therapeutic methods that were derived from both Talmudic studies as well as local folk traditions. Usually, this knowledge was insufficient and the medics who used it were less effective than the doctors who acquired their expertise at Christian universities. The most popular among these academic institutions was the University of Padua, although it was not the only Christian university that opened its doors to Jewish students.[13]

Galen's medicine, which was taught at this university alongside astronomy and mathematics, drew upon observable laws of nature, philosophy and medicine that had been discovered by ancient scholars such as Aristotle, Empedocles and Hippocrates. At the time, the principles derived from Galen's canon seemed rational, coherent and logical. Considering the fact that Galen's medicine prevailed as a scientific dogma in Europe for nearly fifteen centuries, whilst in Poland it prevailed as late as the eighteenth century, it is reasonable to assume that it must have appeared as an attractive discipline for those of rational mind such as our author. His coherent logic is apparent in the fragment quoted above, in which he attempts to present a cause-effect relationship when explaining the rules of digestion with different foods combinations.

The names of Galen and Avicenna appear most frequently on the pages of our book. However, its author also refers to the medical authorities of the Islamic culture, namely Rhazes, Averroes or Avenzoar as well as to the expertise of other most famous ancient physicians: Hippocrates and Dioscorides. He even draws upon the work of the Byzantine Greek physician Paul of Aegina. Most significantly, however, scattered amidst the ideas and opinions of the most prominent scientists of the ancient and early medieval times, we can find the intellectual legacy of Moses Maimonides, a philosopher and physician who was held in the highest esteem by Jewish rational dissidents.

First, as a student of the great Averroes and subsequently as an independent thinker and practising physician, Maimonides remained under significant influence of the prominent Arabic culture and medicine, as well as ancient foundations of philosophy that the Arabic culture had discovered—the works of Aristotle in particular. Notwithstanding these influences, Maimonides not only managed to retain his Jewish identity, but he also had a significant impact on the understanding of the principles of Judaism, skilfully combining Aristotelian thought and scientific knowledge with a profound faith in the Jewish God (Grözinger 2004: 462). His entire philosophy was permeated by questions regarding the relationship between

[13] For more information on Jewish medical students at European universities see Ruderman (2005: 125–132) and Goody (2010: 2. Chapter). For the particular importance of the educational environment of Padua, for its numerous Jewish students see, inter alia, Warchał (1913), Shatzky (1950), Ruderman (1995: 3. Chapter)), Efron (2001: 13–24, 28), Grözinger (2009: 30).

faith and science, the mystery of creation and scientific discovery, as well as the connection between body and soul. These concepts are also examined in his medical treatises which not only discuss theoretical medicine but also provide practical advice in this area. In the field of medical treatment Maimonides is considered to be the first physician who viewed appropriate diet as a remedy for both the body and the soul.[14] Maimonides exerted great influence on our anonymous author, not only in terms of his medical expertise but also in terms of his entire philosophical concept. This influence is clearly demonstrated in the view, which the author expressed in Hebrew on the title page of the book, and which matches the opinion of Maimonides:

> Since maintaining of a healthy body and soul is a weighty obligation of the Torah and to do the opposite means to utterly disobey it [the Torah], for indulging in too much fare and drink and taking unclean food rots the mind. Thence, preserving health precedes the Torah. (*Seyfer derekh eyts ha-khayim* front page)

Our anonymous author is an altogether mysterious character. We cannot exclude the possibility that he was a Jewish man who hailed from the territory of the Commonwealth, and who – as a young man – left the country to study abroad, possibly in Padua. We could further speculate that many years later he returned to his native land with a fresh perspective of mature man equipped with the most modern knowledge and a new point of reference. It is, however, more probable that our author was a Jewish physician who, having received his education abroad, arrived to Poland on the invitation of a local magnate. It is likely that after few years he became thoroughly acquainted with the appalling state of local dietary and medical knowledge. Possibly, in order to remedy this situation, he decided to anonymously publish our book which disagreed with the existing Talmudic dogmas of the time. Incidentally, it is interesting to note, that had it not been for the early publication date, namely 1613, which appears on the title page of the book, the portrait constructed above would perfectly fit the Sephardic physician and thinker, Solomon Delmedigo (1591–1655), who resided in Poland between 1620 and 1625. This, however, is a subject for a separate study, which goes beyond our topic.[15]

---

[14] Compare the opinion expressed by one of the ancient Arab scholars quoted in the encyclopaedic biography of Maimonides in EMG (II, 885): 'Galen's skill remedied only the body, while that of Abu-Amran [Maimonides' name in Arabic] endorsed the healing of both the body and the soul. His knowledge not only distinguished him as the most prominent physician of his times but it also cured [his contemporaries] of ignorance and lack of knowledge' [translated from German by E.G.].

[15] I elaborate on this speculation in Geller (2015: 62–64).

2.7 Summary

It is important to emphasise universal humanism and rationalism displayed by the anonymous author on the pages of his book, a treatise that inherently fits the European mainstream medicine of the time. It is this universal humanism and rationalism that distinguish this book as a rare phenomenon against the background of the then-contemporary medical works, as well as moralistic rabbinical dietary and hygiene guidebooks belonging to the genre of *musar* literature which were addressed to Yiddish readers.[16] It is quite interesting that many of them have survived in the old print library sections in Germany, Austria, Great Britain and Israel, to name but a few, while our Eastern European Yiddish self-help book, a work of universal and all-embracing nature, seems to be a solitary piece of work, preserved in a unique copy rather by accident.[17]

The Yiddish *Regimen sanitatis* is an extraordinary example of transmission and adaptation of scientific medical knowledge that crossed the boundaries between Jews and Gentiles. In his remarkable endeavour, the anonymous writer was undoubtedly inspired by the humanistic spirit of his times but equally by the ideas of such rational Jewish dissidents like Maimonides. Both inspirations appear to be of particular importance especially when we reflect on the fact that medical knowledge and practices employed by the Jews of Eastern Europe, from where the book originates, are considered to have been—at that time—exceptionally superstitious and obscure. *Seyfer derekh eyts ha-khayim* is a proof to the contrary.

References

Adamson, M. W. (1995). Medieval dietetics. In *Food and drink in 'Regimen Sanitatis' literature from 800 to 1400*. Frankfurt am Main: Peter Lang.

Bendowska, M. (2017) 'Sefer derech ec ha-chajim' na tle dziejów drukarstwa żydowskiego w I Rzeczypospolitej. *"Kwartalnik Historii Żydów"/Jewish History Quarterly, Warsaw, 3*, 381–403.

Bylebyl, J. (1979). The school of Padua: Humanistic medicine in the sixteenth century. In C. Webster (Ed.), *Health, medicine and morality in the sixteenth century* (pp. 335–370). Cambridge: Cambridge University Press.

Büchlein der Gesundheit. In Hagenmeyer, C. (1995). Das Regimen Sanitatis Konrads von Eichstätt. Quellen – Texte – Wirkungsgeschichte (pp. 223–254). Stuttgart: Franz Steiner Verlag.

Efron, J. M. (2001). *Medicine and the German Jews*. New Haven/London: Yale University Press.

EMG – Enzyklopädie Medizingeschichte. (2007). W. E. Gerabek, D. H. O Bernhard, K. Gundolf, & W. Wegner (Eds.) (Vol. 1–3). Berlin/New York: Walter de Gruyter.

[16] To compare for example with the Brantspigel (1596) a moralistic treatise addressed to common Yiddish reading public containing also rules of dietetic and hygienic behavior.

[17] A good background for a better understanding of this fact is presented by Ruderman (1995:54): "Thus the homeland of Copernicus, at least at first glance, seems an unlikely setting, for a serious interest in scientific matters, even on the part of Jews living near the University of Cracow".

Geller, E. (2009). A new portrait of early seventeenth-century polish Jewry in an unknown Eastern-Yiddish remedy book. *European Judaism*. London, *42*, 62–79.
Geller, E. (2013a). Knaanic glosses as code-switching phenomenon. In O. Blaha et al. (Eds.), *Knaanic language: Structure and historical background* (pp. 108–129). Academia: Prague.
Geller, E. (2013b). Dziecko jako przedmiot zainteresowania późnorenesansowej medycyny żydowskiej. In A. Jeziorkowska-Polakowska & A. Karczewska (Eds.), *Żydowskie Dziecko* (pp. 21–33). Lublin: Towarzystwo Naukowe KUL.
Goody, J. (2010). *Renaissance. The one or the many?* Cambridge/New York: Cambridge University Press.
Grözinger, K. E. (2004). *Jüdisches Denken. Theologie, Philosophie, Mystik. Bd. 1. Vom Gott Abrahams zum Gott des Aristoteles*. Frankfurt/New York: Campus Verlag.
Hagenmeyer, C. (1995). *Das Regimen Sanitatis Konrads von Eichstätt. Quellen – Texte – Wirkungsgeschichte*. Stuttgart: Franz Steiner Verlag.
Preuss, J. (1911) *Biblisch-talmudische Medizin. Beiträge zur Geschichte der Heilkunde und der Kultur überhaupt* [Reprint 1992 Wiesbaden: Fourier Verlag].
Rosman, M. (2002). Innovative Tradition: Jewish culture in the Polish Lithuanian commonwealth. In D. Biale (Ed.), *Cultures of the Jews. A new history* (pp. 519–572). New York: Schocken.
Rostafiński, J. (Ed.). (1893). Andrzeja z Kobylina Gadki o składności członków człowieczych. *Z Arystotelesa i też z naszych mędrców wybrane* 1535. Kraków: Akademia Umiejętności.
Ruderman, D. B. (1995). *Jewish thought and scientific discovery in Early Modern Europe*. Detroit: Wayne State University Press.
Ruderman, D. B. (2005). *Early Modern Jewry. A new cultural history*. Princeton/Oxford: Princeton University Press.
Shatzky, J. (1950). On Jewish medical students of Padua. *Journal of the History of Medicine, 5*, 444–447.
Shatzmiller, J. (1994). *Jews, medicine and medieval society*. Berkeley/Los Angeles/London: University of California Press.
Steinschneider, M. (1852–1869). *Catalogus librorum hebraeorum in Bibliotheca Bodleiana*, Berlin. http://reader.digitale-sammlungen.de/de/fs1/object/display/bsb10983599_00001.html
Warchał, J. ks. (1913). Żydzi polscy na uniwersytecie padewskim, *Kwartalnik poświęcony badaniu przeszłości Żydów w Polsce*. Warszawa, *I/3*, 38–72.

Chapter 3
'When the Rabbi Meets the Doctor': Differing Attitudes to Medical Diagnosis Among Halakhic Authorities in Eastern and Central Europe in the Sixteenth to Nineteenth Century

Eliezer Sariel

Abstract The purpose of this study is to assess the significance of changes in the attitudes of *halakhic* authorities in Poland to medical diagnosis during the transition from the early modern to the modern period, as reflected in the case of "Attribution of a sighting of blood to a lesion". In the eighteenth century the *halakhic* discussion migrated westward to the German states, where a variety of approaches developed, most of them tending to rely on medical diagnosis. The differences between the attitudes of authorities in different geographic spheres can be explained by the differing status of the physicians: whereas in the German states physicians enjoyed an important status in Jewish society, in early modern Poland the quality of medical training was poor and they did not enjoy the prestige of their German counterparts. In the first half of the nineteenth century, we find a discussion of this matter among the authorities in Posen, a region that passed into the hands of Prussia and experienced an accelerated modernization process. Their approach lending legitimacy to medical diagnosis can be explained on the background of the shift among the authorities in the German states in combination with the increasing connection between Posen and Prussia and the improvement in the professional level of the physicians. The findings can facilitate a clearer understanding of the central processes in the development of Jewish Law and medicine in Jewish society in Poland.

Keywords Halakha · Female period · Rabbinic rulings · Medical diagnosis

E. Sariel (✉)
Shaanan College, Haifa, Israel

Ohalo College, Katzrin, Israel

M. Moskalewicz et al. (eds.), *Jewish Medicine and Healthcare in Central Eastern Europe*, Religion, Spirituality and Health: A Social Scientific Approach 3,
https://doi.org/10.1007/978-3-319-92480-9_3

3.1 Introduction

The purpose of this study is to assess the significance of changes in the attitudes of *halakhic* authorities in Poland to medical diagnosis during the transition from the early modern to the modern period, as reflected in the case of "Attribution of a sighting of blood to a lesion". The findings can facilitate a clearer understanding of the central processes in the development of Jewish Law and medicine in Jewish society in Poland.

According to *Halakha*, i.e. Jewish Law, a woman is forbidden to have any physical contact with her husband during her period. In addition to this fundament of Torah law, rabbinic law requires the woman to examine herself by checking for blood stains on a white cloth for another 7 days after the end of her period. Only if the examinations have come out completely clean over those 7 days, physical contact is permitted. A common problem is when the woman finds a red stain on the white cloth but claims that her period was over and the red stain is the result of an interior lesion or inflammation. The low levels of hygiene during the medieval and early modern periods made this question very common, and we have dozens of responsa (rabbinic rulings) dealing with it. When physicians supported the woman with a medical diagnosis, claiming the blood, which was seen on the cloth, did not come from the uterus, the issue of the reliability of medical diagnosis became woven into the response's fabric.

3.2 The State of Scholarship with Regard to the Attitudes of *halakhic* Authorities to Physicians and Medical Diagnosis

Any discussion of the place of medical diagnosis in the laws of attributing a sighting of blood to a lesion is bound up with the question of the interaction between the *halakhic* realm and the medical realm in the modern era. Most of the scholarly discussion of this interaction is centered on developments in the early modern period. The work of Ruderman, who focused on the central role of the physician in understanding the changes that Jewish society underwent in this period in Central and Western Europe (Ruderman 1992, 1995, 2001, 2007), led to a parallel discussion regarding Eastern Europe as well. Fishman noted the fact that, as opposed to Italy and Spain, in sixteenth to seventeenth century Poland the phenomenon of physicians who were also Torah scholars was unknown (Fishman 1997). He enumerated several reasons for this, including the fact that local universities were closed to Jews, who were obliged to go to Padua in Italy to learn medicine. Also, university studies required knowledge of Latin, which was uncommon amongst Polish Jewry, and the readiness to spend several years at an institution which served not only as a place of academic study but as a Christian theological center (Fishman 1997, p. 572–574). His research is borne out by Gelber's early findings that lack of regulation led to a situation where many, including Jews, served as physicians without obtaining appropriate training (Gelber 1957). Reiner argued that Eastern European Jewry did not generally contend with the challenge of science in the early modern period (Reiner 1997). Efron examined the status of physicians among the Jews of Germany from the middle ages until the twentieth century (Efron 2001). He noted the popularity of the medical profession in Germany from the Middle Ages on, and contrasted the limitations imposed on physicians in Germany serving in key roles in the Jewish community with the absence of such limitations in Jewish communities in Poland (Efron 2001, p. 40–45). Feiner viewed Jewish physicians in the German states in the early eighteenth century as harbingers of the changes the assimilation of enlightenment values would bring into Jewish society in the late eighteenth century (Feiner 2004, p. 21–23, Feiner 2010, p. 30).

The scholarly discussion, then, suggests that the phenomenon of rabbi-physicians was unknown in Eastern European Jewish society in the early modern period, and the influence of medical knowledge on the development of the *halakha* was limited, whereas in Central European Ashkenazic society Jewish physicians may be said to have been the pioneers, over the course of the eighteenth century, of a broader change in attitudes, reflected, as well, in the influence of medical knowledge on the development of the *halakha*. The limited range of cases which informed this body of scholarship fit in comfortably with the conventional description of the attitude of the *halakhic* authorities as part of a defensive Orthodox response, which seeks to preserve the *halakha* against a novel threat (Efron 2001, p. 187–233).

In this study, I will seek to focus on the attitude of Polish *halakhic* authorities to medical diagnosis in the early modern and modern periods, as reflected in the laws of attributing a sighting of blood to a lesion and the conclusions which may be drawn with regard to the character of *halakhic* discourse in Poland and the character of the mitzvah-observant Jewish society in these periods.

3.3 The Attitude of Physicians in Poland to Medical Diagnosis in the Context of the Laws of Attributing a Sighting of Blood to a Lesion in the Early Modern Period

3.3.1 Background

The starting point of the *halakhic* discourse on this topic is in the *Tosefta* (a *halakhic* source from the Tannaitic period, redacted in the third century C.E.). The Sages relied on physicians, who explained that an unusual substance which emerged from the vagina "like red shells" and "like red hairs" was caused by an internal lesion. That is to say, the Sages attributed the sighting of blood to a lesion on the basis of the estimation of physicians that the origin of the blood was caused in an internal lesion (Tosefta Niddah 4:1–2).[1] The Babylonian Talmud cites the *Tosefta* with a highly significant textual variant: according to this variant the Sages[2] were not satisfied with the medical diagnosis and required the substance emerging from the vagina to be immersed in water; if it dissolved in the water it was determined to be blood, and the woman ruled to be in a state of impurity (BT Niddah 22b).[3] This version reverses the situation, implying that the diagnosis of the physicians cannot be relied upon to attribute the sighting to a lesion, although it is still possible to make the attribution when there are additional supports for the medical diagnosis.

Up to this point it may be said that according to the version in the *Tosefta* itself the sighting of blood may be attributed to a blood-producing lesion on the basis of a medical diagnosis on its own, whereas the Babylonian Talmud's variant does not consider the medical diagnosis to be a sufficient basis.

3.3.2 The Approach of the Maharam of Lublin

For more than 1000 years, until the golden age of the Jewish community in Poland, the second half of the sixteenth century and the first half of the seventeenth century, there was no significant progress in the *halakhic* analysis of physicians' credibility on this issue. Maharam Lublin's (1558–1616) responsum centered the attention of the Poskim (*halakhic* authorities) on the question of the physicians' credibility. The importance of the Jewish golden age in Poland on the development of Jewish Law cannot be overemphasized. Not only did this golden age produce incomparable *halakhic* experts like Moshe Iserlish, Yoel Sirkish, David Segal, Shabtai Cohen but also extended the *halakhic* frontier into new areas.

The Maharam of Lublin weighed the reliability of medical diagnosis, and determined that despite the fact that the Babylonian Talmud implies that the Sages did not rely on physicians (BT Niddah 22b), physicians should be accorded the same degree of reliability which the Sages accorded to the woman:

> "Even if a woman clearly sees actual blood but the physicians say that it is clear to them that she has a lesion in the womb which is the source of the emergent blood, or if the woman herself states that she knows she has a lesion even inside the source itself where the blood emerges, we rely upon this and declare her pure." *(Responsa Maharam Lubli*n 111).

[1] This variant is attested both in the Vienna manuscript as well as the first print edition.

[2] Rabbi Asher Ben Yechiel, known as the "Rosh"(1250–1327), cites an explanation that immersing the substance in water was a requirement of the physicians, because they were not convinced that the substance in face was produced by the lesion (*Tosaphot Rosh on Tractate Niddah*, Rabbenu Asher., 2006. "It was stated that this woman has a lesion in her bowels").

[3] Ms. Vatican 113 has a Babylonian variant in the first case, where the woman saw "something like red shells", similar to the variant in the Tosefta, and the requirement to immerse it in water does not appear there, whereas in the second instance, at the sighting of "something like red hairs", the requirement to immerse in water arises (the understanding of the words "to put into the water" is unclear). The Babylonian variant which Rabbi Yosef Karo had before him was consistent with the variant in the Tosefta (Bet Yosef YD 191).

In the event of "a woman who produces something resembling shells" the Sages required immersing the substance in water, as the physicians did not declare that the blood was produced by a lesion, rather they declared that this was a peculiar case of "a unique phenomenon" which was not blood, therefore the Sages required that it be substantiated that it was indeed not blood.

It is worth noting that the Maharam of Lublin referred in a general way to "physicians" without expressly relating to their ethnic identity. Moreover, the Maharam's responsum was written for an observant Jew from a family of physicians, who apparently was himself a physician.[4] What the Maharam wrote served as the opening shot in a skirmish over the laws of attributing a sighting of vaginal blood to a lesion, which may be said to have served as a pretext for a broader battle over the measure of *halakhic* reliability to be accorded to physicians' diagnoses.

3.3.3 The Common Approach Among the Poskim in Poland in the Early Modern Period

There are two other responsa from Poland from that period. In one instance Rabbi Abraham Epstein requests the Maharsha to affirm his ruling permitting a woman who finds a sand-colored red stain on the cloth with which she wiped herself after urinating. He pays a lot of attention to the question of the reliability of physicians. As opposed to the Maharam of Lublin, he sets forth detailed criteria and argues that medical diagnosis can only be relied upon if it is based on the opinion of two Jewish physicians (*New Responsa of the Bach* 36).[5] The Maharsha, in his brief response, affirms Rabbi Epstein's ruling, but avoids directly addressing the ethnic identity or number of the physicians. As opposed to the responsum of the Maharam, the responsa of Rabbi Epstein and the Maharsha did not have a lasting impact on the *halakhic* discourse (*New Responsa of the Bach* 37).[6]

In 1654, Rabbi Joshua Heschel was appointed rabbi of Karkov.[7] His grandson, Rabbi Gabriel Eskles, rabbi of Metz in the first half of the eighteenth century, reports

---

[4]The Maharam addressed his correspondent as "a holy and awesome personage on a par with the angels, the venerable and wondrous master of Torah and piety my teacher and rabbi Shlomo Rofe may his Rock preserve him in life *(Responsa Maharam Lublin*, 111).

[5]In Rabbi Epstein's view, the requirement in the case of Rabbi Tzadok to immerse substance in water came from the physicians. As noted above (note 2) this explanation appears already in Tosaphot Rosh, but Rabbi Epstein was not familiar with that source. In the second part of the responsum Rabbi Epstein follows the Rama and holds that the woman can attribute the sighting to a lesion on the basis of the *halakhic* category of "sfek sfeka".

[6]The Maharsha became involved in this issue while in Lvov, i.e. the question and the responsum were composed between 1614 and 1625.

[7]Not to be confused with Rabbi Joshua son of Rabbi Yosef, author of Responsa Pnei Yehoshua, rabbi of Karkov some years previously, until his passing in 1648. Among his outstanding disciples were Rabbi Shabtai Cohen, author of the Shach, Rabbi Gershon Ashkenizi, author of Responsa "Avodat HaGershuni", and Rabbi Aharon Shmuel Kaidanover, author of "Emunat Shmuel".

that in a case involving his own daughter, Rabbi Joshua Heschel refused to rely on the opinion of physicians in order to permit her to her husband until a diagnosis was obtained from a Jewish physician:

> When a certain young girl was still being raised in the home of my grandfather-in-law, my teacher and rabbi, the rabbi of all of the exile, Rabbi Heschel of blessed memory, it occurred that his daughter, may she rest in paradise, raised a matter similar to the case of the separation mentioned above. My master and teacher my grandfather-in-law of blessed memory could not bring himself to permit her to her husband for a long period of time. Even the physicians said that it was on account of sand in the kidneys and she reported pain caused by this disruption. It took two years until the late pious Rabbi David Rofe of blessed memory took great pains to clarify the matter as clear as a new garment, which the blood was from a wound she bore *(Responsa Shev Yaakov 36)*.

Rabbi Joshua Heschel's reluctance to make a permissive ruling while waiting for a properly justified diagnosis from an observant physician is reflected as well in the *halakhic* dialog which took place in Poland in the latter third of the seventeenth century.[8] Rabbi Mordechai Kahana turned to the Rabbi of Posen, Rabbi Yitzhak, presenting him with his *halakhic* position. In his view, in the Babylonian Talmud, the only reason the Sages relied on the physicians' opinion[9] was that the unusual substance, resembling hairs or shells, proves the existence of a wound, whereas in other cases the medical diagnosis cannot be relied upon (*Be'er Yitzhak* 41). Rabbi Yizhak of Posen responded that non-Jewish physicians may be relied upon on when "their statement does not contradict the tradition of the Sages" (*Be'er Yitzhak* 42); in other words, the non-Jewish physician has no authority of his own.

To this point it may be concluded that, except for the Maharam of Lublin, the *halakhic* authorities in Poland in the seventeenth century expressed an array of reservations about relying on the diagnosis of physicians. Among these authorities were Rabbi Abraham Epstein, Rabbi Joshua Heschel, Rabbi Mordechai Kahana and Rabbi Yitzhak of Posna, who were concerned about inaccurate diagnoses, and demanded clear proof that the source of the blood was, indeed, a lesion, and were prepared to rely only upon an observant Jewish physician (Sariel 2011, p. 233–336).

[8] Both of these responsa were written when Rabbi Yitzhak served as rabbi of Posen, 1667–1685.

[9] Rabbi Mordechai Kahana explains that the requirement to immerse it in water is a function of the laws of purity and impurity, whereas from the point of view of the laws of menstruation, it is agreed by all that the woman is not considered impure.

3.3.4 *The Approach of the Early Modern Authorities in Poland Versus the Approach of the Authorities in the German States in the Eighteenth Century*

Surprisingly, in spite of the fact that the matter of physicians' credibility emerged and was discussed in Poland, and in spite of the fact Jewish population in Eastern Europe was seven times bigger than the Jewish population in central Europe, towards the end of the seventeenth century and the beginning of the eighteenth century the *Halakhic* discourse of physicians' credibility shifted completely to central Europe. While we can find varied positions among the *poskim* of Central Europe, the majority of them tended to give credit to medical diagnosis with the requirement that it be supported by Jewish physicians or with self-examination based mostly on the *posek's* own eyes, common sense and logic (Sariel 2011, p. 236–244).

Making sense of the difference in approach between the *poskim* of Poland and of Germany will require a reexamination of the findings presented above with regard to the difference between the status of physicians in Poland and the German states (section 2). In the early modern period Eastern European Jewish society did not produce the rabbi-physicians and the influence of medical knowledge on the development of the *halakha* was limited. In Central European Ashkenazic society, on the other hand, a significant shift occurred in the eighteenth century, with Jewish physicians heralding a broader change in Jewish society. In light of these circumstances, it is not surprising that in the mid-seventeenth century in Karkov Rabbi Joshua Heschel refused to rely on physicians, even when the consequences for his daughter were quite severe, having to remain in a declared state of impurity for 2 years!. Only a Jewish doctor, referred to as "pious" by the grandson, succeeds in producing sufficiently clear and persuasive evidence that the medical diagnosis may be relied upon.

3.4 The Approach of the *Poskim* in Posen to Medical Diagnosis in the Early Nineteenth Century

The *halakhic* discussion of the reliability of medical diagnosis in the matter of attributing a sighting of vaginal blood to a lesion having migrated in the eighteenth century from Poland to Germany, in the first half of the nineteenth century we once again find *halakhic* authorities in the Polish lands at the center of the discussion. It is not coincidental that these *Poskim* are concentrated in the region of Posen, a region which, following the partitions of Poland, passed, in the latter part of the eighteenth century, to Prussian rule, in the wake of which it underwent a process of accelerated modernization (Sariel 2013). As I have demonstrated elsewhere, there is a correspondence between facing the challenge of modernization and the need of the community for *halakhic* authorities with vision (Sariel 2011, p. 149–152). Posen, which, on the one hand, underwent a process of accelerated modernization and, on the other hand, still held a significant Orthodox community, spearheaded Polish Jewry's struggle with the challenges of modernization.

At the beginning of the eighteenth century Rabbi Gabriel Eskles, rabbi of Metz, informs his correspondent, Rabbi Yaakov Popers, of his *halakhic* approach; in cases where the theoretical conclusion was stringent he customarily turned to physicians to apply his tendency to practical leniency in matters of menstrual impurity: "that the establishment of harmony between a man and his wife and releasing her from the bondage of being lonely for her husband, I say to seek if a remedy may be found for this woman and in fact it is my practice in such matters to find justification from the physicians who are available" (*Responsa Shev Yaakov* 36).

Over a hundred years later Rabbi Akiva Eiger takes the identical tactic, when asked about the case of a woman who suffers from hemorrhoids, and who suffers pain in the anal region when discharging her bowels. The woman found blood on the examination cloth during the examinations of the 7 days, i.e. a severely prohibitive sighting of blood. In addition he defines the hemorrhoids as a wound that "*is capable of* producing blood", and rules in the manner of the Shach and Rabbi David Segal that a sighting of vaginal blood cannot be attributed to a lesion that is "capable of producing blood". In light of these circumstances Rabbi Eiger reaches the conclusion that the woman must be prohibited to her husband: "I am, to my distress, unable to find a path to permit the woman" (*Responsa Akiva Eiger* 61).

Even so, Rabbi Eiger's strong wish to permit the woman leads him on a circuitous quest which reaches as far as Frankfurt. First, he calls out the categorization of hemorrhoids as "a wound capable of producing blood" and raises the possibility of recategorizing it as 'a wound that is known to produce blood': "Prima facie there is a basis for leniency in the matter at hand, as the condition of *Goldiner* [=hemorrhoids – E.S.] may be considered as a wound that is known to produce blood" (*Responsa Akiva Eiger* 61). Yet he is not satisfied with this resolution, because according to the medical knowledge accepted in his day it was possible that the blood of the hemorrhoids had its origin in the uterus: "However, it is yet unclear if there is an artery that is dedicated to the drainage of the blood of the golden vein adjacent to or at the edge of the uterus, or whether there is no dedicated artery but once the drainage channel to the anus is blocked the blood becomes distributed among the organs and intermingles with the blood vessels of the source and become menstrual blood" (*Responsa Akiva Eiger* 61).

To make sense of Rabbi Eiger's uncertainty, it must be noted that hemorrhoids were known as "the golden vein". The medical assumption was that hemorrhoids result from the body's need to drain blood refuse, which, similarly to feces, finds its way to the anal region. Physicians disputed whether this was an actual artery which the blood feeds into and it goes to the anus or an artery without a direct egress which leads to a woman's central gathering place for blood – the uterus. For example Martin Luther suffered from hemorrhoids. He describes the origins of the term "golden vein" thus: "This disease is called golden vein, and it is actually golden, because it is said that with all of this blood all of the unhealthy substances are evicted from the body as though constituting a *porta sterquillinii* for all evils" (Banov 1964). Kassouf discusses the perception of hemorrhoids as a "Jewish" disease in the eighteenth century (Kassouf 1998). A similar formulation of the conundrum may be found in the writings of Rabbi Eiger's son-in-law, the Chatam Sofer:

> The golden vein is a circular surface in the rectum where the blood drains through the anus. Generally this is periodically regulated and at times the path is obstructed and the blood does not drain and this causes great maladies and pains. Sometimes the blood bursts out and its customary path is obstructed, and it is released through the penis. This is what the Shulhan Arukh described as one who produces blood from the penis and sometimes it even comes up and is excreted via the mouth and nostrils, but in a case where the drainage is not obstructed and the blood comes out through the anus it is not common for the blood to come out through the penis for a man or the vagina for the woman. (*Responsa Chatam Sofer, Yoreh Deah* 185).

The Chatam Sofer raises the possibility that blood from the "golden vein" reaches the oral and nasal cavities, in accordance with the medical view that various organs are able to communicate and blood can pass between them. 500 years previously Rabbi Hayyim Or Zarua wrote on the same vein: "It is conceivable that all of the [body's] cavities communicate with one another by the means of pain, and the nether cavities are as the upper cavities, and just as the nasal mucus can drain through the mouth when a person is ill, such is the case in the nether cavities as well (*Responsa Maharah Or Zarua* 8). Westreich sees the Chatam Sofer's attitude to medical diagnosis as an example of an exception to the image of rigidity attributed by Katz to Orthodoxy in general and to the Chatam Sofer in particular. (Westreich 2009). This paper exemplifies how the position of the Chatam Sofer is consistent with the accepted position of the German *Poskim* throughout the eighteenth century.

The medical uncertainty has far reaching consequences for the laws of menstruation. If hemorrhoid blood goes directly to the anal region, then it is considered blood of a lesion and the woman is menstrually pure; if, on the other hand, it passes through the uterus, then it is considered menstrual blood and the woman is prohibited. In order to settle the uncertainty Rabbi Eiger turns to the professional opinion of medical experts: "I have enquired and discovered that the physicians are in dispute over this and I wrote to a professor in the congregation of Frankfurt and he replied that it is not a dedicated artery, rather it flows through the vaginal blood vessels, while another expert physician told me that this is not the case, rather it flows through a dedicated artery, and last Tuesday I discussed this again with yet another physician who flatly agreed with the opinion of the professor" (*Responsa Akiva Eiger* 61).

The contradictory responses prevent Rabbi Eiger from ruling on the basis of medical science, but the very fact that he refers to the physicians indicates a readiness, in principle, to rely upon medical diagnosis in the matter of attributing menstrual blood to a lesion. In this manner he follows the mainstream of *halakhic* authorities in the German states, who, as opposed to the authorities of Poland, put their trust in medical diagnosis. There are several possible explanations for Rabbi Eiger's preference for the approach of the German halakhists: first, the ruling from Poland with regard to the reliability of physicians which is most often cited in the *halakhic* literature is that of the Maharam of Lublin, who was prepared to rely on the opinion of the physicians. It may be presumed that the position of Rabbi Abraham Epstein, the Maharsha and Rabbi Yitzhak of Posna was not even known to Rabbi Eiger (Sariel 2011, p. 236–237). In addition, the "migration" of the discussion from Poland westward over the course of the eighteenth century eroded the dominance of the Polish approach. Thirdly, at the end of our discussion of the differences between Poland and Germany in matters concerning the reliability of physicians, we raised the possibility that, among other things, they stem from differences in the quality of the physicians. The political changes at the end of the eighteenth century created a close connection between Posen and Germany and gave Rabbi Eiger ready access to the German medical system, which was considerably more sophisticated and advanced than the Polish system. This is evidenced by his choice to address his doubts to a German physician in Frankfurt rather than a Polish physician in Karkov.

Rabbi Eiger's ultimate dispensation of the matter is also consistent with the prevailing trend of the eighteenth century which includes broad acceptance of "The Maharil's examination method" (Sariel 2011, p. 214–216). The examination is carried out in the following manner: a piece of cloth is inserted deep into the vaginal cavity in the direction of the uterus, past the point of exit of the urine, and the woman is required to make water. If blood is found in the urine but the cloth remains unstained, this proves that the blood did not emerge from the uterus and the woman, therefore, is considered pure:

> Thus, my opinion is that Your Eminence instruct her to perform the aforementioned examination, if she can do so, as follows: when she senses the urge to discharge, she should thoroughly wash the vaginal region, and fill the entire chamber with cloth covering the entire vagina to the point where the penis reaches at the height of intercourse; subsequently she should urinate, and when she senses the blood dripping, insert the cloth to the uteral labia at the spot where she senses the drip of the blood, and after removing the cloth and finding the blood spot, wash the spot so that she have no concern that any blood remained there, and remove the cloth from inside. If she then finds that it remains clean, this proves that the blood is from the wound (*Responsa Akiva Eiger* 61).

Rabbi Eiger seeks to base the *halakhic* ruling on an empirical inquiry. Placing an obstruction in the birth canal, between the uterus and the vagina, allows one to discover whether the blood is coming from the uterus. With regard to the reliability of physicians and the inquiry into the objective facts, Rabbi Eiger follows the trend of the eighteenth century German authorities. In light of the information available to him he did not view himself as turning from the way of the Polish authorities rather as continuing and augmenting their approach.

Like Rabbi Eiger, Rabbi Gabriel Katz (1783/4–1864), rabbi of Tirschtiegel (1815–1831), also consulted with physicians in order to reach his *halakhic* rulings (*Responsa Teshuvat Gavrei* 24). When asked with regard to a pregnant woman suffering from incessant bleeding due to uterine prolapse, he turned to a medical diagnosis. In contrast, however, to the case of Rabbi Eiger, the medical diagnosis determined that the source of the blood was in the uterus. Despite this medical opinion, he decided to perform the same examination proposed by Rabbi Eiger:

placing an obstruction in the birth canal between the uterus and the vagina in order to determine whether the source of the blood is in the uterus. The examination was performed twice with contradictory results: once the blood was found only on the vaginal cloth but not on the cloth obstructing the passage from the uterus, whereas the other examination found blood only on the cloth obstructing the passage from the uterus. Ultimately, the traditional tendency to be lenient on the woman tipped the scales, relying on the possibility that blood whose source is the uterus but is not menstrual blood does not render the woman menstrually impure. To affirm his permissive ruling he turned to his colleague Rabbi Shmuel Zanvil, rabbi of Wollstein (*Responsa Teshuvat Gavrei* 24). Rabbi Zanvil had a hard time accepting the possibility of declaring blood emerging from the uterus to be pure; nevertheless, he also clearly tended toward leniency. Like Rabbi Eiger, he found the *halakhic* loophole in the empirical examination, and proposed performing a comprehensive examination after the "*Hefsek Tahara*" (self-examination to establish cessation of menstrual flow). The comprehensive examination is to be performed in every place but where the woman felt pain in previous examinations. If the blood was discovered upon a previously clean examination cloth when she examines the places where she felt pains, the woman may be permitted.

Thus far, we can say that Rabbis Eiger, Katz, and Zanvil, are eager to declare the woman pure and try to use medical diagnosis or examinations seeking to clarify the objective reality to assist in their efforts to leniency. This approach is consistent with the prevailing approach of eighteenth century *Poskim*. It may be supposed that the improvement in the professional standards of physicians in Posen also helped in the move towards reliance on medical diagnosis.

3.5 Conclusion

Following the status of medical diagnosis in the *halakhic* discourse of the *halakhic* authorities in Poland in the matter of "attribution of a sighting of vaginal blood to a lesion" indicates that in the early modern period, the golden age of Polish Jewry, the Maharam of Lublin heralded the expansion of the *halakhic* discussion of this matter to include the discussion of attitudes toward the field of medical diagnosis. While his pioneering approach indeed expands the horizons of the *halakhic* discourse, his *halakhic* ruling, which lends a critical status to medical diagnosis, was not accepted by his colleagues in Poland at the time. Most of the Polish authorities at that time discounted the importance of medical diagnosis in the process of *halakhic* ruling. In the eighteenth century the *halakhic* discussion migrated westward to the German states, where a variety of approaches developed, most of them tending to rely on medical diagnosis. The differences between the attitudes of authorities in different geographic spheres can be explained by the differing status of the physicians: whereas in the German states physicians enjoyed an important status in Jewish society, in early modern Poland the quality of medical training was poor and they did not enjoy the prestige of their German counterparts. In the first half of the nineteenth century, we find a discussion of this matter among the authorities in Posen, a region that passed into the hands of Prussia and experienced an accelerated modernization process. Their approach lending legitimacy to medical diagnosis can be explained on the background of the shift among the authorities in the German states in combination with the increasing connection between Posen and Prussia and the improvement in the professional level of the physicians.

References

Banov, L. (1964). Painful proctologic disease of prominent persons: Some famous people who have had anorectal diseases. *Southern Medical Journal, 57*, 328–331.

Ben Avraham, Y. (1982). *Responsa Be'er Yitschak*. Jerusalem: Machon Yerushalaim.

Ben Binyamin, Y. (1742). *Rsponsa Shav Ya'akov*. Frankfurt: Unknown Publisher.

Ben Gedalyah, M. (1881). *Responsa Maharam Lublin*. Warsaw.

Ben Shmuel Sirkish, Y. (1980). *New Responsa of the Bach* (p. 36). Jerusalem: Unkown Publisher.

Ben Yaakov, G. (1826). *Responsa Teshuvat Gavrei* (p. 24). Frankfurt (Oder).

Ben Yechiel, A. (Rabbenu Asher 1250–1327). (2006). In Y. Steinberger (Ed.), *Tosaphot Rosh on Tractate Niddah*. Jerusalem: Mosad Harav Kook, 22b s.v.

Ben Yitschak Ashkenazi, C. (2002). In M. Avitan (Ed.), *Responasa Maharh Or Zarua*. Jerusalem: Unknown Publisher.

Efron, J. M. (2001). *Medicine and the German Jews: A history*. New Haven: Yale University Press.

Eiger, A. ben M. (1987). *Responsa Akiva Eiger* (Vol. 1, p. 61). Bnei Brak.

Feiner, S. (2004). *The Jewish enlightenment, Jewish culture and contexts* (pp. 21–23). Philadelphia: University of Pennsylvania Press.

Feiner, S. (2010). *The origins of Jewish secularization in eighteenth-century Europe, Jewish culture and contexts* (p. 30). Philadelphia: University of Pennsylvania Press.

Fishman, D. F. (1997). Rabbi Moshe Isserles and the study of science among Polish rabbis. *Science in Context, 10*, 571–588.

Gelber, N. M. (1957). Letoldot Harof'im Hayehudim BePolin. In Y. Tirosh (Ed.), *Shai LiYshayahu: Sefer Yovel LeRabbi Yeshyahu Wolfsberg, ben hashishim* (pp. 347–371). Tel Aviv: Hamerkaz Letarbut Shel Hapoel Hamizrachi.

Kassouf, S. (1998). The shared pain of the golden vein: The discursive proximity of Jewish and scholarly diseases in the late eighteenth century. *Eighteenth-Century Studies, 32*, 101–110.

Reiner, E. (1997). The attitude of Ashkenazi society to the new science in the sixteenth century. *Science in Context, 10*, 589–603.

Ruderman, D. B. (1992). Contemporary science and Jewish law in the eyes of Isaac Lampronti of Ferrara and some of his contemporaries. In F. Talmage & B. Walfish (Eds.), *The Frank Talmage memorial volume* (pp. 24–211). Haifa/Hanover: Haifa University Press/University Press of New England in association with Brandeis University Press.

Ruderman, D. B. (1995). *Jewish thought and scientific discovery in early modern Europe*. New Haven: Yale University Press.

Ruderman, D. B. (2001). Medicine and scientific thought: The world of Tobias Cohen. In R. C. Davis & B. C. I. Ravid (Eds.), *The Jews of early modern Venice* (pp. 191–210). Baltimore: The Johns Hopkins University Press.

Ruderman, D. B. (2007). The history of invention: Doctors, medicine, and Jewish culture. In W. Cutter (Ed.), *Healing and the Jewish imagination spiritual and practical perspectives on Judaism and health* (pp. 195–206). Woodstock: Jewish Lights Pub.

Sariel, E. (2011). *Tradition in time of crisis: The history of rabbinic law in the Posen Region during the first half of the 19th century as a case study of the relationship between modernization and the formulation of Halacha*. Dissertation thesis. Hebrew University.

Sariel, E. (2013). The distinctiveness of the Jews of Posen in the first half of the nineteenth century. *In the East Lie My Roots; My Branches in the West: LIYB, 58*, 175–192.

Sofer, M. (2008). *Responsa Chatam Sofer Hashalem* (pp. 1–6). Jerusalem: Machon Lehotsaat sfarim Vechecker Kitvay Yad Al Shem Chatam Sofer.

Westreich, E. (2009). The response of Jewish law to modern science and state laws in German-speaking countries in the nineteenth century. In J. Brunner, & S. Lavi (Eds.), *Juden und Muslime in Deutschland: Recht, Religion, Identität* (Tel Aviver Jahrbuch für deutsche Geschichte 37, pp. 53–62). Göttingen: Wallstein-Verl.

Chapter 4
The Debate over Early Burial Amongst Jews in the Duchy of Mecklenburg-Strelitz in the 1790s

Hans-Uwe Lammel

Abstract Until recently the historiographical focus exploring the early burial practice in Jewish communities was concerned with the community of Mecklenburg Schwerin in the eighteenth century. In the first debate considered as 'the first halachic dispute in the era of emancipation' the Schwerin Jewish community was able to find a compromise settlement between the claim of the pietistic ruler Duke Frederic of Mecklenburg Schwerin and the own religious tradition. Shortly after that, 1787, the Jewish community of Alt Strelitz was demanded by the Duke of Mecklenburg Strelitz Adolf Friedrich IV to reveal their burial practice. In contrast to the conflict in Mecklenburg Schwerin the Alt Strelitz community answered self-confidently the duke and presented the burial practice which they did not want to give up. For more than 7 years they succeeded in continuing their habit inclusively the decided will to avoid early burial. Far away from a progressive narrative of emancipation and assimilation my argument is focused on the hindrances in the process of emancipation and assimilation, the role of the professionalized medicine on the one side and the concepts of the Jewish community concerning tradition and innovation on the other. The community of Mecklenburg Strelitz is an excellent example to show in which way a prosperous community made it to find its own path between Haskalah and German Enlightenment in insisting on tradition and demonstrating own medical expertise as well.

Keywords Jewish burial practice · Jewish community of Alt Strelitz · Duchy of Mecklenburg-Strelitz · Apparent death debate · Haskalah and enlightenment

H.-U. Lammel (✉)
Department of the History of Medicine, University of Rostock, Rostock, Germany
e-mail: hans-uwe.lammel@uni-rostock.de

M. Moskalewicz et al. (eds.), *Jewish Medicine and Healthcare in Central Eastern Europe*, Religion, Spirituality and Health: A Social Scientific Approach 3,
https://doi.org/10.1007/978-3-319-92480-9_4

A collection of documents published in April 1787 in the *Berlinische Monatsschrift*[1] reveals that attempts to align Jewish burial customs with Christian ones were not made entirely without force (Friedländer 1787, p. 318). The assembler of and commentator on these texts, David Friedländer, believed that the issue these 'authentic copies of sovereign decrees' addressed, namely how the dead were to be buried, was a 'matter of interest to humanity' and expressed relief that his 'fellow believers had done away' with this 'old and crude offence' of burial before sunset. He again thanks, on behalf of his 'enlightened confrères', the 'sovereign's prudence' and declares that the ordinance in question was a 'fresh victory over an old, appalling prejudice'. 'This hard won victory', he continues, is 'not amongst the more noble in nature', as it had not been 'secured through compelling argument, but by force from one side, and not without resistance from the other' (Ibid.). The essay set out to examine the role of force in what had come to be called the burial controversy, which had arisen in 1772 in the Duchy of Mecklenburg-Schwerin and lasted over 20 years. What is of interest here is the force that was used to impose the rule from above. Force was wielded first and foremost to prohibit the hitherto existing custom of burying the dead before sunset. Sovereign authority also implied authority over Jewish concepts of time and cleanliness as well as related practices that were singled out and called into question. The concept of force under examination here is a combination of *potestas* and *violentia*, was inherited from the seventeenth century and also implies the element of emotional self-control (*Affektbeherrschung*) (Meumann and Niefanger 1997, p. 12–14).

In this essay I will take a closer look at the Jewish community of Alt Strelitz, in the Duchy of Mecklenburg-Strelitz. In 1787, the members of this community, in a self-assured and self-reliant move, proposed a peaceful solution to a conflict that had been provoked by a Christian incursion into their community, begun 15 years earlier in the form of the first official ordinance of 1772 in Mecklenburg-Schwerin. Their proposed solution differed from official Christian rules that had been enacted as well as from other Jewish attempts at resolving the conflict (Silberstein 1929/30, p. 239).[2] The Alt Strelitz community's proposal was unique in that it was not based on an asymmetrical relationship between Jews and Christians; it was not at all obvious from the beginning what direction the eventual solution would take or which side to the conflict would be pressured to make concessions. Instead, the community presented an entirely self-contained plan for resolving the conflict, taking the matter into their own hands, particularly the key issue: establishing death.

This discussion is not centrally concerned with the Christian medical debates in which an anxiety related to quick burial and reports about the successful reanima-

[1]The *Berlinische Monatsschrift* was the mouthpiece of the Berlin Enlightenment, that rationalist variety of the European Enlightenment primarily embodied by clergymen and teachers, nearly all of them Prussian civil servants. I would like particularly to thank Dr. Małgorzata Anna Maksymiak (Berlin/Rostock) for reading the manuscript with a critical eye and for her helpful suggestions.

[2]The only remark made by the Rostock rabbi regarding the subject of this essay was: "Thanks to a petition submitted by the Jewish community, the matter had remained dormant until 29 November 1793."

tion of people who had been presumed dead reinforced one another, prompting a 'powerful obsession' with apparent death (Ariès 1991, pp. 504–517). This essay focuses on the reactions of the Jewish community, which has received insufficient attention in the existing scholarship. The community members' predominant concern was, as the doctor Salomon Seligmann Pappenheimer (1740–1814) put it, avoiding the impression of being 'descended from cannibals' (Pappenheimer 1798, p. 8). This dispute between Christians and Jews about a traditional Jewish practice warrants more extensive historical consideration, particularly because the aspect of force, which Christians had introduced to dissuade the Jewish community from their practice without directly addressing its use, has not yet been sufficiently examined.

I will do three things in the following essay. First, I summarise the background to the Alt Strelitz debate: the dispute over burial that took place in Mecklenburg-Schwerin in 1772. I then outline the course of the debate in the Duchy of Mecklenburg-Strelitz, which began in 1787. Finally and more briefly, I address the drastic shifts in concepts of time and purity that had been invoked in the official ordinances, shifts that did not come about without the introduction of force.

Deborah Hertz's interrogation into how Jews became German, which she examined in terms of conversion and assimilation (Hertz 2007, cap. 2), provides the conceptual framework for my own investigation. In this essay, I argue that the burial dispute forms not only a part of the broader shifts in the Christian-Jewish conflict that characterised the end of the eighteenth century, but the culmination of it. With Deborah Hertz's argument as its starting point, this essay will investigate the forms and paths taken by 'emancipation' and 'assimilation' in the late eighteenth century, the misunderstandings and obstacles encountered and the cultural meaning of the burial debate.

I shall begin by briefly outlining the 1772 dispute in the Duchy of Mecklenburg-Schwerin regarding early burial amongst Jews, which provided the background to the Alt Strelitz debate. The controversy over the Jewish custom of early burial, the 'first Halachic debate of the Age of Emancipation', as Daniel Krochmalnik recently and fittingly characterised this significant event and the related dispute (Krochmalnik 1997, p. 126), had been precipitated by a university lecturer in Mecklenburg. His name was Oluf Gerhard Tychsen (1734–1815). An Orientalist scholar at the college in Bützow, Tychsen provided, in the form of a *pro memoria* dated 19 February 1772 and addressed to the Duke of Mecklenburg-Schwerin, the impetus for the issuing of a decree that would regulate what were, in his opinion, indefensible Jewish customs. This ruling, issued on 30 April 1772, stipulated that the Jewish community of Schwerin must wait 3 days to bury deceased members of the community (LHAS 1, fol. 1 r and v).[3] In his *pro memoria*, Tychsen not only attacked rabbis, who would be responsible for ensuring the implementation of the rule, but addressed the traditional burial practices directly, ridiculing them with an allusion to the existence of

[3] Reprinted by Silberstein (1929/1930), pp. 278–279. Though this is only a copy, I will still draw attention to orthographic differences between the document and Silberstein's transliteration.

'Jewish devils and ghosts ruling over the night'.[4] He placed particular emphasis on medical rationales that aimed to avoid the accidental burial of the apparent dead. This allowed him to hold a metaphorical knife to the Pietistic and sentimental sovereign's breast in the form of a horrifying and irrational question: 'If he would allow living beings or, at the very least, men and women of whom doubt remained whether they were well and truly dead to be buried under earth belonging to him?' (Silberstein 1929/1930, pp. 278 and 279). Tychsen thought it necessary to enhance his argument with examples from Johann Jakob Schudt's atrocious anti-Semitic book, which purported to report on this practice and its consequences (Schudt 1715, p. 355).

The outraged and troubled Jewish community in Schwerin took two Jewish scholars into their confidence, soliciting advice from Moses Mendelssohn in Berlin and Rabbi Jakob Emden of Hamburg.[5] It should be recalled that the 'custom of early burial had secured legal sanction' in Hamburg in 1710 (Silberstein (1929/30), p. 233, note 1). The new ducal decree created a significant conflict for the community in Schwerin; suddenly, from 1 min to the next, it seemed, Jewish customs that had long been accepted were under Christian attack. Ultimately there was much uncertainty about how they should proceed with these new rules.[6] Mendelssohn, after demonstrating that the conventional custom of early burial did not stand up to a strict examination of Jewish religious and Halachic texts, did not stop at advising the community to defer to the sovereign's demands. In a move that can hardly be overestimated, he appended to his letter a draft of a response to the duke, the most critical aspect of which was the suggestion that a doctor be consulted. The community in Schwerin adopted Mendelssohn's draft and used it. Mendelssohn's proposal conceded to the ducal decree whilst also seeking to preserve existing customs in the broadest terms possible. It acknowledged the Christian debate about apparent death whilst simultaneously avoiding a full adoption of Christian notions and practices. Decisions about religious matters were to remain in the hands of the community.

[4] See also the exceptionally thought-provoking volume by Claire Gantet and Fabrice d'Almeida (Eds.) (2007). *Gespenster und Politik. 16. bis 21. Jahrhundert*. München: Fink.

[5] Other Protestant and Jewish clergymen were also involved, as evidenced in the two letters from Yacov ben Yitzhac Reuven (Sel.) to Tychsen in May and on 5 July 1772, in which Johann Friedrich Vincent Nölting (1735–1806) and Martin Friedrich Pitiscus (1721–1794), both of whom were pastors in Hamburg and connected to establishments for prostelytisation there, were mentioned. Tychsen's 'effective deed' was known to Reuven thanks to Adler. Universitätsbibliothek Rostock, Handschriftenabteilung, Mss. orient. 267a, Nr. 62 and 58. I would like to thank Dr. Małgorzata Anna Maksymiak (Berlin) for informing me about the existence of this correspondence, for her translation of it from Hebrew and her comments.

[6] This debate has been the subject of a number of examinations, most recently by Krochmalnik and Gerda Heinrich: Akkulturation und Reform. Die Debatte um die frühe Beerdigung der Juden zwischen 1785 und 1800. In: Zeitschrift für Religions- und Geistesgeschichte 50 (1998), pp. 137–155. See also Goldberg, S.-A. (2011). *Les deux rives du Yabbok. La maladie et la mort dans le judaïsme ashkénaze*. (pp. 148–154) 2. edition. Paris: Les Éditions du Cerf, and the essay by the Rostock rabbi Siegfried Silberstein cited above. I would like to thank Nathan Wachtel (Collège de France, Paris) for bringing Goldberg's book to my attention and for the patient discussion and kind advice he provided during his stay in Rostock.

The cultural experience of difference was particularly crucial, so it is not at all surprising that converts would play a not insignificant role in the transmission and transformation of cultural models (Lammel and Busch 2017).

It must be kept in mind that the central concern in the Christian sovereign's public reasoning on this matter was promoting the happiness of all of his subjects, which included the protected Jews. And it was this obligation to protect that was at stake here, especially in light of French and English reports of apparent death, which had begun appearing in the 1740s and remained current in the public's imagination thanks to the most convoluted tales (Krochmalnik 1997, pp. 109–120; Patak 1967, pp. 26–54). The issue's religious aspects were of far less interest than the way in which the sovereign's wisdom and enlightenment in fulfilling his duties could be demonstrated by incorporating natural historical knowledge in the design of the commonwealth, as defined by *potestas*. Whilst political calculations clearly prevailed in Schwerin, the statements made by the Orientalist from Bützow indicated that he was instead inspired by the concept of proselytisation he had learnt and practiced at the Institutum Judaicum et Muhammedicum in Halle (Bochinger 1998; Doktór 2009; Veltri 2009). First and foremost, Tychsen saw in this religious conflict a type of 'Jewish project', of which he hoped to be master planner (Lammel and Busch 2017).

The revised Schwerin decree of August 1772 stipulated that, from that point forward, burial of the dead be delayed by 3 days, until unmistakable indications of death, signs of decay, set in. If the deceased's survivors insisted on adhering to the traditional Jewish custom of burial before sunset, then a doctor would have to confirm death and approve the body for interment. The community therefore had two options for handling the conflict: they could either submit to the authority of a doctor in order to maintain the tradition of early burial, or they could consider themselves innovators after the Christian fashion and wait 2–3 days instead.

Having summarised this background, I will turn my attention in the following section to a case involving the Jewish community of Alt Strelitz that took place 15 years later, in 1787. In that same year Friedländer, who had after Mendelssohn's death assumed the leading position in the Berlin Haskalah, submitted his texts to the editors of the *Berlinische Monatsschrift*. During this period, Alt Strelitz was the residence of Duke Adolf Friedrich IV of Mecklenburg Strelitz (1738–1794), a secondary line of the princely house of Mecklenburg.[7] The local Jewish community had officially existed since 1768 and comprised 70 Jewish families that had settled in the area (Hofmann 2007, p. 26),[8] with the first synagogue there having been dedicated in 1763 (Ibid., p. 25). The duke instructed the leader of this community to explain the community's position regarding the burial of its deceased members.

[7] Today Strelitz lies about an hour north of Berlin by train.

[8] To get a sense of the Strelitz community's large size, compare the fact that only slightly more than 500 Jews lived in Paris at this time. Arnold, R. (2007). Integrationsdebatten 1807 und 2007. "Le Grand Sanhédrin" und die Deutsche Islam Konferenz. In: *Dokumente 5/07*, pp. 17–20, here 20a. By the end of the century, every fourth inhabitant of Strelitz was Jewish and the Alt Strelitz community numbered 600 individuals. Hofmann (2007), p. 31.

This demand had been occasioned by the burial of a poor Jew in the afternoon on the same day on which he had died. The instigator on this occasion was not a university professor, but two government officials, H. Seip and A. von Kamptz, who approached the duke regarding this matter on 24 August 1787. They declared the case to be a 'matter of conscience' and a 'matter of concern for an administration with humanity's best interests at heart', a job both religious and political in nature (LHAS 2, draft: fol. 294–295r, last version: fol. 296–297r).[9] They hoped to prevent 'such terrible evil by law'. This is the first indication that Friedländer's message, published just several months previously, had also been received in Strelitz and contributed to a growing awareness. The poor Jew's early burial had, the letter continued, 'renewed our outrage at the Jews' inhumane customs.' It demanded a 'prohibition' of early burial and emphatically stressed that this would not contravene Jewish law. It must be clear by now that they were invoking arguments Mendelssohn had presented in his reply to the worried letter the Schwerin Jewish community had sent in May 1772 and to the Berlin philosopher's delineation of what belonged to Jewish law and what did not. The reference to Mendelssohn is not surprising; originally published in HaMeassef, the letter had been available in German since 1787, when Friedländer included it amongst his collection of documents. In fact, I would suggest that Friedländer published this collection of five thematically related texts in a Christian journal solely as a means of framing the translation of Mendelssohn's letter to the Schwerin community. These two government officials absolutely believed themselves to be in the right and recommended that the duke issue a sovereign decree.

The duke in Strelitz reacted in quite a different manner than had the duke of Mecklenburg-Schwerin some 15 years before. It is not improbable that economic considerations played a not terribly small role given the territory's limited size and the empty ducal coffers (Hofmann 2007, pp. 19, 21, 24). The duke had already allowed 'peddling' in 1772, though visiting people and places where 'illness are rampant' was explicitly subject to 'grave punishment' (Ibid., p. 28). More importantly, Jews in Alt Strelitz were able to buy property and build houses, and their economic success was demonstrated by the workshops they operated (Ibid., p. 27).[10] Meanwhile, Adolf Friedrich refrained from immediately issuing an ordinance based on the submitted information, but instead sent a communication on 5 September 1787 to the Jewish community in Strelitz, requesting they respond to the allegations related to early burial. Seip and von Kamptz were charged with delivering the ducal

[9] 'As is known,' the emperor 'has set an example through his own decrees' that 'such terrible evil' be prevented 'by law'. The demand for a 'prohibition' of early burial, which would not be contrary to Jewish law, 'is related instead simply to the hotter regions where they once resided, where signs of decay set in more quickly, and where it could pose a risk to the living to allow the dead to remain above ground for longer periods' – so went the arguments in Tychsen's *pro memoria* – and indications too that 'the nation's own elders desire such a prohibition' were submitted to the duke, and it was then asked 'whether it would not be agreeable' 'to introduce a sovereign decree.'

[10] In 1788 Fromet Mendelssohn, Moses Mendelssohn's widow, and her children acquired the house in Strelitz that had previously belonged to the court Jew and later financial agent Nathan Meyer, though with the condition that it 'not be sold again to a Jew'.

request, and they made it clear in no uncertain terms that should the community persist in following the custom under question, they might expect 'punishment and cassation of all of their privileges and, if warranted, even corporal punishment' (LHAS 2, fol. 298r and v, letter to the "Judenschaft in Altstrelitz", 5 September 1787). One was no longer prepared to countenance tolerating this practice, and they were ready to take extreme measures. The balance of power between the Jewish community and the Christian sovereign was, up until this point, not at all dissimilar to that in Schwerin in 1772. In this respect, the situations in Schwerin in 1772 and in Alt Strelitz in 1787 were comparable, with one minor exception. Whereas the representatives of authority in Schwerin called for moderation (Lammel and Busch 2017), it was they who favoured a more drastic approach in Strelitz.

The community's elders were instructed 'to make our will and command on this matter known in the synagogue' and to appear 'on the 19th of this month before the privy council', where 'necessary matters will be discussed'. We know neither what was discussed at this meeting nor the outcome. What we do have is the letter, dated 31 October 1787, from the elders and leaders of the Jewish community in Alt Strelitz, signed by Moses Mendel, Lipman Abraham, Simon Jacob, Moses Canter, Gesel Isaac, Joseph Marcus and Philip Joachim (LHAS 2, fol. 299–301v: letter of the "Ältesten und Vorsteher der Judenschaft in Alt Strelitz", 31 October 1787). They declared that they are not motivated 'by ignorance or imprudence' and that their 'teachers forbid leaving the deceased unburied overnight without good cause.' This is their 'genuine belief' insofar that 'the deceased is truly deceased and there remains no doubt that revival is beyond all hope, for we hold it to be a most grave sin to bury someone alive.' With this, the Strelitz community demonstrated that they understood the problem posed by the anxiety surrounding the phenomenon of apparent death and had formulated a position on the question. They justified their stance with the Halachic principle that human life must be preserved above all else: *pikuach nefesh*.

The community leaders then attempted to expand the parameters of the discussion by drawing attention to the creation in Alt Strelitz of a holy guild, the *chevra kadisha*, and reporting on the work done by its 40–50 members, including the head and attendants (Ibid., fol. 299v and 300r).[11] Just like members of the Christian *Caritas*, its members were inspired by *rachmanut* (mercy), *ahava* (love) and *hachnasat orchim* (hospitality), and their female members' typical duties included sewing the graveclothes and washing the female bodies, while the men carried the coffins as well as dug and closed the graves. The duties were even more extensive in Alt Strelitz. It was stipulated that the head must be informed as soon as any community member had been ill for 2 days. The head was then charged with 'visiting twice' daily to check on the ill person. He would direct several members of the

[11] For Berlin, see Jacoby, J. (1989). Anfänge und Entwicklung der jüdischen Krankenpflege in Berlin. In: D. Hartung-von Doetinchem, & R. Winau, (Eds.), *Zerstörte Fortschritte. Das Jüdische Krankenhaus in Berlin 1756–1861 – 1914 – 1989* (pp. 28–67, 250–253, here 29–30, 251). Berlin: Edition Hentrich, and Halévy, M. A. (1970). Die Idee der Caritas in der jüdischen Religion. In: *Zur Geschichte der Jüdischen Krankenhäuser in Europa* (pp. 10–19). Düsseldorf: Triltsch.

group to alternate visits to offer various forms of assistance and advice, including soliciting a doctor's opinion. Should the person's condition deteriorate, they would remain at his or her bedside day and night to keep watch. If death seemed imminent, the entire group would be summoned to attest to the actual death. They emphasised that the group's members possessed detailed knowledge of death, having personally witnessed numerous cases already. The letter subsequently listed and explained the signs for which the attendants would remain alert and which would confirm that death had 'actually occurred'. There were essentially nine points:

1. When the pulse becomes irregular and returns to normal, even if slight and weak, so that there are more than 120 beats per minute,
2. When the person is clammy with sweat,
3. When spasms appear, such that the ill person twitches,
4. When facial expressions change and the light begins to leave the eyes,
5. When the breath in the chest begins to rattle,
6. When one clearly sees that the ill person's limbs stir, particularly the hands and feet, as well as when the nose seems to grow, becoming cold and pointed,
7. When the back or other parts of the body are covered in mottling,
8. When the smell of death is present,
9. And finally death is confirmed when the mouth gapes open, and too when a feather or stem is held up to the mouth or nose to check whether the breath of life is still present (LHAS 2, fol. 300r–300v).

The Strelitz community demonstrated with its reply not only that they were cognisant of and understood the challenges posed by a Christian semiotics of death and that they were willing to engage seriously with these. They also, and above all else, sought to assure their interlocutors that they possessed the requisite knowledge to partake equally in this public discussion, one that aimed to ensure 'that no trace of hidden life has been missed' (Hufeland 1791, p. 44). That the group had access to a community doctor and that the *chevra kadisha*, particularly in smaller communities, acted as a sponsor for the *heqdoshim*, the hospitals, also likely played a significant role. No more can be said here about how or even whether the community doctor and the society cooperated, particularly since the society for visiting the ill was first and foremost an institution dedicated to supporting community welfare, whereas the community doctor was charged with caring for the ill, not the dying.[12] The

[12] For more, see Eliezer Sariel's contribution in this volume, which explores the complex relationship between Halacha and medical advice and the relationship between rabbi and doctor in the early modern period. The independent investigation by the *poskim*, with a probable diagnosis made on the basis of experience, played a critical role in this.

Taking it for granted that a small community would appoint a community doctor reflects the experience of communities in Greater Poland, i.e. the existence of a Jewish community doctor who had passed his exams at a Christian university. Lewin, L. (1912). Jüdische Aerzte in Grosspolen. *Jahrbuch der Jüdisch-Literarischen Geellschaft 9*, pp. 376–420, here 371. Following the death of Markus Moses, a graduate of the college in Bützow who possessed a doctorate of medicine, on 28 March 1786, Isaak Levy, originally from Königsberg and with a doctorate from Frankfurt an der Oder, was appointed the community doctor of the Jewish community of Strelitz. We can assume

Strelitz community's response also signalled that they would not be satisfied with simply accepting the sovereign's decree. Their reply was in fact quite the opposite, offering their own solution towards defusing the problem. Their list of nine points encompassed everything that was known at that time about establishing death and avoiding the phenomenon of apparent death. From this self-assured description it is also clear that the society in Alt Strelitz that was charged with accompanying the dead to their burial had expanded – as in other Jewish communities throughout Europe – into a society responsible for visiting the ill, a *chevra bikur cholim*, or at the very least, the former found that its duties had melded together with those of the latter.

It was the leaders and elders of the Alt Strelitz Jewish community themselves who petitioned the duke to have a doctor examine their practices. This job was assigned to the personal physician Joachim Jaspar Johann Hempel (1707–1788), who was also the court physician to the Pietistic Princess Augusta in Dargun and whose opinions on the matter will be examined below.

Even though the Jewish community's elders believed the new rule to be unnecessary, there was in fact one particular circumstance in which they believed it appropriate to involve a doctor and for which they were prepared to allow a body to remain unburied for as long as was required for a doctor's examination. This exceptional case was that of sudden death. They summarised their stance, which was negative on the whole, by explaining that their religious rationale was reason enough for them to follow, 'without providing the slightest justification', the rules laid down by their teachers. They considered themselves entitled to this stance, since there could well be secret reasons for the rules, reasons that remained unknown to them (LHAS 2, fol. 300v).

The case was handed over to the personal physician Hempel on 1 November in the form of the letter from 31 October 1787 with the request for his opinion attached. He took until January of the following year to provide his response: 64 pages, covered in writing on both sides, 'ex propria auctoritate medica' (Ibid., fol. 304r–358r, Hempel's comment, 8 January 1788). Hempel's greatest concern was the phenomenon of apparent death, the existence of which the 'Jewish community' seemed 'unaware', he wrote, and which could not be sufficiently addressed by the nine points that had been submitted. Hempel's praise for the Jewish community's efforts, however, is more important for the argument I am advancing here. He highlights for the duke how 'cautious' the 'Jewish community' were being, even as he expressed in the very same breath his scepticism about whether the Jews were at all in a position to utilise the nine points as described. He was in fact doubtful whether these points, in the form described, would ever really be seen, 'even if they were to manifest themselves in one or another person caught in the throes of death'. This expert on physiology entirely sidestepped the potentially explosive conflict that could have resulted from confronting a religious question with a medical response. And he did this though he was not himself insensitive to the matter at hand. He provided, in

that he also assumed Moses's responsibilities as a doctor in the Jewish hospital. He left Strelitz in 1808.

great detail, a medical description of how apparent death was possible (Ibid., fol. 320v–332v.), and he laid out in full the medical authorities on which his argument rested (Ibid., fol. 343r–351r).[13] Hempel declined to examine in further detail existing Jewish burial practices, since this would touch 'too closely on their ceremonial nature to permit one to pass judgement on them'. He made explicit reference to Mendelssohn's letter to the Schwerin community from May 1772 that Friedländer had printed, and he pointed out that Moses Mendelssohn had himself lain at rest in his house for 24 h before being buried.[14]

The distinctiveness of the nine-point programme submitted by the Jewish community in Strelitz is now clear: matters of religion and ritual should remain inviolate from the sovereign's desire to regulate his subjects' health. As such, the autonomy and self-government that the large community in Alt Strelitz had enjoyed since the 1760s ought, it was implied, to be extended to cover this complex area of Christian-Jewish coexistence. Thanks to the explication and exchange of facts and opinions regarding jurisdiction, it was possible to delay for several years the use of force that had been addressed in Friedländer's text. Since the personal physician was unable to shift the discussion from one about apparent death to one about lesser degrees of enlightenment, which was ultimately what this argument was about, he settled on a different strategy.

Hempel not only consoled himself with enlightened self-assurance that the 'local Jews, were they guided by reason,' would oppose 'this prejudice', appreciate the duke's 'command' and see it for what it was, an expression of 'sincere grace and supreme sovereign indulgence' (Ibid., fol. 359v). He considered himself to be fulfilling his responsibilities as a personal physician (Lammel 2016), but his actions instead embodied that 'superstition and foolishness' Mendelssohn had invoked to describe the mask of enlightened anti-Judaism that some assumed (Mendelssohn 1989, p. 328). Hempel additionally shared some rules that he had ostensibly developed based on his own experience as a doctor and his son's experience in Neubrandenburg, medically sound practices for handling the recently deceased. He clearly believed this outlet to be a fitting means for their wider dissemination. He presented a 'summary' on two pages, to which he added an extended opinion on the following day (LHAS 2, fol. 359r and v, 9 January 1788). In doing so, he notes that the Jewish community itself had asked that if 'a physician [knows] of more than their 9 points of death', that this information be conveyed to them (Ibid., fol. 359v).

Hempel aimed to demonstrate that his Christian medical approach to handling the dead was the only correct one because he allows for sufficient time for decomposition to begin, eliminating any worry about apparent death. This is an indication

[13] He makes reference here to 'a number of writers', particularly Johann Peter Brinkmann, J. J. Bruhier d'Ablaincourt, Camerer.

[14] An argument that was frequently invoked in the debate: See Friedländer (1787), 329 footnote. The Christian reaction to this 'exception' was derisive and bitter, as evidenced in K. W. Ramler, K. W. (1786). Sulamith und Eusebia. Eine Trauerkantate auf den Tod Moses Mendelssohns. *Berlinische Monatsschrift 1*, 481–489, here 484: '"Bury him late, so he might still wake!" Such were the cries heard throughout the whole city.'

of the force with which the sovereign rule inserted itself, in the guise of medical knowledge, into the Jewish community's understanding of purity. Concepts of corporeality and (religious) purity were invoked that could not have been more opposed to one another. He advised, for example, that the deceased not be taken from the deathbed immediately after breathing has stopped. It would instead be better to leave the body in the bed for another 8 h, so that it could cool 'temperately'. The practices of the Jewish burial society were completely opposed, since its members, who were part of the society that visited and cared for the ill, immediately moved and cleaned the deceased's body, preparing it for burial. 'When [after eight hours] one pulls back the bed's covers' and 'discerns a loathsome smell', this is not necessarily the 'foul odour of death', but rather 'an odious sweat and discharged excrement' that could be washed away with warm urine or warm water. It would be 'a grave mistake' were one to then bring 'the deceased to a cold room' and lay him on a 'cold table or plank', opening 'windows and doors' so that the 'body stiffens'. At this point, there is no way that one could be 'certain that this is a case of complete [death]; a few so-called vital spirits might yet be present which could reanimate the heart and restore life', of which there were examples. Thus 'caution dictates' that 'the body be laid in a temperate room, to best keep the cold at bay'. This had 'the benefit of allowing the genuine, foetid odour of death' to appear 'on the third or, at the latest, fourth day, the true sign of death'. In contrast, 'the putrid stench is much delayed in cold' before it appears (Ibid.). Whilst practices in Jewish communities were designed to avoid exactly this advancing decomposition of the corpse so as to safeguard, insofar as was possible, the contamination of those who took part in these rituals, the Christian-medical discourse betrayed other concerns. Based on a belief in a state of being supposedly between life and death, in which, it was thought, one could discern 'degrees of life' (Herz 1788, p. 10), an 'état intermédiaire', the search for signs of life worked according to its own logic and transformed itself into a search for impurity as a criterion of truth (Ibid., p. 13, 58).[15] This is the profound epistemological reversal, after which it would be death that would determine life, to which Michael Foucault has already made reference (Foucault 1988, pp. 162–210).

In just a few years the discourse appears to have changed entirely; a religious matter had become one concerned with medicine and policing. Marcus Herz[16] and his publications in the late 1780s played a not insignificant role in stimulating this

[15] Herz cites Thiéry, F. (1786). *La vie de l'homme respectée & defendue dans ses derniers momens; ou instruction sur les soins qu'on doit aux morts, & à ceux qui paroissent l'être &c*. Paris, p. 56. German edition: Thiery, F. (1788). *Unterricht von der Fürsorge, die man den Todten, oder denen, die todt zu seyn scheinen, schuldig ist wie von den Leichenbegräbnisse und Begräbnissen*. Lübeck: Donatius.

[16] For more on Herz's scepticism regarding medical questions, see also Münch, R. and Lammel, H.-U. (1997). Versuch und Experiment bei Marcus Herz. In: Michael Hubenstorf et al. (Eds.). *Medizingeschichte und Gesellschaftskritik. Festschrift für Gerhard Baader*, Husum. Matthiesen, pp. 101–122.

debate, which was conducted in both Hebrew and German (Krochmalnik 1997, pp. 137–145; Pelli 1979, pp. 207–211).[17]

Whereas the plan to establish a morgue in Berlin in 1792 under the leadership of Joseph Mendelssohn (1770–1848) led to the founding of the 'Society of Friends' ('Gesellschaft der Freunde') (Krochmalnik 1997, pp. 146–149; Pannwitz 2007), the Mecklenburg community's statement of opinion allowed the matter to remain dormant there until 29 November 1793. A tutor in the house of the chamber agent Nathan Meyer, 'Candidate Michaelis', sent to the charitable society a copy of a work (LHAS 2, fol. 362–365), originally published in Berlin, written by Joel Löwe (1763–1802), professor in Breslau and senior head teacher at the Königliche Wilhelms-Schule (Ibid., fol. 366–402).[18] Michaelis explicitly demanded a prohibition against early burial. For him, this was a 'matter of policing' that had been drawn into the 'sphere of religion' (Ibid., fol. 362v). Michaelis compares the 'morality of Judaism' to the 'absurd and inhumane principle' which 'allows the life of *one* person to be handled with such capriciousness' (Ibid., fol. 363r. emphasis in the original). On 18 December 1793 the Strelitz duke informed the community elders and Rabbi Levin Emanuel by means of a preliminary answer that he himself wished to contribute to the recently submitted proposals towards settling the conflict. He made particular reference to the positions taken by two Jewish scholars – he presumably meant Mendelssohn along with Herz or Emden – and requested an opinion

[17] In 1797 Issac Euchel collected and published materials related to the 1785 debate, which had been conducted in the pages of HaMe'assef and begun with an anonymous essay by Euchel alongside the publication of the documents from Mecklenburg that Mendelssohn had given him, in a volume entitled 'Ist nach dem jüdischen Gesetze das Übernachten der Toten wirklich verboten? In einem Schreiben an den Herrn Professor Löwe in Breslau'. In: Euchel, I. (2001). *Vom Nutzen der Aufklärung. Schriften zur Haskala*, ed., translated and commentary by Andreas Kennecke, Düsseldorf: Parerga, pp. 119–137, Hebrew 159–176. He undertook a philological analysis of Old Hebrew and rabbinical textual evidence, as a result of which he concluded that an 'exact stipulation of time' appeared nowhere and that keeping the deceased overnight, 'which has been of late the cause of much fuss originates in a simple misunderstanding' (135). He furthermore determined that the debate had demonstrated that there were still a good many 'enemies of the *yehudim*' at the end of the eighteenth century (120, passim). His concern was limited to the subject in and of itself, not with undermining the ceremonial Jewish law, as certain rabbis claimed (136). One could not put off mourning three whole days just because there could be a case of apparent death, and there were exceedingly few people who could lie still 3 days whilst in agony (137). In this manner, Euchel engaged in an enlightened discussion, one marked by 'the intent to use facts to refute, [not] to ridicule' (120), in which he integrated historical, philological and medical arguments to modernise this 'custom of our forefathers'.

[18] The text on the title page reads: 'Missive addressed to the honourable members of all worthy and charitable societies in Jewish communities that assume responsibility for caring for the ill and burying the dead, by Joel Löwe. Löwe, J. (1792). Schreiben an die würdigen Mitglieder sämmtlicher löblichen und wohlthätigen Gesellschaften in der [!] jüdischen Gemeinden die sich die Krankenpflege und die Beerdigung der Todten zur Pflicht machen von Joel Löwe. Aus dem Hebräischen mit Anmerkung und Zusätzen von Michaelis zum besten der guten Sache, Berlin: in der jüdischen Freÿschule. Freudenthal, M. (1893). Die ersten Emancipationsbestrebungen der Juden in Breslau. Nach archivalischen und anderen Quellen dargestellt. *Monatsschrift für Geschichte und Wissenschaft des Judentums 37*, pp. 41–48, 92–100, 188–197, 238–247, 331–341, 409–429, 467–483, 522–536, 565–579, here 570–579.

regarding the essay that Löwe had published in Berlin. In a communication to the ministries in Berlin and Schwerin dated 20 December 1793, the Strelitz government made clear that it would inform both in a similar fashion of the decree that was to be issued shortly. It was suggested that similar measures be adopted there as well as part of a coordinated action (Ibid., fol. 411r). Only when this had come to pass were the Strelitz Jews willing to acquiesce to the decree (Ibid., fol. 411v). This was made clear in their statement from 18 December. In a letter dated 22 December 1793 the elders and heads of the community appealed against Michaelis' attempt to intervene and against the opinion issued by Löwe in Breslau (Ibid., fol. 412–425). Schwerin invited the elders and heads of the community on 2 January 1794 to suggest the changes they would make to the Schwerin rule dated 31 August 1772. Though this text has not been found, the Schwerin court agent Michel Ruben Hinrichsen gave his assurances in a letter dated 14 January that the rule would be 'followed to the letter' and spoke of involving a doctor in each and every case of death (Ibid., fol. 429r).[19] Thus Strelitz was informed on 19 January that the ordinance of 1772 was being observed, which the elders further affirmed with the enclosure of a copy of the document (Ibid., fol. 427–428). The government in Schwerin, however, remained suspicious. On the following day, an order was sent to the county physician and *Sanitätsrat* (title conferred on German medical practitioners) Wilhelm Johann Conrad Hennemann (1755–1822) in Schwerin. Even whilst steady 'physiologicisation' had seemingly defused the conflict, at least in terms of the use of force, the communication made reference to the 'unfortunate discovery' that there was no unmistakable criterion for establishing death except 'the beginnings of decay'. One had to rely on 'plausibility' and 'outward appearance' to establish death if one would not wait for the 'onset of actual decay'. But 'plausibility proves nothing'. Duke Friedrich Franz (1756–1837), successor to Duke Friedrich of Mecklenburg Schwerin (1717–1785), made absolutely clear that 'the doctor, on whose opinion everything' is based 'must exercise the greatest caution'. He sought to 'recommend to and impress upon' Hennemann and other doctors a 'caution as profound as humanly possible' (Ibid.).[20]

What most disturbed the Jews of Strelitz was having to depend on and defer to a doctor's opinion (Ibid., fol. 436r–441v).[21] An advisor drew on the Schwerin ruling of 30 April 1772, suggesting that a sentence from it, which stipulated that a corpse remain unburied for 3 days unless a consulting doctor provided written consent for an early burial, be adopted into the regulation meant for the Jews of Alt Strelitz, Mirow and Fürstenberg. One hoped that this rule, thanks to the dearth of doctors, would effect a form of cultural conditioning (Ibid., fol. 438v).[22] The regulation applied to all forms of death (Ibid., 242). The advisor furthermore thought the establishment of a 'separate morgue' was desirable, as had been suggested by Mendelssohn, Herz, Löwe and others (Friedländer 1788, p. 329; Herz 1788,

[19] Letter of Friedrich Franz to Hennemann. In 1808 Hennemann became a personal physician.

[20] Silberstein (1929/30), 241.

[21] Declaration of the Strelitz'Jewish Community, 28 January 1794.

[22] Silberstein (1929/30), 241.

pp. 55–58; Silberstein 1929/30, p. 242). In an interesting turn, Michaelis' translation of Mendelssohn's letter, which he had sent to the elders in Schwerin, was enclosed with the copy, made expressly on this occasion for this purpose, of the 'Rescript addressed to the protected Jews in the Ducal Lands of Mecklenburg regarding the burial of their dead', dated 30 April 1772 (LHAS 2, fol. 492–495). The plan was not included. Archival sources additionally demonstrate how an opinion issued by three royal Prussian state councillors to the privy councillors of Mecklenburg-Strelitz would determine the future direction of the debate. From Berlin, it appeared that the issue in Strelitz could be defused if the problem of being buried too early were simply reframed as one affecting all people in general. It was suggested that it would be far better 'to hold off coercing' the Jews and, aside from that, to support initiatives such as the burial society (Ibid., fol. 497r and v).

On 13 February 1794 Duke Adolph Wilhelm of Mecklenburg-Strelitz finally issued a ruling on the matter under contention, including the clause that was decided on in Schwerin (Ibid., fol. 485 r und v).[23] In what might seem an allusion to the image of the Jew that appeared in Tychsen, Joel Löwe expressed thanks to Adolph Friedrich on 16 March 1794 for the new rule and emphasized that behind it lay the belief that a 'better and more sensible treatment [...] of my poor nation' would shape 'them into better and more valuable citizens'. And, he added, it must never be assumed that neither 'religion nor a complete corruption of morals and the heart' could render them 'thoroughly insusceptible to virtue, forever irredeemable' (Silberstein 1929/30, p. 243).

As far as the Mecklenburg-Strelitz government and the duke were concerned, a politically feasible solution had been found for the religious problem. The same can be said for Mecklenburg-Schwerin. The Enlightenment's insistence on tolerance was surrendered in favour of a pragmatic resolution to a religious conflict. At the same time, it was believed that the forcible nature of this intervention in the community's ways had been mitigated, as the second generation of *maskilim* were as confident as the Christian followers of Enlightenment of the fundamental significance of natural science and medicine for their own religion.[24]

It is not terribly surprising that the influence of the Jewish community in Alt Strelitz was significant enough that 7 years were spent engaged in conflict resolution and that the use of force could be reduced to a minimum until the sovereign ordinance resolving the Christian-Jewish conflict over burial was issued in 1794. The elders and leaders of the Strelitz community petitioned the duke 2 days before Christmas 1793 to be allowed to maintain their old customs. They quoted Michaelis, who wrote that every person who aspired to 'such a resurrection' ought to be allowed to stipulate in his last will and testament 'how long his expired body would lay in this state of expectation'. They thus pointed to the conflict's religious dimension and

[23] Ibid., 242.

[24] See the headword in Hufeland, Chr. W. (1808): Der Scheintod, oder Sammlung der wichtigsten Thatsachen und Bemerkungen darüber, in alphabetischer Ordnung mit einer Vorrede, Berlin: Buchhandlung Matzdorff, Reprint Bern 1986, p. 128–132, ‚murderous burial of Jews', "mörderische Judenbeerdigung".

warned against an escalation of force. They invoked the events of the French Revolution and asked whether, if things continued apace, the 'Jewish religion in its entirety would be buried' and made reference to 'more recent writers' who ventured 'to ransack their old customs' and sought to 'destroy Jewish law' (LHAS 2, fol. 412r–424r., hier 422v, 423v; Silberstein 1929/30, p. 239).[25] Those in Strelitz were not keen to lead the way. They hoped to delay their compliance until after Berlin, Hamburg and Schwerin had taken the first step, which is to say adopted the new ordinance issued by the duke in Strelitz. 'Enlightened, well-read Jewish men' ought to join with the rabbis, especially the rabbi in Berlin, who was 'famed for his erudition in matters both theological and mundane' (Ibid., fol. 421v– 422r).[26]

What is of interest here is not simply that this debate and its reverberations sought to replace the fear of the corpse with a fear of apparent death, a hypothesis that ought also to be examined in the Christian context. This debate took the discursive direction it did as a result of the attempt to draw a connection between apparent death and Jewish customs regarding burial, which predicated not only Jewish backwardness when it came to burial practices, but also a serious bias based on 'prejudices'. Both benefited the eventual amalgamation of the Jewish and Christian Enlightenments. What has remained overlooked, however, are the multiple additional layers of Jewish self-conception and Christian reservations that lay behind the Jewish reticence regarding the burial question. Friedländer had noted earlier that 'on this thorny stem *prejudice*, many a rose' grew 'that deserves the safeguarding, if not active care, of moral betterment' (Friedländer 1788, p. 318; Emphasis in original). This is nothing more and nothing less than the recognition that prejudices are vitally important in determining the direction of thought and action in all cultures.

The first 'rose' that was crushed by the debate over premature burial amongst Jews was related to Jewish conceptions of time, which had been profoundly irritated (Goldberg 2000, 2001, 2004). I am specifically referring to the notion that the new day begins on the evening of what others, including Christians, would conventionally hold to be the day before (Heschel 1981). It was the unimpeded course of the Jewish year that had made life in the diaspora, subject as it was to Christian concepts of time, possible in the first place, and each disturbance of this balance between contrast and commonality on the road to providence had profound consequences for the communities' eschatological conceptions, in which apparent death could became a 'terrible notion of non-existence' (Mendelssohn 1989, p. 277). The important ethical argument advanced by the Strelitz community is in this same vein, namely that every person ought to be able to decide himself how long 'his expired body' should 'lay in expectation [of resurrection]'.

The second 'rose' was notions of purity (Burschel and Marx 2011; Burschel 2014), Christian as well as Jewish, that were completely turned on their head as a result of this cultural negotiation. Peter Burschel has drawn attention to the foundational cultural importance of the discourse on purity in understanding the early

[25] Letter of the Eldest and Heads of the Strelitz'Jewish Community to the Duke, 22 December 1793.

[26] Silberstein (1929/30), 240.

modern period, particularly religious conflicts. This examination has further substantiated many of his assertions: that purity requires impurity, that purity serves 'to reconcile perceptions and harmonise experiences' with the goal of 'homogenising interpretations of one's self and of the world', and that purity draws 'boundaries between religious, social and ethnic groups' (Burschel 2011, pp. 16–18). But the most interesting result of our considerations regarding the cultural meaning of this debate about burial is that Christian notions of physical impurity, which purported to be an expression of religious purity in that one waited until signs of decay appeared before death could be established, sought to call into question Jewish religious 'impurity', interring the body in as unaltered a state as possible, and succeeded in doing so. This could not have occurred without force, on both sides, and injury to a 'well-established sentiment'. Montaigne had already demonstrated how absurd it was to throw around allegations of cannibalism; the debate in Alt Strelitz revealed how difficult it is to defend oneself against such allegations, even when one has a voice.

It is evident that as far as the debate about the custom of early burial amongst Jews is concerned, the permissibility of force was, if not exactly praised by those representing both the Christian Enlightenment and the Haskalah, at least regarded as being legitimate and sensible in terms of putting new ideas into practice. This moment, in which force and injured feelings were confronted with one another, deserves closer examination. This certainly was a case in which a sovereign decree touched on a solution to a religious problem, thereby prompting a political regulation, as had been demanded by Johann Peter Frank (Frank 1788, pp. 624–631)[27] and Marcus Herz (Herz 1788). But the argument was much more complex. The use of force was legitimated in language through the choice to depict this as a belligerent dispute, and the argument was advanced that the Christian side in the burial dispute could gain the upper hand by assuming a human, supra-religious perspective. This would avoid any association with a 'religious war'. Force was further legitimated through the repetition of the stereotype of Jewish obstructiveness, which was contrasted with the figure of the enlightened Jew, and through recourse to an aesthetic argument: Jewish 'prejudice', according to the Christian argument summarised by Friedländer, outraged sentiments. The question of whether the writer was here reproducing an enlightened Christian opinion or, rather, referring to an 'old [Jewish] sentiment' that was offended by 'prejudice' is one that can be illuminated with reference to the example of Mecklenburg-Strelitz and the 'obstructiveness' on the part of the Jewish community there. These reciprocal charges of offence shed new light on the debate and on the depth of wounded feelings, which were to be a subject for discussion, and on the dynamic between the use of force (*violentia*) and emotional control. This essay asks what type of wounded feelings these were and who was the injured party. By clarifying the Christian 'bias' that it was in possession of a 'new semantics of death', it becomes apparent that wounded Jewish feelings are also at issue here if decay is the only remaining unmistakable sign of death that may serve

[27] "Von der Gefahr lebendig begraben zu werden, und von allzuspätem Begräbniß."

as a solution to the unresolved public problem (Hufeland 1791, p. 23, footnote).[28] This involves those effects of force that touched on the conception of time and practices related to purity. It was still within the logic of the Christian argument to refer, on the one hand, to the presumption that enlightened Jews had offended unenlightened Jews in order to explain why it was believed that one could take such unrestrained action – and with Christian support.[29] On the other hand, the aesthetic affected by the 'wounded feelings' only obfuscated that which was actually of central concern: insufficient religious tolerance and religious intervention, hope on the part of the second generation of *maskilim* that the Christian side too would exhibit a willingness to accommodate. Though the description presented here focuses on a writer for whom, at least at first glance, an objective point of view was of the utmost importance in juxtaposing force and obstructiveness, there is an unmistakeable bias in his text and the collection of documents. An insufficient ability to reason through the question of early burial was transformed into a deficit and a sign that enlightened thought was found wanting amongst Jews. On the other hand, Judaism's cultural-religious difference primarily served to help formulate Protestant Christianity's self-definition, in terms of what it was and what it ought to be, as well as what it did not want to be, namely heading towards religious intervention (Heschel 1999, pp. 63–64).

When reversing death could be added to the list of a doctor's duties, when a distinction was drawn in the *Encyclopédie*'s entry on 'mort' between 'incomplete', i.e. reversible, and 'absolute', i.e. irreversible, death and its author, Ménuret de Chambaud (1739–1815), labelled resuscitations as 'réssurections naturelles' and supernatural 'resurrections', including the one in the Bible, as natural reanimations, as 'miracles de la Médecine', then we can start to see clearly the hubris[30] that lay behind the cultural upheavals in the second half of the eighteenth century that we have come to call the (medical) Enlightenment. The debate on apparent death placed the Christian stance towards death in an uncomfortable position, one in which the trend towards secularisation is absolutely clear. The erection of buildings for the dead in Christian cemeteries, methods of ensuring that no hints of reanimation would be missed, the medical concept of 'vital spirits', which could remain present in the body even after the soul has escaped: all of these innovations and new ideas could be considered, even from the Christian perspective, to be the stimulating influence of recent discoveries and knowledge on one's own religious rituals. But they could also be understood to represent a threatening situation, a forcible shift in existing beliefs and practices. Turning their attention to Jewish burial customs would help them to find their way out of this challenge. Not only were religious-

[28] 'None of these signs are unmistakable.' The essay was previously printed in *Der Teutsche Merkur*, 1790, Nr. 5.

[29] Von Braun, Chr. (2001). *Versuch über den Schwindel. Religion, Schrift, Bild, Geschlecht*. Zurich: Pendo, pp. 477–495, identified the interaction between simulation and fiction during the phase of the 'biologicisation of anti-Judaism' at the end of the nineteenth century (491).

[30] I have followed Krochmalnik (1997), 116, in calling this process 'hubris' and have borrowed his examples.

cultural differences revealed, but the exercise allowed them to witness how adherents of another religion, the backwardness of which was a matter of unshakable conviction for many, managed, in dealing with the consequences of the debate on apparent death, to participate in or at least acquiesce to much more profound shifts and changes. Their adaptation to that which they believed to be uncanny was meant to strengthen Christians' ability to handle their own culturally and religiously challenging predicament. Anti-Semitic stereotypes were certainly helpful in this matter, as demonstrated by Johann Jakob Schudt with his assertion that Jews were incapable of any and every form of martyrdom (Schudt 1715, p. 25; Mosse 1985, p. 190). Present-day interpretations posit that apparent death took the place in Christian discourse of the 'fear of death and not wanting to admit the reality of death, of natural feelings of disgust towards the corpse' (Krochmalnik 1997, p. 110). These do not stand up to examination if one does not include the Jewish counterpart. The enlightened Jewish discourse reveals that the Christian Enlightenment and Jewish Haskalah coalesced at this moment in the dispute over burial after both sides adopted a human perspective. These ideas, inspired by notions of equality and equity, drew their authority from their association with the overcoming of stereotypes – a common cause that was foremost in mind. Later, the French Revolution would provide support, to no small degree, to a similar aim. Nevertheless, these very same ideas proved to be, in relation to the practice of early burial common amongst the Jews, a dead end. What was found to be especially problematic was the force wielded by sovereign authority that replaced a moral, community-building obligation with the power of natural scientific, physiological facts, the core substance of which – and this is what is most striking – was on very unsteady footing indeed.

Friedländer's characterisation of what happened over the course of the last 20-odd years of the eighteenth century, using the events in Alt Strelitz as an example, bears witness to a cultural and psychological shift that centred around the issue of force. Forceful action taken to overcome obstructiveness in the face of enlightenment was believed justified and morally sanctioned if one was in the position to plausibly make the case that such action was for the benefit of all humankind (Arendt 2000; Eisenstein-Barzilay 1956).

References

Arendt, H. (2000). Aufklärung und Judenfrage (1932). In H. Arendt (Ed.), *Die verborgene Tradition. Essays* (pp. 117–135). Frankfurt/Main: Jüdischer Verlag.

Ariès, Ph. (1991). *Geschichte des Todes (1978)* (H. H. Henschen, & U. Pfau, Trans.) (5th ed., pp. 504–517). Munich: Deutscher Taschenbuch-Verlag.

Bochinger, C. (1998). J. H. Callenbergs Institutum Judaicum et Muhammedicum und seine Ausstrahlung nach Osteuropa. In J. Wallmann & U. Sträter (Eds.), *Halle und Osteuropa. Zur europäischen Ausstrahlung des hallischen Pietismus* (pp. 331–348). Verlag der Franckeschen Stiftungen Halle im Max Niemeyer Verlag: Tübingen.

Burschel, P. (2014). *Die Erfindung der Reinheit. Eine andere Geschichte der frühen Neuzeit.* Göttingen: Wallstein.

Burschel, P., & Marx, C. (2011). *Reinheit*. Wien/Köln/Weimar: Böhlau.

Doktór, J. (2009). Georg Widmann und die Anfänge der Pietistischen Judenmission. *Kwartalnik historii Zydów, 4*, 449–490.

Eisenstein-Barzilay, I. (1956). The ideology of the Berlin Haskala. *Proceedings of the American Academy for Jewish Research, 25*, 1–37.

Foucault, M. (1988). *Die Geburt der Klinik. Eine Archäologie des ärztlichen Blicks* (1963). Frankfurt/Main: Fischer Taschenbuchverlag.

Friedländer, D. (1787). Ueber die frühe Beerdigung der Juden. Ein Brief an die Herausgeber, nebst einigen Dokumenten. *Berlinische Monatsschrift, 1*, 317–333.

Goldberg, A.-S. (2000). *La clepsydre*. tome 1: *Essai sur la pluralité des temps dans le judaisme*. Paris: Albin Michel; tome 2: *Temps de Jérusalem, temps de Babylone*. Paris: Albin Michel 2004. German edition: Goldberg, A.-S. (2009). *Zeit und Zeitlichkeit im Judentum* (M. Mühlenberg, Trans.). Göttingen: Vandenhoeck & Ruprecht.

Goldberg, A.-S. (2001). Les jeux du temps dans la culture ashkénaze. *Revue des Études juives, 160*, 155–168.

Hertz, D. (2007). *How Jews became Germans. The history of conversion and assimilation in Berlin*. New Haven: Yale University Press.

Herz, M. (1788). *Über die frühe Beerdigung der Juden. An die Herausgeber des hebräischen Sammlers. Zweyte verbesserte und vermehrte Auflage*. Berlin: Christian Friedrich Voß und Sohn.

Heschel, A. J. (1981). *The Sabbath. Its meaning for modern man* (7th ed.). New York: Farrar, Strauss & Giroux. German edition: Heschel, A. J. (2001). *Der Schabbat. Seine Bedeutung für den heutigen Menschen* (R. Olmesdahl, Trans.). Berlin: Jüdische Verlagsanstalt.

Heschel, S. (1999). Revolt of the colonized: Abraham Geiger's Wissenschaft des Judenthums as a challenge to Christian Hegemony in the academy. *New German Critique, 77*, 61–85.

Hofmann, P. (2007). *Jüdisches Leben in Mecklenburg-Strelitz*. Friedland/Mecklenburg: Steffen Verlag.

Hufeland, C. W. (1791). *Ueber die Ungwißheit des Todes und das einzige untrügliche Mittel sich von seiner Wirklichkeit zu überzeugen, und das Lebendigbegraben unmöglich zu machen nebst der Nachricht von der Errichtung eines Leichenhauses in Weimar*. Weimar: C. J. L. Glüsing.

Krochmalnik, D. (1997). Scheintod und Emanzipation. Der Beerdigungsstreit in seinem historischen Kontext. *Trumah, 6*, 107–147.

Lammel, H.-U. (2016). Leibarzt und Professor. Zum Verhältnis von Hofmedizin und Universitätsmedizin am Beispiel von Mecklenburg-Schwerin, 1776–1800. *Mecklenburgische Jahrbücher, 131*, 75–102.

Lammel, H.-U., & Busch, M. (2017). Haskala, Pietismus und der Rostocker Orientalist Oluf Gerhard Tychsen (1734–1815). *Aschkenas. Zeitschrift für Geschichte und Kultur der Juden, 27*(1), 195–238.

LHAS 1 = Mecklenburgisches Landesarchiv 2.26-1, Großh. Kab. I, 7832.

LHAS 2 = Mecklenburgisches Landesarchiv 4.11-16 Judenangelegenheiten (Acta judaeorum) Mecklenburg-Strelitz, Nr. 57.

Mendelssohn, M. (1989). Manasseh Ben Israel: Rettung der Juden. Vorrede. In M. Mendelssohn, & M. Thom (Ed.), *Schriften über Religion und Aufklärung*. Berlin: Union Verlags.

Meumann, M., & Niefanger, D. (1997). Für eine interdisziplinäre Betrachtung von Gewaltdarstellungen des 17. Jahrhunderts. Einführende Betrachtungen. In M. Meumann & D. Niefanger (Eds.), *Ein Schauplatz herber Angst. Wahrnehmung und Darstellung von Gewalt im 17. Jahrhundert* (pp. 7–23). Wallstein: Göttingen.

Mosse, G. L. (1985). *Nationalism and sexuality. Respectability and abnormal sexuality in modern Europe*. New York: Howard Fertig, Inc. German edition: Mosse, G. L. (1985). *Nationalismus und Sexualität. Bürgerliche Moral und sexuelle Normen*. München/Wien: Hansa.

Pannwitz, S. (2007). *Die Gesellschaft der Freunde 1792–1935. Berliner Juden zwischen Aufklärung und Hochfinanz*. Hildesheim: Olms.

Pappenheimer, S. S. (1798). *Deduktion seiner bereits heraußgegebenen Apologie für die frühen Beerdigung der juden*. [Breslau].

Patak, M. (1967). *Die Angst vor dem Scheintod in der 2. Hälfte des 18. Jahrhunderts* (Med. Thesis University Zurich. Zürich 1967).

Pelli, M. (1979). *The age of Haskala. Studies in Hebrew literature of the enlightenment in Germany*. Leiden: Brill.

Schudt, J. J. (1715). *Jüdische Merkwürdigkeiten*. part 2. (libr. VI, cap. 37). Frankfurt/Main and Leipzig.

Silberstein, S. (1929/30; 1930/31). Mendelssohn und Mecklenburg. *Zeitschrift für Geschichte der Juden in Deutschland, 1*(3), 233–244, (4), 275–290; 2(2), 165–166.

Veltri, G. (2009). Die Diarii des Callenberg-Instituts: eine Quelle zur jüdischen Kulturgeschichte in der ersten Hälfte des 18. Jahrhunderts? *Kwartalnik historii Żydów, 4*, 652–661.

Part II
Modern Jewish Healthcare: Community and the State

Chapter 5
German Medicine, Folklore and Language in Popular Medical Practices of the Eastern European Jews (Nineteenth to Twentieth Century)

Marek Tuszewicki

Abstract Medical customs of the Yiddish-speaking Jewish communities in Eastern Europe consisted of various elements, only some of which, mainly those associated with the Rabbinic tradition, could be described as idiosyncratic. Ashkenazi folk medicine was a complex heterogeneous system, to a large extent dependent on its social, geographic and historical milieu. It interacted with other systems: the official medicine and local folklore(s). In the following article several examples of German influences on the Jewish folk medicine will be indicated, as they appear in the sources written or published in the Russian Empire and Galicia in the late nineteenth and early twentieth centuries. Its intention is not only to present the visible impact of such works as Christoph Wilhelm Hufeland's *Makrobiotik oder die Kunst, das menschliche Leben zu verlängern* or Heinrich Paulizky's *Anleitungen für Landleute zu einer vernünftigen Gesundheitspflege*, and not only to enumerate excerpts from the early modern German-Yiddish medical literature, but also to shed some new light on the presence of such influences in the Yiddish folklore.

Keywords Jewish medicine · German medicine · History · Folklore · Eastern Europe · Nineteenth century · Twentieth century

The traditional Jewish community in Eastern Europe – a community that spoke Yiddish, was deeply religious, and cherished the values of Ashkenazi culture – held a wide range of beliefs associated with health. At a time when ethnographic research flourished, at the turn of the twentieth century, the folk medical beliefs and practices of Jews fascinated scholars to the same extent as analogous customs of other groups of the population. This interest yielded results in the form of numerous collections of folk *curiosa*, printed in separate volumes or in ethnographic periodicals in

M. Tuszewicki (✉)
Institute of Jewish Studies, Jagiellonian University, Kraków, Poland
e-mail: marek.tuszewicki@uj.edu.pl

M. Moskalewicz et al. (eds.), *Jewish Medicine and Healthcare in Central Eastern Europe*, Religion, Spirituality and Health: A Social Scientific Approach 3,
https://doi.org/10.1007/978-3-319-92480-9_5

German, Russian, Polish, and later in Yiddish and Hebrew as well.[1] Similar to those of Christian people, these beliefs were objected to by doctors and social activists engaged in the promotion of modern medicine. Such principles were perceived as manifestations of backwardness and superstition, harmful on all accounts, and had to be eradicated as soon as possible. The misunderstandings which arose at that time in the discussion of Jewish folk medicine took root and manifest themselves to this day. However, one should remember that beliefs concerning health constitute an inseparable element of the broad world view of a given community, also providing testimony of the relations of its culture with the neighbouring cultures (Yoder 1982, pp. 191–193; Libera 1995, pp. 11–12). Comparative research, conducted on the source material, proves that there are numerous relations among the folk medicine of Jews and that of Polish, Belarusian and Ukrainian peasants. Research also indicates the permanence of beliefs (promoted by medical authorities at least since the Middle Ages) that are based on obsolete theories associated with human anatomy regarding, among others, the place of man in the world, and the relations between him and other beings.

For many centuries, the Jews of the Slavic areas had close contacts with their fellow believers who inhabited Germanic countries. It was actually until the eighteenth century that these groups considered themselves as members of one community – of the community of Ashkenazi Jewry – the representatives of which had common roots, cultivated the same customs, and spoke a language which was generally referred to as *yidish-taytsh* (the Jewish German language). A clear separation the Ashkenaz into western and eastern was brought about by dynamic growth of the Jewish population in the Polish-Lithuanian Commonwealth, and, especially, by the rapid acculturation and assimilation of the German Jews caused by the progress of the Enlightenment. After Poland was partitioned, the culture of the East European Jews developed in an uninterrupted manner in Galicia, the western gubernyas of the Russian Empire, and a number of neighbouring regions (northern Hungary, Romania). Relations between this culture and the western Ashkenaz became looser. Nevertheless, the Jewish inhabitants of Kraków (Cracow), Lwów (Lviv), Warszawa (Warsaw) and Wilno (Vilnius), as well as the inhabitants of hundreds of minor cities and towns, cultivated, to a considerable degree, the Jewish-German cultural heritage. One way this aspect manifested itself in the area of health was in giving infectious diseases names that were deeply rooted in the Yiddish language (*pokn, mozlen, pest* i.e. 'smallpox', 'measles', 'bubonic fever'), as well as the names of certain other ailments and medications. At the beginning of the twentieth century, modern German medicine also exerted an influence upon the opinions of the traditional Jewish community.

The present article will demonstrate the scale of the influences of German medicine, folklore and language upon the traditional Jewish community in Eastern Europe. These influences had a highly multifarious, not always direct, nature, many of which constituted testimony of progressive modernisations. However, other influences proved that attachment to the early, pre-Enlightenment opinions regarding health remained.

[1] On the history of Jewish ethnography in Eastern Europe see: Gottesman 2003.

5.1 Traditional Society and Modern German Medicine

In 1891, in Lviv, the book *Ta'amei ha-Minhagim*, a collection of detailed explanations of rituals and customs, a part of the canon of books of customs popular at that time (*sifrei minhagim*), was published for the first time. It was written by Avraham Itzhak Sperling, a schochet and supporter of Hasidism and the Zaddik of Olesko. This publication enjoyed great success and over the course of subsequent decades expanded to four volumes in the case of the Hebrew edition, and to two volumes of the concise translation into Yiddish. Apart from religion-related content, the the second volume also featured a chapter devoted to traditional charms (*segulot*) and natural medications (*refuot*) (Sperling 1924/1925, pp. 39a–56a; 113–128). The latter chapter contains formulae for preparation of such concoctions as zinc ointment (*tsink-zalb*), spirits of camphor (*kamfer shpirtus*), and remedies for headache and diarrhoea. Sperling also included (in the Yiddish edition) an extremely interesting formula which he ascribed to the Viennese professor Josef Škoda (1805–1881): "Used to treat coughing. The book *Be'erot ha-Mayyim* features a formula of Professor Škoda, which helped many people who suffered from coughing. First it must be shown to a doctor" (Sperling 1909, p. 115). This remark is followed by a description of ingredients, information about the proportions and application of the medication thus prepared.

Be'erot ha-Mayyim is an anthology of Hasidic commentaries and homilies published for the first time in Przemyśl in 1888 as a supplement to a sixteenth-century ethical work entitled *Ta'amei Mitzvot*. The author and publisher of the book, Abraham Abbele Kanarvogel, was descended from a well-to-do Jewish family from Rzeszów. He lived in Rymanów where he visited the local Zaddik's court. The last part of the anthology bears the name *Be'er Mayyim Hayyim*. It consists of three chapters devoted to the problems of health and treatment: traditional remedies, pharmacist's medications, and remedies recommended by doctors. Kanarvogel, who for many years was undergoing treatment for a lung disease, included in his book recommendations and formulae derived from the Austrian doctors Heinrich Bamberger (1822–1888) and Adalbert Duchek (1824–1882) as well as Škoda.[2] A more in-depth analysis of his book indicates that the information that he had acquired directly from Viennese professors could have been supplemented by the author with material derived from printed sources, above all from the handbook *Klinisches Rezept-Taschenbuch für Praktische Ärzte*, published in Vienna (1).

Both Sperling and Kanarvogel were representatives of Galician orthodoxy in its Hasidic perspective. They treated with equal seriousness the traditional remedies and charms whose use was attested by the authority of the Zaddikim and the medications available in pharmacies. Hasidic literature provides ample evidence that in the second half of the nineteenth century even the most conservative representatives of the Jewish community underwent treatment by doctors. These people eagerly

[2] Bamberger's and Duchek's medical advice may be found in Sperling's Hebrew edition of his second volume.

embarked on trips to major urban centres in hope of receiving help from *Profesorn* (professors), the representatives of medicine who were held in high regard.[3]

Trips made in order to seek consultation of doctors usually constituted one of the final stages of the treatment. The journey itself was a considerable financial and logistic effort, and the cost of treatment was also considerable. However, throughout the entire nineteenth century, the Jews in Eastern Europe had at their disposal literature which reflected the ideas of the developing medicine, written in Hebrew and in Yiddish.[4] These were above all foreign publications. The task of translation of these works was undertaken by the advocates of the Jewish Enlightenment (the Haskalah). As early as before the final fall of the Polish-Lithuanian Commonwealth, Moses Marcuse's book *Sefer Refuot ha-nikra Ezer Israel* (Poryck 1790), an extensively revised translation of a well-known medical handbook of a Swiss professor, Samuel August Tissot, *Avis au peuple sur sa santé* (Lausanne 1761), was published. Written in robust Yiddish language, the book remained both modern and well-established in the reality of the Polish-Lithuanian province. Despite the great value of Marcuse's work, it is difficult to unambiguously indicate its influence on the medical practices of the subsequent generations (Efron 2001). The second translation of Tissot's work, *Refuot ha-Am* (Żółkiew 1794), published in Hebrew owing to the inspiration of Moses Mendelssohn, was more successful.[5] The author of the translation, Menachem Mendel Lefin of Satanów (1749–1826), was considered a brilliant master of Hebrew style. This work saw even a second edition, published in 1851 in Lviv owing to the initiative of a Warsaw physician, Mateusz Studencki (1805–1875). The translation was extended by remarks concerning motherhood, drawn from his Hebrew booklet *Rofe ha-Yeladim* (Warszawa 1847).[6]

The translation of *Anleitung für Bürger und Landleute* (first edition published in 1793) by a German doctor, Heinrich Felix Paulizky (1752–1791), was an even greater success. The work appeared as *Marpe le-Am* in 1834–1842 in two volumes, translated into Hebrew by Yehuda Bezalel Eliasberg (1800–1847) on the basis of the Polish edition *Medycyna dla ludu wiejskiego* in Walenty Szacfajer's translation (Wilno 1818). Until the late 1870s, another edition was published, this time in Żytomierz (1868). Extensive fragments of this work appeared in a small booklet entitled *Imrei Israel* (Berger 1911/1912) – a chapter devoted to pharmacist's medica-

[3]"He was the neighbourhood pharmacist, which practically made him a doctor, which was as far as one's imagination could stretch. Unless it was to a "professor". In dire medical cases, the victim was sometimes operated on by a "professor" – a sign of both privilege and despair" (Gay 2001, p. 50).

[4]Despite the obvious limitations, the sources also provide evidence that Jews consulted books in non-Jewish languages, e.g. Russian, Polish or German.

[5]The initial chapters were published under the title *Moda le-Bina* (Announcer of Wisdom) in Berlin in 1789.

[6]The booklet reached many representatives of the Jewish community; it was quoted, *inter alia*, by the author of the collection of *segulot* titled *Mare ha-Yeladim* (View on Children, Jerusalem 1900).

tions for convulsions.[7] References to this work also recurred in manuscript sources.[8] The book, written in a way that did not encroach upon the tradition and built a bridge between biomedicine and the writings of Moses Maimonides, became a true best-seller and an essential item in the libraries of Jewish feldshers and other practitioners of medicine. In the memoirs from the Polesie town of Motol, which date back to the second half of the nineteenth century, we may read the following words: "Father, who was a weak and very sickly man, visited respectable physicians, would go to trips to Vienna and Berlin, and he himself perused the popular book of medications entitled *Marpe le-Am*, he loved to hold longer and shorter conversations with Moishe-Yehuda [a local paramedic – MT]" (Chemerinski 1927, p. 33).[9] The supporters of the Haskalah attacked Eliasberg's translation for its excessive sanctimoniousness and concessions to tradition. Nevertheless, it is these aspects that enabled this work to enter the libraries of many rabbis and learned Talmudists (Epstein 1995, p. 154).

Although the aforementioned translations of German medical writings represent the state of knowledge of the turn of the nineteenth century, their relevance on the bookshelves of Jewish libraries did not decrease even before the beginning of the twentieth century. However, these works were soon to face competition on the dynamically developing publishing market. Aaron Bernstein's *Naturwissenschaftliche Volksbücher* lexicon, which devoted some space to medical questions, was among the books translated into Yiddish (Bernstein 1908–1914) and Hebrew (Bernstein 1895–1900). Jewish doctors who directly addressed Jewish patients by means of the written word also appeared. Eventually, shortly before the outbreak of the First World War, the first Yiddish-language periodical devoted exclusively to medical questions, which featured original and translated articles of doctors of medicine, was published in Warsaw.[10] Among those doctors there were obviously also representatives of German medicine. After all, as early as in the 1920s and 1930s Jewish social organisations, including the Towarzystwo Ochrony Zdrowia Ludności Żydowskiej [Society for Protection of Health of the Jewish Population], furthered popular medical knowledge owing to translations from western European literature. Special significance was attributed especially to Gottfried von Rittersheim's *Die Gesundheitspflege des jüngeren Kindes*, publicised in the TOZ "Folksgezunt" periodical.[11]

The image of the influence of contemporary German medicine on the medical opinions and practices of the Jews would be incomplete if we omitted a number of

---

[7] *Imrei Israel* (Words of Israel) is a very interesting example of a publication which combined the advice of folk medicine, which did not fail to refer to magical activities, with modern medicine. Both categories of advice were arranged in neighbouring columns, whereby the natural-modern part is filled mainly by quotations from *Marpe le-Am* (Medicine for People).

[8] Many excerpts from *Marpe le-Am* are featured in the London manuscript MS 9862 from the collection of The Library of the Jewish Theological Seminary.

[9] *Vide* Rozowski 1947, p. 18.

[10] i.e. the periodical *Der Yidisher Hoyz-Doktor* (The Jewish House Doctor) which was published as a supplement to the Warsaw daily *Haynt* (1912–1914).

[11] The Yiddish edition bore the title: *Der Arumgang un Flege fun a Zoygkind* (Preparing and Feeding an Infant, 1926).

concepts rejected by modern authorities. Over the course of the nineteenth century, traditional Jewish communities came to be interested in the ideas of mesmerism and hypnosis (Reiser 2014, 2015). However, the community which respected educated physicians, and was willing to accept their recommendations based on a worldview different from the traditional, had no possibility to verify the pseudo-scientific theories and basically did not distinguish them from the entire body of "medical knowledge". Therefore, in the medical handbooks of the Jewish Orthodox people, both the representatives of biomedicine and the pioneers of competing systems were simply referred to as 'doctors'. Sometimes the facility with which the latter reached the traditional Jewish communities seems even surprising. There were two important reasons for this state of affairs: the accessibility of publications in Jewish languages and a certain extent of compatibility which characterised these concepts in reference to the medical worldview, which for centuries underwent no particularly profound transformations.

In Jewish sources, a definitely strong position is occupied by Christoph Wilhelm Hufeland (1762–1836). This advocate of macrobiotics, a theory which strictly associated one's health with daily diet, was quoted repeatedly by Hebrew and Yiddish authors. His most important work, *Makrobiotik oder die Kunst das menschliche Leben zu verlängern* (Polish edition: *Sztuka przedłużania życia ludzkiego*, Warszawa 1828) was translated in Hebrew as *Ruah Hayyim*, published in Lviv in 1831. Another of Hufeland's works, the *Enchiridion Medicum,* was translated into Hebrew as *More Darkei ha-Refua*, published in Żytomierz in 1869. A much less popular reference work, it did not perform as great a role "on the Jewish street".[12]

5.2 Long-Lasting Traces of Pre-modern Medical Ideas

A considerable share of the rich medical literature preceding the onset of the Enlightenment was constituted by the collection of books known as *segulot u-refuot,* which discussed practical remedies of various kinds, frequently of a magical character, and medications recommended by medical authorities (Matras 1988, 1997; Zinger 2009). In this group, one may classify manuscript and printed collections of information on health, diseases, and pharmacist's medications, which in passing engaged such questions as unlucky days, protection against highwaymen, identification of a thief, glimpses of the invisible, and the extraction of wine from a wall. The *segulot u-refuot* books were written both in Hebrew and in Yiddish. Sometimes researchers are familiar with the authors of these books to the same extent as with

[12] The popularity of Hebrew translations of Hufeland's works is not the only instance in which the traditional communities consulted these Enlightenment publications which described the world of nature. Among others, *Reshit Limudim* of Baruch Lindau (Berlin 1788) was a work based to a great extent on a German textbook entitled *Naturgeschichte für Kinder* (Göttingen 1778). *Vide* Kogman 2009.

the educated doctors, other times as *baalei-shem*. Works published anonymously were also not rare occurrences (Matras 1988).

In the second half of the seventeenth century, print editions of a few books deeply rooted in the contemporary medical opinions and directed at readers unfamiliar with the holy language were published. Issachar Ber Teller (born ca. 1607), who originated from Bohemian Prague, a student of Josef Salomon Delmedigo, published a handbook of slightly over 100 pages entitled *Be'er Mayyim Hayyim* ca. 1650. It contained a brief description of the four elements of which the human body was thought to consist, a description of the humoral theory and a collection of natural remedies against various ailments. Moreover, towards the end of the book, the author included a Hebrew translation of the aphorisms of Hippocrates, although this part was not included in the second edition.[13] More than two decades after Teller, two books by Moses ben Benjamin Wolf of Kalisz were printed: *Yerushat Moshe* (Frankfurt am Mein or Wilhermsdorf 1677) and *Yarum Moshe* (Amsterdam 1679, Frankfurt am Mein 1710). This doctor, educated in Germany and Padua and taking up practice in his native city, provided Yiddish-language readers with access to information concerning the means of improving health recommended by contemporary doctors. The former of these titles was a sort of a *pharmacopoeia*, while the latter was a collection of advice on combatting a number of ailments, including considerations about the treatment and diet of a sick person. In this work, the author makes reference to the authority of Hippocrates, Galen and the "doctors of the holy community of Padua".

A work which influenced to an even greater extent the popular medical collections, thus consolidating the position of early medical opinions, was the quasi-encyclopaedia *Ma'ase Tuviya* by Tobias Kohn (1652–1729). It was published for the first time in Venice in 1706 or 1707. The author of this work, the son of the rabbi of the French Metz, lived in Cracow during his youth, before setting out to Frankfurt an der Oder and then to Padua to study medicine. The knowledge that he acquired and his considerable capacities qualified him to hold the position of a doctor at the sultan's court in Istanbul for many years. *Ma'ase Tuviya*, a two-volume collection on astronomy, botany, and the principles of the structure and functioning of the world, attests that this man was well-versed in the theories on which contemporary medicine is based. Similarly to many Jewish religious thinkers, he remained sceptical toward the Copernican model, although he maintained a keen interest in the discoveries being made in his original field of study. It would not be an exaggeration to say that he belonged to the elite of European doctors, and the work that he left to posterity attests to an equal extent his gift as a writer.[14]

Ma'ase Tuviya became one of the most important encyclopaedic publications, and its popularity extends far beyond the boundaries of the Ashkenazi diaspora. Most importantly, this work became well-established in the Jewish culture of the Polish lands. It was re-published many times in places including Lviv (1867,

[13] The reprint bore the title *Refuot u-segulot* (Medicines and Remedies, Prague 1694), *vide* Heller 2011, p. 745.

[14] For more information about the author and his work see: Schipper 1930; Sadowski 2011.

1874/1875) and Cracow-Podgórze (1907/1908). Hand-written copies were made as well, and transcripts of the work in whole and in part gained the status of an essential item in the libraries of Jewish barbers and feldshers very early on – as early as the eighteenth century.[15] The work of the "proficient and renowned doctor [*ha-rofe mumche ha-mefursam*], our teacher reb Tuviya" (Halpern 1858, p. 34b) was referred to by a considerable number of Jewish medical handbooks. The second volume of Sperling's book of customs *Ta'amei ha-Minhagim* (Sperling 1909, p. 50b) even features an illustration derived from Tuviya's work which presents "the form of a worm seen through a microscope". It is nothing else but a copy of the image of a parasite identified as *crinones* (German: *Mitesser*) in the late 17th c. by a German doctor, Michael Ettmüller (1644–1683),[16] author of the empirical testimony of the existence of parasites which supposedly fed in the skin of infants and caused pulmonary consumption.

From the early decades of the eighteenth century publications of acknowledged Kabbalists-healers began to be published. The *baalei shem* were the intermediaries between Heaven and Earth who used the "holy" names of God and the angels.[17] Their works combined elements peculiar to early medicine with mystical remedies to a greater extent than in the case of doctors educated at universities. This was done to such an extent that for a long period of time these works were considered above all collections of magic lore and the sources of amulets, copied by Eastern European Jews in the subsequent centuries. The Enlightenment criticism perceived these books as sources of ignorance, whereas the *baalei shem* frequently belonged to the intellectual elites of their communities.[18] Moreover, recent research proves the existence of close ties between the authors and publishers of this kind of book with modern pharmacy, not only regarding textual aspects, but also as far as personal contacts and shared interests are concerned (Petrovsky-Shtern 2011, pp. 43–44).

The activities of the printing press in Żółkiew was of crucial importance for the emergence of literature written by practicing Kabbalists. The first editions of Yoel Halpern's *Toldot Adam* (1720) and *Mifalot Elokim* (1724) left this press. Both works contained rich troves of medical advice and recommendations of a natural and Kabbalistic-magical nature. *Zewach Pesach* (1722), strictly devoted to the subject of the plague, was also published in Żółkiew. Many other titles were published in German centres: Hamburg (*Sefer Zechira*, 1709), Schulzbach (*Shem Tov Katan*, 1706) and Wilmersdorf (*Amtachat Binyamin*, 1716; *Menachot Yaakow Solat*, 1731). It is worthwhile to add that not all of the books belonging to the *segulot u-refuot* collection dating to that period were printed (Petrovsky-Shtern 2004).

The *segulot u-refuot* literature, either that written by doctors or by the practitioners of the Kabbalah, left an indelible mark on the medical opinions and practices

[15] Moses Marcuse derisively remarked that the Jewish barbers in the republic do not study medicine: only "one or two of them got a smattering of *Ma'ase Tuviya*" (Marcuse 1790, p. 82a).

[16] *Vide* Ettmüller 1682a, b.

[17] For information about the elements of shamanism in the *baalei shem* practices see: Rosman 1996, pp. 13–19.

[18] See the chapter devoted to the *baalei shem* and their works in: Etkes 2005, pp. 7–45.

of Eastern European Jews. The most important works in this category were republished many times in the subsequent centuries. Until the outbreak of the First World War the *Sefer Zechira* was published no less than 32 times, *Mifalot Elokim* – 17 times, *Shem Tov Katan* – 16 times, and *Toldot Adam* – 7 times.[19] One should also mention a number of manuscripts which are extensive copies of the aforementioned works. Among the medical handbooks which appeared in Ashkenazi communities at the turn of the twentieth century there are hardly any which failed to refer to the *segulot u-refuot* works. Testimony to this is furnished by the materials gathered by the pioneers of Jewish ethnography.[20]

5.3 Folklore of the Germanic-Slavic Borderland

Modern scholarship no longer maintains a direct link with ethics or demonology. Owing to the great efforts of a number of generation of doctors, the pre-Enlightenment ties linking the physical and metaphysical aspects of human speculation were severed. However, the history of European culture contains numerous instances of explaining diseases by the influence of the heavens or the world of demons. Folk culture also contained such instances, therefore in our discussion of the problem of Jewish medical practices we are forced to consider these practices from the perspective which involves magical activities, sorcery and demonology – referred to by nineteenth-century doctors as manifestations of superstition – and also religious rituals. The presence of themes peculiar to the folklore of German-language areas in the opinions of Eastern European Jews is complex in nature. Not all opinions of the Jewish people paralleled in the cultures of German or Polish people should actually be considered borrowed opinions. Many of these opinions are universal in nature, such as the anthropological fear of death. Other opinions, such as belief in the evil eye (*ayin-hore*), were widespread among the inhabitants of Europe. Still, other opinions, although their sources may indeed be perceived in German folklore, were profoundly transformed over the course of centuries. In any case, in our discussion of the influences of German culture upon the medical customs of the eastern Ashkenaz, we must devote some more attention to these aspects.

Many testimonies indicate that Jews shared opinions about witches and witchcraft popular among Germans and Slavs. However, Jews imparted a different character to these opinions. The Talmud already featured speculations about the magical dispositions of women, equivalents of which are furnished by the pagan image of the "knowing woman" (Pol. *wiedźma*, Ger. *Hexe*). A medieval book entitled *Sefer*

[19] The majority of the re-published works appeared in the Polish lands. See also the tables in the appendices to Matras 1997.

[20] *Inter alia* in the article entitled *O lecznictwie i przesądach leczniczych ludu żydowskiego* which appeared in issue No. 48 of "Izraelita" of 1898 (p. 505). In the footnotes the author included the transcription of titles, which shows influences of the Ashkenazi pronunciation: *Tołdos-odom* and *Mifałos-elokim*.

Hasidim, one of the most important sources, which enables us to familiarise ourselves with the mentality of German Jews of the Middle Ages, devotes some paragraphs to witches. In reference to the latter, it applies the word *estrie*, which is related to the Latin *strix* or *striga*, and presents them in a similar way to that of the image featured in non-Jewish descriptions. According to Polish ethnographers, opinions about witches held by the population in Eastern Europe until the beginning of the twentieth century are based on two sources: the Slavic and the Graeco-Roman, which reached these areas along with other influences from the West (Kolczyński 2003). These beings were thought to feed on human blood and meat, usually by preying on infants or lone travellers. Folk culture attributed to them the ability of rapid movement from one place to the other and of assuming the form of animals, especially a black cat or an owl. The witch thus represented in Jewish culture assumed the form of Lilith. In Hebrew manuscripts and on amulets Lilith appears very frequently as the cause of misfortunes attributed by the Christian population to witches. She appeared with various variants of the name Astarihu, which supposedly shares its roots with *estrie*; she was also known simply as 'witch' (Hebr. *machshefa*). A Biblical quotation against her is frequently alluded to (Exodus 22, 17) (Tuszewicki 2015).

In German and Polish folklore witches were accused of casting spells (German: *Geschoss*[21]), an alleged consequence of which was a sudden, ripping pain referred to as *Hexenschuss*[22] or *postrzał*. Ashkenazi Jews shared this opinion, and the Hebrew- and Yiddish-language sources feature practices and spells which served to counteract such misfortunes. A severe headache, sometimes referred to as *heyptgeshpar* or *heyptshayn* (German: *Hauptschein*) was treated by winding the patient's belt around his or her head and uttering a magical formula in which Elijah – a frequent protagonist of Jewish folklore – chases away the sinister Lilith, thus preventing the witch from perpetrating further harm (Goldberg, Eisenberg 1880/1881, p. 7b).

Sometimes folk culture explained diseases not as the result of a spell, but as a manifestation of the presence of demons in the human body. In the beliefs of Eastern European Jews there are many similar concepts, which frequently bear German names. The disease referred to as *ripkuchn/ripkichn* was explained by the Jewish sources as a swollen abdomen or a swollen spleen (less frequently a swollen liver) which accompanies rickets (Halpern 1858, p. 44a; Chotsh 1703, p. 9; Lilientalowa 2007, p. 65; Rosenberg c.1920, p. 23; Alfabet 1924, p. 119). A similar ailment was known as *Rieb-Kuchen* in early German medicine, and its main symptoms included a feeling of "hardness in the side", Crato 1690, p. 199[23]). In one treatment of this disease a fresh porcine or bovine spleen was applied to the swollen part of the body,

[21] *Geschoss* in Hoffmann-Krayer and Bächtold-Stäubli 1974, p. 756.

[22] *Hexenschusz,* in: Deutsches Wörterbuch von Jacob Grimm und Wilhelm Grimm, http://www.woerterbuchnetz.de/DWB?lemma=hexenschusz [24 August, 2016].

[23] See: *Kuchen, Kuche* in: Deutsches Wörterbuch von Jacob Grimm und Wilhelm Grimm, http://woerterbuchnetz.de/DWB/?sigle=DWB&mode=Vernetzung&lemid=GK15294#XGK15294 [24 August, 2016].

or was drawn from head to toe on the patient's body and hung above the chimney so that it dried off along with the disease (Lilienatlowa 2007, p. 64; Rosenberg 1911, p. 37; c.1920, p. 23). However, the most popular means of combating *ripkuchn* consisted in the beating of a swingle, board or a kneading trough covering the child lying on the threshold of the house with a cleaver (Yiddish: *hakn a ripkuchn*). The whole procedure, repeated over 3 days, had to be accompanied by a magic formula (Alfabet 1924, p. 63; Lilientalowa 2007, p. 65; Lew 1897, p. 407). It is worthwhile to note that the Slavic inhabitants of Belarus employed similar methods in the case of *hryzia*, explained as a hernia caused by a disease that "bit into the body" (Wereńko 1896, p. 151; Moszyński 1934, p. 201).

Jews also shared the belief, dating back to antiquity, of the 'wandering womb'. At the end of the nineteenth century this belief assumed the form we are familiar with thanks to non-Jewish ethnographic sources. After a baby was born the womb was thought to be anxious, looking for the baby and causing labour pains by its movement. The womb was equally anxious in women who had not been pregnant for a long time or in women who had never been pregnant at all (Lilientalowa 1905, pp. 160–161). This should be interpreted in the context of the belief in hysteria (Greek: *hystera* 'womb'), a peculiar mood which affects women and bears the traits of a mental disease (King 2014). However, hysteria is only one aspect of the belief in the wandering womb. An equally important symptom of its movement within the body allegedly had to do with various ailments affecting the members which the womb reached. The remedial measures applied in such cases frequently had the nature of magical incantations. In one medieval source there is a Jewish-German spell which begins with the words *Ber muter lig dich* (be at ease, o womb) which besought the womb "for nine generations, and nine pure scrolls of the Torah and for nine angels" (Perles 1887, p. 28).[24] Whereas the Yiddish-language incantation by means of which the traditional healer conjured up "*heyb muter* in a woman", so that the womb could return to the appropriate region of the body dates back to the early twentieth century. A traditional healer should utter the spell, holding his right hand on the omphalos of the female patient.[25] The ailment, caused by the movement of the womb, was also identified to a considerable extent with stomach pain and colic. Such an explanation is featured in the works of researchers into German folklore like Jacob Grimm, who identified the biting *Hachmutter* or *Bärmutter* as colic.[26] This term appears in a similar meaning in the translation of the work of Tissot, *Refuot ha-Am* (Lefin 1851, p. 84b unpaginated). One of the stories attributed to

[24] See Chotsh 1703, p. 11.

[25] MS 9862, pp. 42–44. See a similar incantation in Yiddish which also referred to a woman and was uttered while holding a hand on the omphalos of the sick woman in: Goldberg, Eisenberg 1880/1881, p. 2a. However, it is marked by a peculiar fragment "do not drink her blood, do not break her bones, do not stab her sides, do not weaken her members". On the same page there is also another incantation referring to the womb, which entreated the sickness by: "three Patriarchs and three Matriarchs, the seven seraphim, the seven heavens, the seven shofarot, the sun and the moon with seven planets and the Name of God".

[26] Deutsches Wörterbuch von Jacob Grimm und Wilhelm Grimm, http://woerterbuchnetz.de/DWB/?sigle=DWB&mode=Vernetzung&lemid=GB00933#XGB00933 [24 August, 2016].

Zaddik Motel z Czarnobyla furnishes an example of a man who suffered from *heyb muter*, which the author explained as "abdominal fever" (Hebrew: *kadahat ha-beten*) (Bodek 1927, pp. 22–23). In such cases the displaced womb was treated with the use of a large amount of natural remedies.

As far as skin is concerned, the *miteser* was considered one of the most dangerous parasites, one that attacked mostly children. The original version of this belief, also featured in Polish ethnographic sources, stipulated that minute worms infest the skin of an infant (in its face and under the shoulder blades) which in the course of time become a source of consumption. It is also in this case that sugar or honey was used to remove those worms. These substances, along with hot baths, were supposed to encourage the parasites to "stick their little heads up". The heads were then cut off by means of a knife or another tool with an elongated, sharp edge. At the beginning of the twentieth century these parasites were no longer referred to by this name, whose source lay in German folklore and contemporary medicine. In everyday speech there was only the expression "to eat as if with a *miteser*",[27] to refer to a child with an insatiable appetite (Tuszewicki 2013).

At this point, it is worthwhile to devote a certain amount of attention to two interesting cases of diseases believed to be caused by the moon. The first ailment was thought to have its source as early as in the foetal period, whereby the *monkalb*/*munkalb* (German: *Mondkalb*, English: *mooncalf*), a monster referred to as the "mooncalf", was a victim of the sinister influence of moonlight. The roots of this belief date back to the folk culture of Germanic lands. However, in the Polish lands it appeared relatively early – at the latest in the late sixteenth century.[28] According to various sources, the *monkalb*, although born from a normal relationship of a woman with a man, had the physical properties of a calf or a piglet. It appeared in the world when a man passed water by moonlight before intercourse. The monstrous appearance was not all, though, for this creature was thought to pose a particular threat to its mother, killing her with its shrill squeal (Lilientalowa 1921, p. 1; 2007, p. 20).[29] As late as at the turn of the nineteenth century the concept of the *mondkalb* appeared in handbooks for midwives (e.g. *Grundlegung zur Hebammenkunst für die Wehmütter und für Frauen, die Wehmütter werden wollen*, Flensburg-Leipzig 1793), medical texts, and encyclopaedias. Literature used this term in reference to the so-called false birth, which for a long time was not distinguished from a normal pregnancy, concluded by stillbirth of a formless body.[30] An analogous use of the term *monkalb* may be found in the context of Jewish culture. Yiddish literature of the later period used this term to refer to cases of miscarriage which revealed defects of the foetus. Itzhak Yoel Linetski eloquently described the

[27] The word *miteser* was translated as "table companion" (Lilientalowa 2007, p. 68).

[28] Information about it appears in the legal interpretation of rabbi Samuel ben Fayvush of Przemyśl (16th/17th c.), *Teshuvot ha-gaon rabi Shmuel mi-Premesla* (Responses of the Gaon Rabbi Samuel of Przemysl) In: *Piskei ve-shelot ve-teshuvot Maharash mi-Lublin* (Decisions, Questions and Responses of Mahrash of Lublin), p. 254.

[29] Cf. Grimm 1854, p. 1111.

[30] Lat. *Mola hydatidosa*, *vide* Bandtkie 1827, p. 498.

"ignorance" of the Hasidim who debated about the sainthood of the pregnancy of the Zaddik's wife: "Didn't the great reb Hile say about his old wife that she carried two scrolls of the Torah inside her when she was pregnant... (He referred to two sons, twins) And as she gave birth to a child and a *monkalb*? Well, tough luck" (Linetski 1897, p. 98). In colloquial speech and in folklore this term still refers to an abnormal creature, a monster (Stutshkov 1991, p. 574).

Another example of a negative transformation of the body of a baby due to the influence of moonlight was the *wasermun*, described a number of times by the ethnographer Regina Lilientalowa as "moon water". The symptoms of this condition included diarrhoea, blurred vision and yellowness of complexion. The ethnographer adduced a number of examples of remedies which were supposed to cure this disease by washing or sprinkling the sickly body with water prepared in an appropriate way (Lilientalowa 2007, p. 39–40; 1921, p. 1). However, it is worthwhile to note that the term *wasermun/waserman,* which signifies 'a drowned man' (or 'a water sprite'), functioned in the folklore of the Germanic-Slavic borderlands (Dźwigoł 2004, p. 175). Moonlight carried associations with humidity and coolness, which in turn brought to mind the basic attributes of water. Moreover, in the Slavic tradition the moon was sometimes a significant element of stories about drowned bodies which were revived – a theme which recurs in Jewish folklore as well. Moreover, the symptoms of the disease referred to as *wasermun* may be classified as properties of the appearance of a man who drowned. Any doubts in this respect are eventually dispelled by a quotation from the collection *Sefer Refuot* derived from the Pale of Settlement, published in the "Mitteilungen" of Max Grunwald in 1906. It contains a description of a disease known as "a drowned man or a moon" (*fun waserman un fun lewone*), which was characterised by the following symptoms: blurred vision, diarrhoea, and excessive thirst (Grunwald 1907, p. 145).[31]

5.4 Recapitulation

Over the course of centuries German medicine exerted an influence on the medical culture of Ashkenazi Jews, including that part of the Jewish community which was surrounded by the Slavic population. Before the Enlightenment this influence was exerted by early medicine, the graduates of German universities, physicians, barbers, and the *baalei shem* who printed their books in the German-speaking countries. In the nineteenth century the trends of modern perception of health and its care reached the traditional communities of Jews in the Russian and Galician provincial towns not only due to the works translated from German by the advocates of the Haskalah, but also thanks to the success of modern medicine in fighting diseases (particularly epidemics) and the role played by German universities. Two

[31] See a different incantation related with the disease known as *waser man*, along with the magic activities (pouring water into a bucket, inserting a "ring without a stone" into the said bucket") in: Chotsh 1703, p. 8.

archetypes which existed in the culture of Eastern European Jews at that time are a good representation of the phenomena in question. The first type is the demonic *Daytsh*, the German, who frequently assumes the form of an assimilated Jewish physician, hostile to folk culture and traditional customs. This figure of a clearly negative characterisation, sometimes identified with the devil, appeared in Yiddish stories. The second type is the *Profesor*, a representative of the medical elite, who gives hope of curing any disease. One was supposed to "travel" to such a hero of modern medicine, and renowned physicians from such centres as Vienna or Berlin were particularly respected among the representatives of the Jewish Orthodox community. At the turn of the twentieth century physicians no longer filled people with dread, and the *Daytsh* lost its significance, yielding ground to the positive image of the *Profesorn*.

The presence of German themes in the folk medicine of Jews tells us a great deal about the relationships between this community and European culture. It reveals its Central European dimension, the ignorance of which would inevitably lead to misunderstanding of the essence of the heritage of the Ashkenaz. This presence also attests to the keen interest in modern medicine on the part of those Jewish communities which supported and realised conservative attitudes. The rabbis and Hasidic leaders, even if they initially mistrusted the doctors, with the course of time became more open to the advancements of science. The latter developed particularly rapidly in German centres until the early twentieth century.

References

(1888). *Klinisches Rezept-Taschenbuch für Praktische Ärzte* [Clinical prescriptions pocket-book of practical medicine] (IX edn.). Vienna/Leipzig: Urban & Schwarzenberg.

(1988). *Piskei ve-Shelot ve-Teshuvot Maharash mi-Lublin* [Rulings and inquiries and responses of Maharash of Lublin]. New York: Hershkovich.

Alfabet, F. (1924). Materialn tsu idiomatik fun shtetl Piosk (Piaski), lubliner krayz [Sources for idioms from the town of Piaski, Lublin county]. *Jidisze Filologie, 1*, 61–72.

Bandtkie, J. S. (1827). *Nowy słownik kieszonkowy polsko-niemiecko-francuzki* [New pocket-dictionary Polish-German-French]. Breslau: Korn.

Berger, M. (1911/1912). *Imrei Israel* [Word of Israel]. Kézdiszászfalu: Meir Klausner.

Bernstein, A. (1895–1900). *Yediyot ha-Teva* [Information about nature]. Warsaw: Zaks.

Bernstein, A. (1908–1914). *Bernstein's Natur-Wisnshaftlekhe Folks-Bikher* [Bernstein's popular books on natural science], London: L. Friedman.

Bodek, M. M. (1927). *Mifalot ha-Tsadikim* [Wonders of the Righteous]. Lublin: Publisher unknown.

Chemerinski, C. (Reb Mordkhele). (1927). Ayarati Motele [My town Motele]. *Reszumot, 2*, 7–124.

Chotsh, Z. H. 1703. *Segulot we-refuot* [Remedies and medicines]. Amsterdam: Publisher unknown.

Crato, J. (1690). *Johannis Cratonis (...) ausserlesene Artzney-Künste vor alle des menschlichen Leibes Zufälle, Gebrechen und Kranckheiten* [Johannis Cratonis]. Frankfurt am Main.

Dźwigoł, R. (2004). *Polskie ludowe słownictwo mitologiczne* [Polish folk mythological vocabulary]. Kraków: Wydawnictwo Naukowe Akademii Pedagogicznej.

Efron, J. M. (2001). *Medicine and German Jews. A history*. New Haven/London: Yale University Press.

Epstein, L. (1995). *Caring for the soul's house. The Jews of Russia and health care 1860–1914*. PhD dissertation, Yale University.

Etkes, I. (2005). *The Besht: Magician, mystic, and leader*. New York: Brandeis University Press.

Ettmüller, M. (1682a). De crinonibus seu comedonibus infantium. *Acta Eruditorum, 1682*, 316–317.

Ettmüller, M. (1682b). De sironibus. *Acta Eruditorum, 1682*, 317–318.

Gay, R. (2001). *Unfinished people. Eastern European Jews encounter America*. New York/London: W.W. Norton.

Goldberg, I. Y., & Eisenberg, A. A. (1880/1881). *Sefer lachashim u-segulot ve-goralot u-mazalot* [Book of incantations, remedies, lots and constellations]. Jerusalem: Publisher unknown.

Gottesman, I. N. (2003). *Defining the Yiddish nation. The Jewish folklorists in Poland*. Detroit: Wayne State University Press.

Grimm, J. (1854). *Deutsche Mythologie* [German mythology] (Vol. 2). Göttingen: Dieterichsche Buchhandlung.

Grunwald, M. (1907). Aus Hausapotheke und Hexenküche II (Forts.) [From home pharmacy and witches' kitchen]. *Mitteilungen zur Jüdischen Volkskunde, 4*, 118–145.

Halpern, Y. (1858). *Mifalot Elokim* [Wonders of God]. Lemberg (Lviv): D.H. Schrenzel.

Heller, M. J. (2011). *The seventeenth century Hebrew book*. Leiden: Brill.

Hoffmann-Krayer, E., & Bächtold-Stäubli, H. (Eds.). (1974). *Handwörterbuch des deutschen Aberglaubens* [Dictionary of German superstions] (Vol. III). Berlin: De Gruyter.

King, H. (2014). Once upon a text: Hysteria from hippocrates. In: *Hippocrates' woman: Reading the female body in ancient Greece* (pp. 205–246). London/New York: Routledge.

Kogman, T. (2009). Baruch Lindau's *Rešit Limmudim* (1788) and its German source: A case study of the interaction between the Haskalah and German *Philanthropismus*. *Alef, 9*(2), 277–305.

Kolczyński, J. (2003). Jeszcze raz o upiorze (wampirze) i strzygoni (strzydze) [Once more on upiór (vampire) and strzygoń (striga)]. *Etnografia Polska, 47*, 211–246.

Lefin, M. M. (1851). *Refuot ha-Am* [Remedies for the people]. Lemberg (Lviv): J. Schnayder.

Lew, H. (1897). O lecznictwie i przesądach leczniczych ludu żydowskiego [On healing and medical superstitions of the Jewish people]. *Izraelita, 42*, 406–407.

Libera, Z. (1995). *Medycyna ludowa. Chłopski rozsądek czy gminna fantazja?* [Folk medicine. Common sense or communal fantasy?]. Wrocław: Wydawnictwo Uniwerystetu Wrocławskiego.

Lilientalowa, R. (1905). Wierzenia, przesądy i praktyki ludu żydowskiego [Beliefs, superstitions and practices of the Jewish people]. *Wisła, XIX*, 148–176.

Lilientalowa, R. (1921). Kult ciał niebieskich u starożytnych Hebrajczyków i szczątki tego kultu u współczesnego ludu żydowskiego [The cult of heavenly bodies among ancient Hebrews]. *Archiwum Nauk Antropologicznych, I*(6).

Lilientalowa, R. (2007). *Dziecko żydowskie* [Jewish child]. Warsaw: Midrasz.

Linetski, I. Y. (1897). *Dos khsidishe yingl* [The Hasidic lad]. Vilnius: Bletnicki.

Marcuse, M. (1790). *Sefer Refuot ha-nikra Ezer Israel* [Book of medicines named The Help for Israel]. Poryck: Meshutafim.

Matras, H. (1988). Wholeness and holiness in Sifrei Segulot. *Koroth, 9*, 96–107.

Matras, H. (1997). *Sifrei segulot ve-refuot be-ivrit – tehanim u-mekorot* [Books of remedies and medicines in Hebrew – contents and sources]. PhD thesis, Hebrew University of Jerusalem.

Moszyński, K. (1934). *Kultura ludowa Słowian. Cz. II. Kultura duchowa* [Folk culture of the Slavs. Part II. Spiritual culture]. Kraków: Polska Akademia Umiejętności.

Perles, J. (1887). Die Berner Handschrift des kleinen Aruch [The Bern manuscript of the small Aruch] In *Jubelschrift zum siebezigsten Geburtstage des Prof. Dr. H. Graetz*. Breslau: S. Schottlaender.

Petrovsky-Shtern, Y. (2004). The master of an evil name: Hillel Ba'al Shem and his *Sefer ha-Heszek*. *AJS Review, 28*, 217–248.

Petrovsky-Shtern, Y. (2011). You will find it in the pharmacy. Practical Kabbalah and natural medicine in the Polish-Lithuanian Commonwealth, 1690–1750. In G. Dynner (Ed.), *Holy dissent* (pp. 13–53). Detroit: Wayne State University Press.

Reiser, D. (2014). Idea nieuświadomionego a chasydyzm w Galicji i w Polsce [The idea of the unconscious and Hasidism in Galicia and Poland]. *Cwiszn: Żydowski Kwartalnik o Literaturze i Sztuce, 1–2*, 42–55.

Reiser, D. (2015). *Mesmerism, hypnosis and Jewish mystics in Vienna in the early twentieth century*. European Forum at the Hebrew University working Papers 139.

Rosenberg, Y. Y. (1911). *Rafael ha-Malach* [The Angel Raphael]. Pietrkov: Rosenberg.

Rosenberg, Y. Y. (c. 1920). *Segulot u-refuot* [...] *ibergezetst oyf zshargon fun heyligen seyfer Rafael ha-malakh* [Remedies and medicines... translated into Yiddish from a holy book]. Łódź: I. Szlomowicz.

Rosman, M. (1996). *Founder of Hasidism. A quest for the historical Ba'al Shem Tov*. Berkeley/Los Angeles/London: University of California Press.

Rozowski, M. (1947). *Geklibene szriftn* [Chosen writings]. Buenos Aires: Grupe Khaveyrim.

Sadowski, D. (2011). Jupitermonde und »verschlossene Gärten«: Tuvija Cohens Enzyklopädie der Naturwissenschaften und der Medizin Ma'ase Tuvija (1707) [Jupiter and "closed gardens": Tuviya Cohen's encyclopaedia of nature and medicine Ma'ase Tuviya]. *Yearbook of the Simon Dubnov Institute, 9*, 247–277.

Schipper, I. (1930). Tsu der Eltster Geshikhte fun Yidishe Doktoyrim in Poyln [To the oldest history of Jewish physicians in Poland]. *TOZ-Yediyes, 5–6*, 142–146.

Sperling, A. I. (1909). *Ta'amei ha-Minhagim oyf ivre taytsh* [Reasons of customs in Yiddish] (Vol. 2). Lemberg (Lviv): Avraham I. Sperling.

Sperling, A. I. (1924/1925). *Ta'amei ha-Minhagim* [Reasons of customs] (Vol. 2). Lemberg (Lviv): David Roth.

Stutshkov, N. (1991). *Der Oytser fun der Yidisher Shprakh* [Thesaurus of the Yiddish language]. New York: YIVO.

Tuszewicki, M. (2013). O pasożytach wywołujących suchoty. Na styku żydowskiej kultury ludowej i dawnej medycyny [About Suchote-Causting parasites. The crossroads of Jewish popular culture and old medicine]. *Kwartalnik Historii Żydów, 4*(348), 613–634.

Tuszewicki, M. (2015). Demonology at the crossroads. The presence and significance of non-Jewish beliefs within Ashkenazi Folklore. *Scripta Judaica Cracoviensia, 13*, 93–112.

Wereńko, F. (1896). Przyczynek do lecznictwa ludowego [Contribution to folk medicine]. *Materyały Antropologiczno-Archeologiczne i Etnograficzne, I*, 99–228.

Yoder, D. (1982). Folk medicine. In R. M. Dorson (Ed.), *Folklore and Folklife* (pp. 191–216). Chicago/London: The University of Chicago Press.

Zinger, N. (2009). Who knows what the cause is? Natural and unnatural causes for illness in the writings of Ba'alei Shem, doctors and patients among German Jews in XVIII c. In *The Jewish body. Corporeality, society, and identity in the renaissance and early modern period* (pp. 127–155). Leiden: Brill.

Chapter 6
Jewish Bodies and Jewish Doctors During the Cholera Years of the Polish Kingdom

Katharina Kreuder-Sonnen

Abstract Epidemics challenge the social, ethnic, cultural or national cohesion of a society. When cholera reached the Polish Kingdom in 1892 medical debate about the disease produced a specific understanding of Jewishness. The category was not only considered to be a religious one but a socio-economic one as well. Furthermore, first ideas about a different Jewish biology emerged. However, medical 'othering' during the cholera epidemic concerned unprivileged Jews only. Jewish doctors, for instance, were not considered to be members of a 'different' group. The paper will trace this tension between Jewish 'othering' and religious/national/ethnic indifference. It will show that in the beginning of the twentieth century this indifference became more and more contested. It ended when Jewish members were excluded from Warsaw's Medical Society in 1907.

Keywords Cholera · Jewishness · Othering · National indifference

In 1892 cholera was ravaging Europe[1]. From Baku to Hamburg thousands of people died of the disease. The epidemic also spread to the Polish Kingdom, the Russian partition of the former Polish-Lithuanian Commonwealth. The first case occurred in August in the small town of Biskupice near Lublin. From there, cholera came to Lublin and then to the smaller cities of the region. Other than expected Warsaw remained mostly unaffected (Bujwid 1892, 742; Bujwid 1894, 812; Biernacki 1892).

Cholera outbreaks and other epidemics have been analyzed as sources of growing social tension as well as of efforts to create social order. A prominent example is Michel Foucault's description of the "plague town" in which he shows how

---

This contribution is based on my article "Grenzen ziehen und überschreiten. Ärzte und das Jüdische im Königreich Polen während der Choleraepidemie 1892/93". *Zeitschrift für Ostmitteleuropa-Forschung* 64 (2015), 330–355.

[1] I am grateful to Mikołaj Golubiewski for providing a first draft of an English translation of the text.

K. Kreuder-Sonnen (✉)
Department of Contemporary European History, University of Siegen, Siegen, Germany
e-mail: kreuder-sonnen@geschichte.uni-siegen.de

M. Moskalewicz et al. (eds.), *Jewish Medicine and Healthcare in Central Eastern Europe*, Religion, Spirituality and Health: A Social Scientific Approach 3,
https://doi.org/10.1007/978-3-319-92480-9_6

the epidemic spurred the development of techniques and practices which laid the foundation for a disciplinary society (Foucault 1976, 251–254). Studies of individual cholera outbreaks have interpreted them as historical moments in which the social, ethnic, or national cohesion of societies has been tested, and as situations in which communities disintegrated along these lines (Briggs 1961; Evans 1990; Stolberg 1994; Arnold 2002; Sahadeo 2005; Henze 2011). As Christopher Hamlin put it: "Since the mid-nineteenth century, cholera has been an axis of difference, a means for distinguishing places and races as clean or dirty, and a vehicle for despising others" (2009, 56).

In what follows, I will discuss if and how the 1892/93 cholera epidemic in the Polish Kingdom inspired the formation of boundary lines between Jews and Non-Jews. I will focus specifically on discourses and practices of the medical community. I will show that medical publications on cholera created distinct boundaries between Jews and Non-Jews. Jews were described as an alien and pathogenic group within Polish society. Their being 'other' was defined mostly by social and cultural character traits. Only individual authors defined Jewishness in biological terms. However, a social and cultural understanding of Jewishness at the beginning of the 1890s also implied a national 'othering' from non-Jewish Poles.

In a second step, I will analyze the daily work of Jewish and non-Jewish physicians in the Polish Kingdom. I will thus show that at the end of the nineteenth century the socially, culturally and nationally different Jew was still an unstable and malleable category for the Polish medical community. The socio-cultural differentiation of Jews in the cholera texts did not encompass Jewish doctors. Within the medical profession it was the intellectual attitude which determined whether you were in- or excluded: If Jewish doctors subdued to the prevailing medical thought collective, their religion, culture, nationality, or even race became invisible in daily work and medical debate. Such an approach to Jewishness can be described with Tara Zahra as "indifference" (2010; Zahra and Judson 2012).

The 1892/1893 cholera epidemic occurred in a period which has been described as one of growing tensions between Jewish and Christian Poles. The emergence of ethno-nationalism, anti-Semitism, and racism made it increasingly hard to overcome the boundaries between the two groups (Polonsky 2010, 96, 112; Weeks 2006, 86; Michlic 2006, 51). Medical writing about the cholera epidemic allows for a precise analysis of the formation of these boundaries. My analysis reveals that the process of differentiation was heterogeneous and fractured. Even during a late nineteenth century epidemic when medico-political difference production was at its height there (still) existed the possibility that people did not act according to national or racist boundary lines.

6.1 Drawing Boundaries

The small town of Siemiatycze, today located in Podlaskie Voivodship in the Polish East, was one of the Polish Kingdom's cities which was severely affected by cholera. The disease broke out in August 1893. Among the 6500 inhabitants there were at least 379 who fell ill and 192 who died from the disease (Czarkowski 1894, 114). Siematycze's physician, Ludwik Czarkowski, treated a considerable part of them and reported extensively about his experiences in the medical journal *Medycyna*. Starting with his article, I will examine the medical discussion about the disease and analyze if and how it produced social, cultural, national, and biological dimensions of Jewishness.

Czarkowski's report on Siemiatycze was part of an intensive medical debate on cholera in the 1890s. The questions of how people got infected, how the disease spread, what preventive measures were effective and how to cure cholera were still unresolved.[2] Publications of practicing physicians about how the epidemic had proceeded in their towns, which sanitary measures they had taken and which medication they had applied were all contributing to this open field of research. Czarkowski's article can be considered one of them.

His article starts as follows:

> The small town of Siemiatycze lies partly on a clayey-sandy hill, but mainly on its two hillsides, of which one runs down to the small river Kamionka, that cuts the city in two uneven parts from the east; the other, gentler hillside in the North runs down towards a small brook that joins Kamionka River in the town. A bath is situated at the border of the brook and a mikveh is attached to it. Dirty water from the bath runs directly into the brook. Four lavatories are also situated at the brook's bank... In the town's center, the cloth factory, employing 40 people, has created a pond from Kamionka River.... The pond serves all possible purposes: they water cattle in it, bathe horses, wash linen and woolen yarn; a tannery situated close to the factory throws all its waste into the pond; finally, four Jews take water from this pond each day and deliver it for domestic use... Siemiatycze counts 6500 inhabitants, including 4000 Jews; the population lives in 480 houses, mostly wooden and thus in darkness and dirt... The town was once a wealthy trade town but since the fire in 1863 it has declined steadily and the population impoverished. 2/3 of the Jews are paupers and do not know if they will live to see the next day.
>
> Cholera appeared in our town on the 20th of August last year, at once in 3 locations: at the bath on the brook, in two Jewish streets, and in the suburb of Łojka, inhabited only by Christians (Czarkowski 1894, 113–114).

Czarkowski's introductory paragraphs reveal a whole range of factors that he considers epidemiologically relevant and therefore noteworthy: Siemiatycze's topography and soil structure, its watercourses, the town's economic structure, and the living conditions of its inhabitants. His approach to the epidemic is thus a multifactorial one. In contrast to strict bacteriologists of the time, Czarkowski did not

[2] In 1884 on an expedition to Egypt and India Robert Koch had identified the causative agent of cholera and the "comma bacillus" became a broadly recognized "scientific fact". The role of the microorganism in the course of the disease, the way it spread as well as its relationship to climatic-geographical conditions or individual constitution remained open to debate, however (cf. Harrison 1996; Worboys 2007).

reduce the epidemic to the size of a "comma bacillus". Instead, he presented a variety of possibilities that may have contributed to the development and spread of the disease.[3]

Czarkowski also includes the religious affiliation of Siemiatycze's inhabitants into his list of possible epidemiological factors. The fact that there were Jews and Christians living in Siemiatycze seems to be essential information to him. And already in these first paragraphs he gives some insight into why this is the case. Jews and Jewish ritual infrastructure are situated within a context clearly related to the emergence of cholera. He mentions the mikveh from which dirty water flows through the Kamionka right into the city. He also explicitly states the Jewish religion of those who feed water from the dirty pond into the town's drinking water supply. Furthermore, he writes about the precarious living conditions of the Jewish population and he points out that two of the places where cholera broke out were Jewish (the ritual bath and two streets). Thus, from the very beginning Czarkowski's cholera narrative suggests that Jews play an important role among the different epidemiological factors. He also underlies this specific role by numbers. He states that 334 diseased out of 379 were Jews, only 45 were Christians. In relation to the overall number of inhabitants, this resulted in a 4.5 times higher cholera morbidity among Jews compared to Christians. The cholera mortality rate was even seven times higher among the Jewish population (ibid. 1894, 114).[4]

However, the author does not find the higher Jewish morbidity and mortality rates to be self-evident. Rather, in the remainder of his text he relates this finding to the numerous other epidemiological factors and reasons about their interrelation. In doing so he connects Jews to specific characteristics.

The strongest link Czarkowski creates between Jewishness and another epidemiological factor is between Jews and poverty. It is the poorest part of the population which suffers from cholera and this part is mostly Jewish (Czarkowski 1894, 115). By focusing on such social aspects, Czarkowski follows the common nineteenth century idea that cholera is the disease of the poor (Briggs 1961; Evans 1990, 539; Stolberg 1994; Hamlin, 87).

A second epidemiological factor Czarkowski examines is the profession of the diseased. He states that cholera mostly appeared in specific occupational groups: tailors, shoemakers, grocers, lump collectors, glaziers, butchers, and dyers. Most of these jobs, he states, are carried out by Jews (Czarkowski 1894, 115, Ref. 3). Thus Jews stood out, not only by being exceptionally poor but also by specific occupations and both seemed to make them susceptible to cholera.

However, the high Jewish cholera morbidity and mortality rate could not be explained by poverty and profession alone. To his astonishment, Czarkowski finds that these two characteristics of Jewishness do not necessarily go hand in hand. The many Jewish butchers who contracted the disease could not be regarded as poor.

[3] Regarding the difference between an open but powerless hygiene and specific yet strong bacteriology, see Latour 1988, Ch. 1.

[4] Eastern European Jews had been regarded to play a special if nuanced role already in the cholera epidemic of 1830/1831 (Wolff 2000).

They had access to meat and therefore enjoyed a specifically rich diet. On the other hand, there were very poor Jewish occupational groups such as Siemiatycze's gardeners who did not come down with cholera at all (ibid.).

The diverse epidemiological possibilities competed in yet another way: The topography of Siemiatycze also thwarted an uncomplicated link between poverty, Jewishness and cholera. Czarkowski notes that Siemiatycze's poorest Jews lived in the higher parts of the town, far away from the river and the pond. Despite serious poverty, there were no cases of cholera among the Jews from this area (ibid. 114).

The link between Jewishness, poverty, and cholera, however, seems so strong to Czarkowski that he regards a deviation from this rule as remarkable. In the conclusion of his article, he therefore deduces particularly significant epidemiological factors from this observation. He states that topography plays an important role in the spread of the disease (lower areas close to watercourses are more susceptible than higher areas). Furthermore, he establishes the susceptibility of specific occupational groups to cholera: butchers and dyers were at high risk while being a gardener protected you from the disease (ibid. 137).

This confusing mixture of pathogenic factors meant that the relationship between Jewishness and cholera could not be clearly defined. On the one hand, the triad 'poor – Jewish – infected' was *ex negativo* strengthened by Czarkowski's report. Only by taking it for granted he could derive further decisive epidemiological factors from it. On the other hand, it were just these factors which thwarted the linear connection between cholera, poverty and Jewishness. Occupation and topography also played a role. What was more, Czarkowski had to state that the majority of Christian sick were also particularly poor (ibid. 115).

So in all this messiness, how could Czarkowski explain the higher Jewish cholera morbidity and mortality? In his final remarks, he comes to the following conclusion: "The Jews appear to be of a disposition [*usposobienie*] which makes them more susceptible to the disease. Therefore, the course of the disease is usually more severe among them which can be seen in the higher mortality rate" (ibid. 137). By assuming a Jewish disposition Czarkowski introduces a category which alludes to a biological dimension of Jewishness. As the social understanding of Jewishness no longer helps to draw a boundary line he resorts to the vague concept of *usposobienie*. The term denotes a person's 'nature' but also his character traits. Czarkowski does not use a term which naturalizes in a more explicit way, such as 'physiology', 'biology', or simply 'race'. Nevertheless, with the notion of "disposition" he creates a residuum of the Jewish Other which is clearly not social and which implies a different physicality. Precisely because of the vagueness of the term it created a Jewish difference which was hard to circumvent.

Thus, Czarkowski's cholera narrative developed the category of Jewishness as one which was first of all socially defined: Jews were poor and therefore susceptible to the disease. This social categorization, however, was challenged by the factors 'occupation' and 'topography'. Therefore, Czarkowski adds the Jewish disposition to his list of differentiating dimensions explaining Jewish cholera morbidity and mortality. He thus insinuates a biological definition of Jewishness.

The idea of a Jewish *usposobienie* to cholera appears in no other medical text about the epidemic in the Polish Kingdom. Nonetheless, Jewishness played a crucial role here as well. It was the social categorization of Jewishness which was most prominent. The triad 'poor – Jewish – infected' became a particularly popular motif in cholera narratives.

The Warsaw physician and bacteriologist Odo Bujwid, for instance, explained the first cholera outbreak in the Polish Kingdom in September 1892 in Biskupice near Lublin as follows: An infected traveler from Rostov-on-Don had brought the disease to Biskupice. Then, the district where "many poor Jews" lived had been the breeding ground for the epidemic. Further cases among Biskupice's Jewish residents are never mentioned without the attribute "poor" (Bujwid 1892). The metaphor of the poor Jewish district as "a good breeding ground for the development of cholera" also appears in Edmund Biernacki's description of cholera in Lublin (Biernacki 1892, 838). A Jewish woman and her sick child had brought cholera to the city in the end of September 1892. It then spread among the "dark [*ciemne*] and poor" Jewish population. At first, the Christian population had therefore been unconcerned and thought cholera to be a purely Jewish issue. However, the epidemic did come to the Christian districts too. But it mainly spread in areas which had similarly poor sanitary conditions as the Jewish quarter. Concluding, Biernacki states that cholera afflicted especially such areas which were densely populated, unhygienic, with dirty streets, polluted drinking water, and unsanitary lavatories (ibid. 840–41). Hence, also in Biernacki's writing Jewishness is primarily a social and habitual category and not a biological one. Yet, he explains the attribute of poverty in even more detail: Jews were not only poor but they also lived an unhygienic life which testified to their civilizational "backwardness" (ibid. 838). This coupling of poverty, poor hygienic standards and a certain level of civilization is present also in Czarkowski's article when he describes the wooden houses of Siemiatycze's poor as dirty and "cramped" or when he depicts how Jews take water from the dirty pond.

With their socio-cultural categorization of Jewishness in the context of cholera, Polish doctors joined into a broader European discourse on the disease. It interlinked the conceptual fields of 'cholera', 'orient', and 'Jews'. Between 1840 and 1870 in British, French, and German discourse the disease had become "Asiatic Cholera" whose origin was situated in Bengal (Hamlin 2009, 35–48). In contrast to the mild form of *cholera nostras* the disease had thus become the antithesis of Western civilization (ibid. 57–78).

As a phenomenon of Asian anti-modernity European discourse had connected cholera to the socio-cultural categorization of Jewishness. Representations of traditional Jewry as the counter-image to modernity originated in European Enlightenment and were often closely linked to the construction of women as the Other of modernity (Braun 1992; Hüchtker 2002; Hödl 1997). In the nineteenth century, the linking of Jewishness and anti-modernity was increasingly ascribed to Eastern Jewry. The German *Kaiserreich* therefore introduced a strict hygienic regime at its Eastern border which Eastern European emigrants had to endure. What was more, also during the 1892 cholera epidemic in Hamburg and Paris Jews were blamed for spreading the disease (Weindling 2007, 366–7; Hödl 1997, 64).

However, European cholera discourses did not encompass a biological categorization of Jewishness – even though the creation of a 'Jewish body' imagined as diseased and as spreading disease had begun already in the nineteenth century (Gilman 1995). Yet this discourse fully evolved only after 1900. Furthermore, infectious diseases did not play an important role in constructing the Jewish 'race'. While a supposedly strong Jewish immunity against tuberculosis raised some interest, the main focus of the debates was on anthropometry and mental diseases (Lipphardt 2008, 123).

Thus, a Jewish 'disposition' which did not only refer to socio-cultural dimensions but also to a specific Jewish physique was not broadly discussed in medical debate about cholera in 1892/94, neither in the Polish Kingdom nor in other countries. Yet, Czarkowski's idea of a Jewish *usposobienie* shows that a vaguely defined physical difference of Jews could already be thought and talked about in the 1890s.

Even if a socio-cultural framing of Jewishness created more malleable boundaries than a biologistic one, it still had far-reaching differentiating functions within Polish society. As I will show in the following section, during the last third of the nineteenth century hygiene, modernity and civilization had become central values of the Polish nation.

6.2 Socio-cultural Jewishness and the Polish Nation

Czarkowski, Bujwid, and Biernacki differentiated socio-cultural Jewishness from 'Christians'—not from Poles. On a linguistic level the authors thus remain within religious categories. Yet, during the last decade of the nineteenth century socio-cultural Jewishness could be easily understood as a national category as well. Already in the 1860s and 70s Polish positivists 'sociologized' the concept of Polish nationality. A 'nation' encompassed a homogeneous unity tied together by a common language as well as common cultural and social norms. The positivists found that the Polish one needed severe modernizing efforts. The Polish nation was thus defined as a working group thriving towards progress (Porter 2000, Ch. 2). Such a group necessarily excluded those who did not subdue to the cultural and material values of rational modernity. For the positivists of the 1860s and 70s these were the 'uneducated masses' of Jewish and Christian faith alike and both groups had to be treated the same. Education and enlightenment would merge "dark" and uncivilized Polish peasants as much as traditional Jews within a modern Polish nation. Such were the ideas of the liberal (Christian and Jewish) proponents of Jewish 'assimilation' into the hegemonic Polish culture which was supposed to be Western and civilized (Weeks 2006, 58; Cała 1989, 218). However, Brian Porter points out that – despite all positivist religious tolerance – the disciplining dimension of the 'social nation' already laid the foundations for chauvinist and radical Polish nationalism (Porter 2000, 44).

The medical descriptions of Jews as poor and unhygienic during the cholera epidemic of 1892/93 reveal a similarity to this nationality discourse. In both cases

Otherness is created by the concept of backwardness. Hence, socio-cultural Jewishness in the cholera texts can also be understood as national Othering.

This becomes even more plausible when we consider the nationally loaded hygienic discourse in the Polish Kingdom during the last third of the nineteenth century – a discourse which was prominently framed by leading members of the Warsaw positivists.[5] As in many European countries the context of industrialization and urbanization had brought about a Polish crises discourse which deplored the pauperization of the urban population and predicted the degeneration of Poles by disease – especially by alcoholism and venereal diseases (Gawin 2003, 52–61).

The promotion of public and individual hygiene was thus transformed into a strategy of national self-assertion. This happened in a twofold sense: Health, hygiene, and cleanliness were regarded to be an expression of civilization and modernity which, according to the positivists, were to make a strong nation. What was more, striving for public health was also linked to resisting Tsarist rule. After the January Uprising of 1863 St. Petersburg withdrew public health matters from the realms of Polish self-administration. They were now to be the competence of the Tsarist Ministry of the Interior which, however, remained largely inactive in the field.[6] Many Poles in the Kingdom regarded this reluctance towards public health policies as part of anti-Polish Tsarist politics, especially when they compared their experiences with the social legislation of the German Reich and the Habsburg Monarchy in the two other Polish partitions.

In the Kingdom, public health became the playing ground for local politics and philanthropists. In Warsaw, for instance, in the mid-1880s the local (Tsarist) administration and Polish hygienists managed to erect a sewage system and introduced water filters. Industrials and wealthy nobles funded hospitals, milk stations, sanatoria, recreational parks, etc. The great majority of funds spent on public health matters in the Kingdom came from municipal and private budgets (Hanecki 1968; Więckowska 1984; Więckowska 1985; Caumanns 2006). The Russian administration did perceive these activities as an expression of Polish national self-assertion. This was especially true for the formation of a hygiene movement in Warsaw around the physician Józef Polak. The movement had managed to publish its own journal *Zdrowie* (Health) since 1885 and it had successfully organized hygiene exhibitions in Warsaw. Yet, it was only in 1898 that the Russian authorities allowed for the creation of a proper hygiene society. It had to call itself Warsaw Hygiene Society. "Polish" had not been permitted as a title (Caumanns 2000, 52).

Thus, in the 1890s the Polish nation defined itself also by an idea of hygienic modernity. The Jews which the cholera texts described as sick and unhygienic, providing a "good breeding ground" for the epidemic endangered this Polish project of

[5] One of the outstanding advocates of positivism was the author of *Kroniki* and the famous novel *Lalka*, Bolesław Prus, who repeatedly made individual and public health the subject of discussion. The same was true for Adolf Dygasiński, renowned for his *Nowe tajemnice Warszawy* (cf. Gawin 2003, 58).

[6] In Russia itself, *zemstva* were actively engaged in this field (Hutchinson 1990; Strobel 2013).

self-preservation through modernization and could hence be understood as a nationally alien element.[7]

The socio-cultural and national boundary line between Poles and Jews was not drawn in every social realm, however. Within the professional group of physicians the cooperation between Jewish and Christian doctors continued unperturbed during the epidemic. In the 1890s male physicians[8] still shared a space defined by professional ethics and intellectual values which left little room for socio-cultural, national or even biological differentiation among them. The socio-cultural Jewishness created by cholera narratives dissolved to a certain degree when Jews and Christians came together in the *common* socio-cultural space of a medical elite. In the last section of this article I will show how the Jewish Other was negotiated, practiced – or ignored within Warsaw medical circles.

6.3 Crossing Borders – Christian and Jewish Physicians in the Late Nineteenth Century Polish Kingdom

Medical writing about cholera had linked Jewishness to poverty and hygienic backwardness and had thus made it the national Other. How then did Christian physicians behave towards their Jewish colleagues? Were they included into the Jewish Other or was the category of Jewishness open enough to allow for different dimensions of being Jewish? In order to answer these questions let me first shed some light on the relation between Jewish and non-Jewish doctors in a broader time frame.

Jewish doctors and medical students were an integral part of the medical community in the Polish Kingdom. The medical profession was one of the few possibilities for Jews in the Tsarist Empire to climb the social ladder. Being a physician they could live wherever they wanted, not only in the pale of settlement. It permitted the acquisition of land, at least to some extent and throughout the nineteenth century it promised increasing social prestige. For Jewish women, becoming a physician meant financial independence from their parents or husbands (Epstein 1995, 90–91). Thus, in 1878 58.2% of all matriculated Jewish students in the Tsarist Empire studied medicine. Their share among all medical students was highest in Warsaw. Between 1878 and 1880 it amounted to 17.3% (ibid. 63). After the 1881 pogroms possibilities for Jewish doctors to work in Tsarist public office became restricted. Furthermore, a quota of 5% was introduced to Jewish physicians

[7] The broader medical debate about hygienic modernization also saw the poor, sick, and unhygienic Polish Christians as a group who had to familiarize itself with the ideal of modern Polish hygiene. In the cholera accounts, however, the Christian population did not play a prominent role.

[8] In the time period examined in this paper medical studies were accessible to women only in Zürich between 1872 and 1882. From 1891 onwards women could study medicine in St. Petersburg. Jewish female students, however, received a limited admission quota in St. Petersburg only in 1897.

serving in the military. Additionally, like Christian-Polish physicians they were only appointed to work in Siberia or Turkestan (ibid. 65). In Warsaw, however, employment opportunities came from university clinics, communal and Jewish hospitals as well as private clinics. Moreover, there was always the possibility of opening a private practice.

Warsaw Jewish doctors did not only work in Jewish hospitals or care facilities of the Jewish community. On the contrary, they were also active in university and communal Catholic hospitals. After graduation many Jewish students found employment as medical assistants in university clinics or as research assistants of former professors. Thus, religion was of little importance in the occupational structure of physicians in the Polish Kingdom. Jewish and non-Jewish doctors went through similar career stages and we may assume frequent and manifold contact between the two groups.

This applies to the area of professional associations as well. Jewish physicians did not form any separate medical organization. Instead, they were members of the Warsaw Physicians' Society (Towarzystwo Lekarskie Warszawskie, TLW). There are at least 23 Jewish members to be found on the Society's membership lists of the last 20 years of the nineteenth century. In total, the TLW had between 86 and 155 members in this time period.[9] Many of those Jewish members exemplify the work of Jewish doctors in non-Jewish medical institutions: Albert Rosental (1857–1921), after having studied at Warsaw University, worked at the Holy Jan Boży Hospital for Psychiatry (Szarejko 1995, 335–36). Antoni Elsenberg (1852–1910) (ibid. 130–31) and Ludwik Wolberg (1857–1904) (ibid. 516–17) graduated from Warsaw and Dorpat University respectively and were then employed at the Department of Pathological Anatomy with Professor Włodzimierz Brodowski. Rafał Aleksander Feilchenfeld (1858–1932) (ibid. 127–28) worked at the Hospital of the Holy Spirit in Warsaw, while Wolf Władysław Freudenson (1849–1910) practiced in the clinics of the Warsaw professors Tytus Chałubiński and Karl Lambl (ibid. 143).

An outstanding example of religious indifference in the medical milieu is Jakub Szwajcer (1851–1910). He worked at the Warsaw Emergency Hospital for Infectious Diseases (*Szpital Zapasowy*) from 1887 to 1897 and then became its director. In 1902 he was appointed head of the new Jewish hospital in the Warsaw district of Czyste. Szwajcer was also a very active member of the TLW. He assumed the role of the treasurer and, from 1916 onwards, became member of the society's directorate. In 1924, the TLW made him an honorary member. Szwajcer's esteem and the special role he played in the TWL are certainly related to his extraordinary commitment to public health questions. As a director of *Szpital Zapasowy* he enlarged it by four so-called "pavilions". While he was at the Jewish hospital the number of beds

[9] The lists are available in each yearbook of the *Pamiętnik Towarzystwa Lekarskiego Warszawskiego*. For this article, I have examined all the names suggesting a Jewish background in biographical reference works for Polish doctors. Religion is never explicitly mentioned there. Whenever there was a hint about the Jewish faith of a doctor (such as his activity in a Jewish hospital, involvement in the Jewish community, or a burial on a Jewish cemetery), I have assumed a Jewish background.

also increased considerably (ibid. 367–68). Ludwik Natanson (1822–1896) was a public health activist from an earlier generation. He also served as the head of the Warsaw Jewish community. Between 1863 and 1865 he was president of the TLW (Kośmiński 1888, 341–43). In all of these cases, Jews were not the Other of a nationalized hygienic discourse but active participants of the medical modernization efforts in the Polish Kingdom.

A condition for not making Jewishness an issue within the medical profession was that Jewish doctors joined the majority thought collective. This was of course heterogeneous. Yet, during the second half of the nineteenth century medicine in the Polish Kingdom, as in other parts of Europe, was defined more and more as a science and tied to university education. By studying medicine Jewish doctors partly turned their backs on traditional Jewry. Their openness towards secular forms of knowledge meant their approval of the project of Jewish enlightenment, the Haskalah. Hence, they became part of the endeavor to integrate Jews into the bourgeois middle class and confined Jewry to a solely religious practice. For some, pursuing a medical degree resulted in a total rupture with their orthodox families or with the traditional Jewish world altogether (Epstein 50–51; for the German context, see Wolff 2014).

Next to their commitment to the hygiene movement, a study trip abroad was another element integrating Jews into the usual course of medical academic training and the medical thought collective.[10] After the January Uprising of 1863, the Tsarist government reduced medical training at Warsaw University to a bare minimum. Large parts of Polish students therefore completed their medical studies outside the Polish Kingdom. Estimates assume that during the entire nineteenth century half of all Polish medical students from the three partitions studied abroad (Brzeziński 1991, 84; Nieznanowska 2008, 132).[11] Even though these numbers are insecure and the definition of 'Polish' and 'abroad' sometimes questionable we can state that there existed a culture of academic travel among students in the Polish Kingdom in the nineteenth century (Konopka et al. 1987, 387–88). In order to get a good medical education it was not necessarily requested to complete your entire university degree abroad. You could also go on a shorter trip to study with a specific professor or to learn in a certain laboratory (Kreuder-Sonnen 2012).

Such diverse scientific and cultural interrelations made Warsaw an important location in the transnational European space of knowledge.[12] Contemporaries

[10] For the concept of the thought collective in medicine and its proximity to ideas of a cultural or tribal community, see Fleck 1980, 54–65.

[11] Scholars usually assume those doctors to be 'Polish' who are listed in the biographical reference works by Franciszek Giedroyć and Stanisław Kośmiński or in the medical bibliography of the nineteenth century by Stanisław Konopka. 'Abroad' also encompasses universities within the realms of the empires the Polish partitions belonged to themselves.

[12] Ruth Leiserowitz is conducting a research project about the transnationality of Warsaw students: http://www.dhi.waw.pl/de/forschung/forschungsprogramm/nationale-identitaet-und-transnationale-verflechtung.html#c223 (17.11.2016). On the transnational biographies of Polish scientists and their meaning for Poland in the interwar period, see the works of Katrin Steffen: Steffen 2008, 2013.

therefore regarded the medical thought collective in the Polish Kingdom as a mixture of diverse national, regional, and local forms of knowledge. The Warsaw histologist Henryk Hoyer, for instance, having studied in Breslau and Berlin, characterized Warsaw medicine around the middle of the century as a blend of different medical schools:

> [Warsaw doctors] received their education at different moments in time and from different schools: Some came from the first University of Warsaw, others from Wilno, Kraków, Petersburg, Moscow, Dorpat (now Jurjewo), Paris, and different German universities. The youngest generation, though still considerably small, was already familiar with the latest trends in medicine, but the majority remained attached to the more or less outdated views (Hoyer 1903, 834).

Jewish medical students and doctors were part of this lively border-crossing transfer of knowledge to and within Warsaw. Almost every Jewish member of the TLW had gone on a longer academic trip or had studied a couple of semesters abroad: Maksymilian Flaum (1864–1933) studied in Warsaw, Berlin, and Munich (Szarejko 1995, 181–82), Samuel Grossglik (1860–1931) visited Berlin after completing his medical education in Warsaw (ibid. 87–89), Hipolit Oberfeld (1860–1914) went to Bern (ibid. 210–11), Albert Rosental traveled to Vienna, Munich, Zürich, Basel, Achern, Heidelberg, Bonn, and Paris to study in different psychiatric clinics (ibid. 335–336), Antoni Elsenberg went on academic trips to Paris, Vienna, and Berlin and attended a bacteriology course with Robert Koch (ibid., 131), Zygmunt Kramsztyk (1849–1920) studied in Warsaw, Vienna, and Berlin (Löwy 1990, 121).

Jewish doctors thus clearly contributed to the transnational context of Polish medical knowledge production. Their individual contributions were just as heterogeneous as the entire landscape of medical knowledge in the Kingdom. Elsenberg for instance propagated laboratory medicine in Warsaw. He thus advanced the very idea of modern scientific medicine which wanted to transfer medical practice from the patient's bedside to the laboratory glass tube. In 1886, two years after being named head of the department for skin and venereal diseases at the Jewish Hospital in Warsaw he opened an anatomical-pathological laboratory in his clinic. After returning from Koch's bacteriology course in Berlin he also established a bacteriological laboratory and organized bacteriology courses for colleagues (Szarejko 1995, 130–31). On the other hand, Kramsztyk was a prominent representative of those skeptical towards laboratory medicine. To him, medicine meant first and foremost the art of healing and not a science (Kramsztyk 1895; for a broader discussion of the topic see Sturdy 2011). In order to promote this program he founded the journal *Krytyka Lekarska* (Medical Critique) in 1897 which created a forum for debates in medical philosophy and a publishing home for the so called "Polish school of medical philosophy" (Löwy 1990).

Jewish voices in Polish medical debate were thus as diverse as Christian ones. There was no *Jewish* position in medical discourse in the Polish Kingdom at the end of the nineteenth century. I was able to identify but one article in the TLW yearbook which addressed an explicitly Jewish contribution to medicine: In "Medycyna w Talmudzie" (Medicine in the Talmud) Adolf Grünbaum (1849–1906) explained how certain ritual-hygienic prescriptions of the Talmud generated medical and anatomical knowledge which was still valid (Grünbaum 1884).[13] Grünbaum thus juxtaposed the link 'Jews – poverty – illness' with the image of a specifically hygienic and medically enlightened Jewry. Here, the Jewish religion becomes a catalyst of medical progress. Grünbaum's attempt to ascribe Jewry a special role in the development of modern medicine remained a lonely one, however. There are no comparable articles in other Polish medical journals. Jewish doctors integrated into the medical thought collective of the Polish kingdom without reference to their Jewishness.

It is therefore hardly surprising that the 1884 TLW "Guidelines on the Physician's Rights and Duties" do not contain a single remark about their members' faith. The Guidelines constituted a professional code and defined who was allowed to practice as a proper member of the medical community. These were those men (and rarely women)[14] who were holding a medical degree from university. Excluded from the community were feldshers, healers and "medical sects". The categories 'Jewish' and 'Christian' do not appear in the code (Zasady 1884, 219; Caumanns 2006, 370).

Hence, within the realms of 'modern' and academic medicine Christian and Jewish male doctors in the Polish Kingdom closely cooperated. It was here – among others – where the integrative ideal of enlightened Jewry confined solely to religion was realized. Working in a Jewish hospital was naturally a clear sign of belonging to the Warsaw Jewish community. Yet, medical ideas and practice in these institutions did not differ from Christian ones.

The 1892/1893 cholera epidemic did not sharpen the blurry boundaries. Poor and dirty Jewishness which offered a "breeding ground" for cholera microbes in urban Jewish districts did not encompass Jewish physicians. They were part of the Polish 'social nation' and had abandoned their Jewish Otherness. As academically trained doctors they remained a natural part of the Warsaw medical community, even in 1892/93. So, physicians at Jewish hospitals closely cooperated with their Christian colleagues. Edmund Biernacki, for instance, had reported about cholera in Lublin as a specifically Jewish problem. But he visited the city's Jewish hospitals to exchange information with the local physicians (Biernacki 1892, 843). In Biernacki's perspective, Jewish doctors were not part of the hygienically other Jewry. The bacteriologist Odo Bujwid received stool samples from Jewish colleagues to test them for the cholera bacillus (Bujwid 1892). Exchange across Jewish-Christian border lines continued here as well. Jakub Szwajcer published on one of the few cases of cholera in Warsaw which he treated at *Szpital Zapasowy*. When reporting about his close collaboration with his colleagues from Warsaw University, Przewoski and Janowski, he never calls them Catholic. Together with Janowski Szwajcer had cultivated the

[13] The idea that traditional Jewry already encompassed key elements of modern medicine and hygiene was also conveyed by the Galician social hygienist and Zionist Alfred Nossig (Hart 1995).

[14] The extent to which the TLW excluded women is especially telling in the 1878 case of Anna Romaszewicz-Dobrska who held a medical degree from the University of Zurich and whose membership application to the TLW was rejected (Caumanns 2006, 381). The first female TLW member was Elżbieta Downarowicz who appears on the yearbook's membership list as late as 1896.

cholera bacillus from "his" case and proudly reports that Janowski had been able to present the culture at a TLW meeting (Szwajcer 1892, 661, 667; Janowski 1892).

In 1892–1893, we can thus conclude, Jewishness was not yet an all-encompassing and stable category. Not every Jew was sorted into a specific grid of social, cultural, habitual or biological characteristics. Rather, by adapting to the values of the Polish 'social nation' Jewish doctors had the possibility to reduce their Jewishness to religious realms and to merge into a professional and intellectual community with their Christian colleagues.[15] For the large part of the Jewish population which was disregarded for their backwardness this meant at least an opportunity to discard their socio-cultural ascription. Yet, this possibility was of course bound to the subordination under a hegemonic idea of Polish culture. Since ideas of a distinct Jewish biology remained isolated Jewishness remained malleable. The border line of the Polish social nation could still be transgressed if you were willing to undergo cultural and social transformation.

This changed at the turn of the twentieth century. Jewishness became a much more stable and also a biological category which ascribed Jews specific physiological and psychological characteristics (Lipphardt 2008). European anti-Semitism helped produce these ideas and used them as a basis for its politics. In the Polish Kingdom emerged a political landscape with parties dividing also along Christian-Jewish lines. The campaigns for the Duma elections in 1906 and 1907 provided a first opportunity for an open confrontation of the different political groups and bore clearly anti-Jewish tones. The National Democrats perceived the Jewish candidates as a serious threat to the representation of Polish national interests in St. Petersburg (Trees 2007, 289–293).

In the Warsaw medical community the progressing separation and biologization of Jewishness put an end to cooperative practice and research – at least on an organizational level. In 1907 the TLW decided not to accept any more Jewish members. Therefore, Zygmunt Kramsztyk left the Society in protest (Löwy 1990, 132; for other societies, see Guesnet 1998, 84).

However, even after 1907, there was no total rupture between Christian and Jewish medical colleagues. Jakub Szwajcer, for instance, was admitted to the directorate of the TLW in 1916 and appointed honorary member in 1924. Many Jewish doctors were active in the Second Polish Republic, among others at the National Institute of Hygiene. Yet, in the interwar period medical associations clearly divided along Jewish – Christian lines. In 1921, the Society for the Protection of Jewish Health (TOZ) was formed. It established its own health care facilities and held a series of congresses aimed only at Jews, thus clearly separating itself from the Christian-Polish realm of medicine and public health.

Translated by Mikołaj Golubiewski

[15] This is by no means a plea to regard the sciences as an objective and universal space which would automatically unite diverse groups. Sciences are always laden with cultural, social, political etc. interest. At the end of the nineteenth Century in Central Eastern Europe they were closely intertwined with projects of nationalization (Ash, Surman 2012).

References

Arnold, D. (2002). *Colonizing the body: State medicine and epidemic disease in nineteenth-century India*. Berkeley: University of California Press.

Ash, M. G., & Surman, J. (Eds.). (2012). *The nationalisation of scientific knowledge in nineteenth-century Central Europe*. Basingstoke: Palgrave Macmillan.

Biernacki, E. (1892). Cholera w Lublinie. Sprawozdanie z wycieczki, odbytej w dniu 21, 22, 23 września r.b [Cholera in Lublin. Report from a visit in the city on 21, 22. and 23. September of this year]. *Gazeta Lekarska, 12*, 837–850.

Briggs, A. (1961). Cholera and society in the nineteenth century. *Past & Present, 19*, 76–96.

Brzeziński, T. (1991). Medyczne studia Polaków na uniwersytetach niemieckich [Medical studies of poles at German universities]. *Archiwum Historii i Filozofii Medycyny, 54*(1–2), 83–89.

Bujwid, O. (1892). Dalszy ciąg wiadomości o epidemii cholery w Lubelskiem. Rzecz czytana w Tow. Lek. Warsz. na posiedzieniu nadzwyczajnem w d. 26 sierpnia 1892 r. [Further notice about the cholera epidemic in Lublin. Talk on extraordinary conference of the Warsaw Medical Society on 26 August 1892]. *Gazeta Lekarska, 12*, 741–743.

Bujwid, O. (1894). Kilka uwag i spostrzeżeń dotyczących cholery w Królestwie Polskiem i w Warszawie w r. 1892/3 [Some remarks and thought about cholera in the Polish Kingdom and Warsaw in 1892/93]. *Medycyna, 22*(41), 811–814.

Cała, A. (1989). *Asymilacja żydów w Królestwie Polskim (1864–1897). Postawy, konflikty, stereotypy* [The assimilation of Jews in the Kingdom of Poland (1864–1898): Opinions, conflicts, stereotypes]. Warszawa: Państwowy Instytut Wydawniczy.

Caumanns, U. (2000). Miasto i zdrowie a perspektywa porównawcza. Uwagi metodyczne na przykładzie reform snitarnych w XIX-wiecznej Warszawie [The city and health in comparative perspective. Methodological remarks using the example of sanitary reform in 19th-century Warsaw]. *Medycyna Nowożytna, 7*(1), 45–62.

Caumanns, U. (2006). Modernisierung unter den Bedingungen der Teilung. Überlegungen zur Frage strukturellen und kulturellen Wandels in Warschau am Beispiel öffentlicher Gesundheit. In C. Goehrke, & B. Pietrow-Ennker (Eds.), *Städte im östlichen Europa. Zur Problematik von Modernisierung und Raum vom Spätmittelalter bis zum 20. Jahrhundert* (pp. 365–391). Zürich: Chronos.

Czarkowski, L. (1894). Epidemia cholery azyatyckiej w miasteczku Siemiatyczach (gub. Grodzieńska) w roku 1893 [The asiatic cholera in Siemiatycze (Gouv. Hrodna) in 1893]. *Medycyna, 6*, 113–117, 135–138.

Epstein, L. R. (1995). *Caring for the soul's house. The Jews of Russia and health care, 1860–1914*. PhD dissertation, Yale University.

Evans, R. J. (1990). *Tod in Hamburg: Stadt, Gesellschaft und Politik in den Cholera-Jahren 1830–1910*. Reinbek bei Hamburg: Rowohlt.

Fleck, L. (1980/1935). Entstehung und Entwicklung einer wissenschaftlichen Tatsache. Einführung in die Lehre vom Denkstil und vom Denkkollektiv. Frankfurt: Suhrkamp.

Foucault, M. (1976). *Überwachen und Strafen: Die Geburt des Gefängnisses*. Frankfurt: Suhrkamp.

Gawin, M. (2003). *Rasa i nowoczesność. Historia polskiego ruchu eugenicznego, 1880–1952* [Race and modernity: The history of the Polish eugenics movement, 1880–1952]. Warszawa: Neriton.

Gilman, S. (1995). Der „jüdische Körper". Gedanken zum physischen Anderssein der Juden. In J. Schoeps, & J. Schlör (Eds.), *Antisemitismus. Vorurteile und Mythen* (pp. 167–179). München: Piper.

Grünbaum, A. (1884). Medycyna w Talmudzie [Medicine in the Talmud]. *Pamiętnik Towarzystwa Lekarskiego Warszawskiego, 80*, 192–200.

Guesnet, F. (1998). *Polnische Juden im 19. Jahrhundert. Lebensbedingungen, Rechtsnormen und Organisation im Wandel*. Köln: Böhlau.

Hamlin, C. (2009). *Cholera: The biography*. Oxford: Oxford University Press.

Hanecki, M. (1968). Z dziejów warszawskiej służby zdrowia w latach 1863–1900 [History of Warsaw Public Health Service, 1863–1900]. In S. Kalabiński, & R. Kołodziejczyk (Eds.), *Warszawa Popowstaniowa. 1864–1918* [Warsaw after the uprising. 1864–1918] (pp. 99–158). Warszawa: Państwowe Wydawnictwo Naukowe.

Harrison, M. (1996). A question of locality. The identity of cholera in British India, 1860–1890. In D. Arnold (Ed.), *Warm climates and western medicine: The emergence of tropical medicine, 1500–1900* (pp. 133–159). Amsterdam: Rodopi.

Hart, M. (1995). Moses the microbiologist: Judaism and social hygiene in the work of Alfred Nossig. *Jewish Social Studies. New Series, 2*(1), 72–97.

Henze, C. E. (2011). *Disease, health care and government in late imperial Russia: Life and death on the Volga, 1823–1914*. Abingdon: Routledge.

Hödl, K. (1997). *Die Pathologisierung des jüdischen Körpers. Antisemitismus, Geschlecht und Medizin im Fin de Siècle*. Wien: Picus.

Hoyer, H. (1903). Uwagi nad piśmiennictwem lekarskim polskim [Comments about Polish medical periodicals]. *Gazeta Lekarska, 23*, 833–836, 857–861, 878–882.

Hüchtker, D. (2002). Der „Schmutz der Juden" und die „Unsittlichkeit der Weiber". Ein Vergleich der Repräsentationen von Armut in Stadt- und Reisebeschreibungen von Galizien und Berlin (Ende des 18./Mitte des 19. Jahrhunderts). *Zeitschrift für Ostmitteleuropa-Forschung, 51*, 351–369.

Hutchinson, J. F. (1990). *Politics and public health in revolutionary Russia, 1890–1918*. Baltimore: Johns Hopkins University Press.

Janowski, W. (1892). Badanie bakteryjologiczne pierwszych dwóch przypadków cholery w Warszawie [Bacteriological examination of the first two cholera cases in Warsaw]. *Gazeta Lekarska, 12*, 852–856.

Konopka, S, Podgórska-Klawe, Z., Dzierżanowski, R. (1987). Medycyna [Medicine]. In B. Suchodolski (Ed.), *Historia Nauki Polskiej* [Polish History of Science] (pp. 383–415), Vol. 4: 1863–1918, Part 3, ed. by Z. Skubala-Tokarska, Wrocław: Zakład Narodowy iminia Ossolińskich.

Kośmiński, S. (1888). Natanson Ludwik. In S. Kośmiński (Ed.), *Słownik Lekarzów Polskich* [Dictionary of Polish physicians] (pp. 341–343). Warszawa: Gebethner i Wolff.

Kramsztyk, Z. (1895). Czy medycyna jest nauką, czy sztuką? [Is medicine a science or art?]. *Gazeta Lekarska, 15*, 1099–1103, 1126–1129.

Kreuder-Sonnen, K. (2012). Wie die Mikroben nach Warschau kamen. Wissenstransfer in der Bakteriologie in den 1880er Jahren. *NTM. Zeitschrift für Geschichte der Wissenschaften, Technik und Medizin, 20*(3), 157–180.

Latour, B. (1988). *The pasteurization of France*. Cambridge: Harvard University Press.

Lipphardt, V. (2008). *Biologie der Juden. Jüdische Wissenschaftler über „Rasse" und Vererbung, 1900–1935*. Göttingen: Wallstein.

Löwy, I. (1990). *The Polish School of philosophy of medicine: From Tytus Chałubiński (1820–1889) to Ludwik Fleck (1896–1961)*. Dordrecht: Kluwer Acadmic Publishing.

Michlic, J. B. (2006). *Poland's threatening other: The image of the Jew from 1880 to the present*. Lincoln: University of Nebraska Press.

Nieznanowska, J. (2008). Polsko-niemiecka wymiana myśli medycznej w XIX wieku – założenia metodologiczne projektu badawczego. Udział polskich autorów w dziwiętnastowiecznym niemieckim czasopiśmiennictwie medycznym na przykładzie periodyków wydawanych w Berlinie [Polish-German exchange in 19th-century medicine: Methodological thoughts about a research project. The presence of Polish authors in German medical periodicals published in Berlin]. In M. Sachs, B. Płonka-Syroka & F. Dross (Eds.), *Współpraca na polu medycyny między niemcami i polakami/Austausch in der Medizin zwischen Deutschen und Polen* (pp. 131–141). Wrocław: Arboretum.

Polonsky, A. (2010). *The Jews in Poland and Russia. Volume II: 1881 to 1914*. Portland: Littman Library of Jewish Civilization.

Porter, B. (2000). *When nationalism began to hate: Imagining modern politics in nineteenth-century Poland*. New York: Oxford University Press.

Sahadeo, J. (2005). Epidemic and empire: Ethnicity, class and „civilization" in the 1892 Tashkent cholera riot. *Slavic Review, 64*(1), 117–139.

Steffen, K. (2008). Wissenschaftler in Bewegung. Der Materialforscher Jan Czochralski zwischen den Weltkriegen. *Journal of Modern European History, 6*(2), 237–261.

Steffen, K. (2013). Experts and the modernization of the nation. The arena of public health in Poland in the first half of the twentieth century. *Jahrbücher für Geschichte Osteuropas, 61*, 574–590.

Stolberg, M. (1994). Public health and popular resistance: Cholera in the Grand Duchy of Tuscany. *Bulletin of the History of Medicine, 68*, 254–277.

Strobel, A. (2013). Die Gesundung Russlands. Hygienepropaganda in Russland um 1910. *Jahrbücher für Geschichte Osteuropas, 61*, 531–551.

Sturdy, S. (2011). Looking for trouble: Medical science and clinical practice in the historiography of modern medicine. *Social History of Medicine, 24*(3), 739–757.

Szarejko, P. (1995). *Słownik Lekarzy Polskich XIX Wieku* [Dictionary of Polish physicians in the 20th century] (Vol. 3). Warszawa: Towarzystwo Lekarskie Warszawskie.

Szwajcer, J. (1892). Pierwszy przypadek cholery azyatyckiej w Warszawie [The first case of asiatic cholera in Warsaw]. *Medycyna, 20*(41), 657–662, 676–680.

Trees. P. (2007). *Wahlen im Weichselland: Die Nationaldemokraten in Russisch-Polen und die Dumawahlen 1905–1912*. Stuttgart: Franz Steiner Verlag.

von Braun, C. (1992). „Der Jude" und „Das Weib". Zwei Stereotypen des „Anderen" in der Moderne. *metis, 1*(2), 6–28.

Weeks, T. R. (2006). *From assimilation to antisemitism: The „Jewish question" in Poland, 1850–1914*. DeKalb: Northern Illinois University Press.

Weindling, P. J. (2007). Ansteckungsherde. Die deutsche Bakteriologie als wissenschaftlicher Rassismus 1890–1920. In P. Sarasin, S. Berger, et al. (Eds.), *Bakteriologie und Moderne. Studien zur Biopolitik des Unsichtbaren, 1870–1970* (pp. 354–374). Suhrkamp: Frankfurt.

Więckowska, E. (1984). Formy opieki zdrowotnej na ziemiach polskich w końcu XIX wieku do lat 30 wieku XX [Forms of health care in the Polish lands from the end of the 19th century to the 1930s]. *Wiadomości Lekarskie, 37*(13), 1049–1055.

Więckowska, E. (1985). Początki opieki zdrowotnej nad ludnością Królestwa Polskiego na przełomie XIX i XX w. [The beginnings of public health care for the population of the Polish Kingdom at the turn to the 20th century]. *Zdrowie Publiczne, 5–6*, 227–237.

Wolff, E. (2000). Juden als Verkörperung von Armut und Unsauberkeit in ärztlichen Berichten über die Choleraepidemie in Osteuropa 1830/31. In S. Jersch-Wenzel, F. Guesnet, et al. (Eds.), *Juden und Armut in Mittel- und Osteuropa* (pp. 123–148). Böhlau: Köln.

Wolff, E. (2014). *Medizin und Ärzte im deutschen Judentum der Reformära. Die Architektur einer modernen jüdischen Identität*. Göttingen: Vandenhoeck & Ruprecht.

Worboys, M. (2007). Was there a bacteriological revolution in late nineteenth-century medicine? *Studies in History and Philosophy of Biological and Biomedical Sciences, 38*, 20–42.

Zahra, T. (2010). Imagined noncommunities: National Indifference as a category of analysis. *Slavic Review, 69*, 93–119.

Zahra, T., & Judson, P. (2012). Sites of indifference to nationhood. *Austrian History Yearbook, 43*, 21–137.

Zasady obowiązków i praw lekarzy. (1884). Przyjęty przez Warszawskie Towarzystwo Lekarskie na posiedzeniu w dniu 29 Kwietnia 1884r. [Principal duties and rights of the physician. Accepted by the Warsaw Medical Society on its session of 29 April 1884]. *Pamiętnik Towarzystwa Lekarskiego Warszawskiego, 80*, 213–227.

Chapter 7
Work of Jewish Medical Community and the Health Culture at School in the Second Republic of Poland (1918–1939)

Beata Szczepańska

Abstract This article is an attempt to more closely examine health problems in the area of education in Poland during the interwar period while taking into account the actions of the Jewish medical community. Because of the large number of Jewish doctors in this professional group it is possible to refer to the issues in question by looking at their activity within Jewish social organisations such as TOZ the Society for Safeguarding the Health of the Jewish Population and CENTOS [*Central Office of the Union of Societies for Care of Orphans and Abandoned Children*] that focused on private schools for Jewish children. The activity of representatives of the Jewish medical community especially advocating of school hygiene in public education is also considered in the analysis. Considerations concern in particular these elements of school health culture that are connected to hygiene and medical care of pupils' health as well as to the evaluation of the condition of children's health, and to the investigations initiated in this respect. One can notice here some attempts to show otherness of Jewish pupils observed against the general background. There is also some information about organized activity in respect to health prevention in schools, and of issues related to hygiene education. Conceptual, organizational, and practical activities of Jewish doctors for school hygiene have been presented in the article. Special achievements of Jewish doctors in treatment and prevention of tuberculosis were reflected in creation of forms of healthcare for pupils suffering from or threatened by this social disease, as well as in the search for an explanation of the incidence of this disease. These considerations are based on the image that had been created in the Polish inter-war medical and pedagogical magazines. The following examples were taken into consideration: Warsaw Medical Magazine [*Warszawskie Czasopismo Lekarskie*], Child Care [*Opieka nad Dzieckiem*]. Also, the contents of Jewish press published in Polish by TOZ and CENTOS, such as Social Review [*Przegląd Społeczny*], Health Almanac [*Almanach Zdrowia*], and the magazine containing Polish abstracts TOZ-Jedies/Social Medicine [*Medycyna Społeczna*]were analysed.

B. Szczepańska (✉)
University of Lodz, Faculty of Educational Sciences, Lodz, Poland
e-mail: beata.szczepanska@uni.lodz.pl

M. Moskalewicz et al. (eds.), *Jewish Medicine and Healthcare in Central Eastern Europe*, Religion, Spirituality and Health: A Social Scientific Approach 3,
https://doi.org/10.1007/978-3-319-92480-9_7

Keywords School hygiene · Health promotion · Health protection · Health education · Jews in Poland · Doctors · Magazines · Interwar period

7.1 Introduction

School health culture is an issue closely related to school hygiene in interwar Poland. This issue was widely understood as "[…] the entirety of health issues concerning school […], in other words a collection of school obligations focused on improvement of pupils' health". (Kacprzak 1933). Health culture was intended to be a basic task for school. The schools should institutionally focus on counteracting any threats to pupils' health and on reinforcing their health (Sosnowski 1920). The elements of school culture can be distinguished by establishing a common area of school hygiene in the structures of the ministry's agencies created at the beginning of the interwar period, such as the Division of School Hygiene and Physical Education in the Ministry of Religious Denominations and Public Enlightenment [*Wydział Higieny Szkolnej i Wychowania Fizycznego w Ministerstwie Wyznań Religijnych i Oświecenia Publicznego*, hereinafter called: MWRiOP] and the Division of School Hygiene in the Ministry of Public Health *[Wydział Higieny Szkolnej w Ministerstwie Zdrowia Publicznego]*. The elements of school culture were as follows: material school environment, medical and hygienic care of pupils' health, hygiene education, medical pedagogy, care and social assistance for pupils, hygiene education and hygiene promotion (Szczepańska 2014).

A strong interest of the resurgent Polish State in health promotion through the school system and at school resulted mainly from the dramatic demographic and health situation of Poland's society after the World War I. However, it took into consideration the achievements of Polish thought and practice in the area of school hygiene from the period when Poland was partitioned. The Jewish medical community contributed to creating this new educational reality which was aimed at building healthy society via school. The presence of the Jewish minority, a group which was setting up organisations for health protection, in the demographic structure of the second Republic of Poland as well as a large percentage of Jews respecting the religious tradition in the professional group of doctors, were advancing health prevention before the treatment. This constituted the basis for Jewish medical communities' participation in building and developing a culture of healthy schools.

Research presented here focuses on the activities of the representatives of the Jewish medical community which concerned school hygiene in public and private education in Poland after its winning independence, that is in the period of 1918–1939. We can distinguish three fundamental directions of doctors' activities: organizational, practical, and conceptual. Within these activities, we can find references to catalogued fields of school hygiene. This paper is structured according to chronology and problem raised.

Medical and pedagogical magazines 'constitute' the source for deliberations presented here. Periodical magazines published by Jewish social organisations were especially researched. These were: 'Social Medicine' [*Medycyna Społeczna*], 'Health Almanac' [*Almanach Zdrowia*], 'Social Review' [*Przegląd Społeczny*].

'Social Medicine' was the central press organ of the Society for Safeguarding the Health of the Jewish Population [*Towarzystwo Ochrony Zdrowia Ludności Żydowskiej; TOZ*] devoted to the issues of social medicine and hygiene among Jews, edited by Ludwik Wulman. This bimonthly magazine was first published in 1928 in Yiddish with the contents partly in Polish. For the first 8 years 'Social Medicine' was being published in Vilnius, and in 1932 the editorial office was transferred to Warsaw. At that time, Polish summaries of articles began to appear. Intentions were to publish this periodical magazine in two languages, in Yiddish and in Polish, but the outbreak of the Second World War made it impossible.

'Health Almanac' was being published as local press organ TOZ in Lvov. In that city, The Society for Care of Jewish Orphans was publishing the 'Social Review', edited by Józef Kohn. 'Social Review' was a monthly magazine devoted to the issues of social work and childcare and, as an organ of the Union for Care of Jewish Orphans in the Republic of Poland, was first published in 1927.

Also, the author of the paper took advantage of the materials from 'Warsaw Medical Magazine'[*Warszawskie Czasopismo Lekarskie]* and 'Child Care'[*Opieka nad Dzieckiem*]. 'Warsaw Medical Magazine' began its publication by the Polish Society for Social Medicine [*Polskie Towarzystwo Medycyny Społecznej*] since 1924 as a monthly magazine; Zygmunt Srebrny being its editor.'Child Care' was a magazine devoted to the protection of motherhood and care of children and youth published by the Polish Committee of Child Care. This bimonthly magazine was published from 1922 to 1932 when it was transformed into "Child Life" and next into "Youth Life". It was edited by, among others, Bronisław Krakowski and Władysław Szenajch.

Newspapers, as a specific historical source, are characterised by internal limitations. They convey the past in a way that is not fully satisfying, as the problems are often presented in a facile and fragmentary way. Some facts are taken out of the context while the contents are sometimes biased. So, it must be emphasized that, with such methodological assumptions made, these considerations do not aspire to create an overall vision of contributions to the public health made by doctors coming from the Jewish community, but their aim is just to bridge the gap existing in contemporary discussions on school hygiene, which in fact did not take into account Jewish doctors' participation in creation of the health culture of Polish schools.

7.2 The Organizational, Practical, and Conceptual Activity of Jewish Doctors for the Benefit of Children and School Youth's Health in the Interwar Period

A multi-national character of the Republic was manifested by the fact that every third citizen was not Polish. In the doctors' professional group in seven central and eastern voivodeships (while the total was 16), the majority were of Jewish nationality (Więckowska 2004). Such participation of Jewish doctors in this professional group affected their work in the medical care both in public as well as in private schools. That is why school doctors seem to be important in building the health culture in public schools. The tasks of a school doctors as defined in the *'Instruction'* published in 1918 by MWRiOP were as follows: supervision over the sanitary condition in a school building, supervision over pupils' health, care of hygiene education, professional counselling and health education. (Instrukcja 1918). The school doctor was a member of the teachers' meetings. His care covered pupils' health from the beginning of the school admission process, through their stay at school, and, also, prior to leaving school when the doctor was helping to define a pupil's predisposition for a given profession.

In urban and multicultural Łódź, where, in the interwar period, Jews constituted 83% of doctors and Germans: 6% (Mroczka 1997) – the activities of the Jewish medical community in the process of building and reinforcing the school health culture can be considered, from the perspective of the city's social situation, as particularly important. It is worth noting here some of the most prominent Jewish doctors who were working in this area: Bronisław Handelsman and Stanisław Gutentag. For many years, both of them were working as school doctors and teachers of hygiene in Łódź schools, both public and private.

Handelsman (1859–1935), a doctor, pedagogue and a social worker, is known to be the initiator of establishing city medical clinics for child outpatients from public schools as well as of clinics for children suffering from trachoma. He was managing these institutions. He was also very active in the following organisations: The Society for Promoting Education [*Towarzystwo Krzewienia Oświaty*], Polish Society for Research on Children [*Polskie Towarzystwo Badań nad Dziećmi*], The Society Supporting Schools for Beginners and Craft Schools for Jews [*Towarzystwo Popierania Szkół Początkowych i Rzemieślniczych dla Dzieci Wyznania Mojżeszowego*], The Society for Care of the Sick [*Towarzystwo Pielęgnowania Chorych „Bykur Cholim"*]. He also wrote and translated some scientific works and manuals on school hygiene (Kempa and Szukalak 2001).

Gutentag (1866–?) – a paediatrician and a manager of the school doctors' section (the supreme doctor – hygienist of public schools) in Łódź city hall, was an active co-organizer of the Jewish education in Łódź; co-founder and a president of Łódź Jewish Society for Care of Orphans [*Łódzkie Żydowskie Towarzystwo Opieki nad Sierotami*]. He was publishing articles in the Journal of the Municipality of Łódź (1924/34, 1927/16) (Kempa and Szukalak 2001).

The social Jewish movement for improving school health can be traced back to the activity of the first branches of TOZ in Vilnius and Lvov. The Society for Safeguarding the Health of the Jewish Population [*OZE*] supervised sanitary condition of school buildings and dealt with treatment of ill children since 1919 (Kronika 1923). Then, the Society for Promoting Hygiene Among Jews [*Towarzystwo Szerzenia Higieny wśród Żydów*] separated in 1919 from Lvov Sanitary Section of the Rescue Committee [*Sekcja Sanitarna Komitetu Ratunkowego*]. It first concentrated on promoting hygienic knowledge via organized lectures. In 1924, the first organizational action for summer camps for Jewish school children began.

The importance of improving children's health by school was popularised in 1928 as a result of the First Common Meeting of TOZ and the resolutions passed during its plenary sessions and during The School Medical and Hygienic Section.. The members decided that the problem of 'physical recovery' of the Jewish population had to be started from the young generation, and school was its indispensable basis. The detailing of the provisions adopted at the general forum took place during the discussions of The School Medical and Hygienic Section. At the beginning of the session a statement was delivered saying that, generally, school buildings of Jewish religious schools did not satisfy the developing need to comply with the following recommendations concerning hygiene in education:

1. Proper distribution of breaks between lessons and holidays.
2. Separation of school rooms from the private area of teachers.
3. Isolating teachers who are ill of contagious diseases.
4. Creation of classrooms consisting of pupils of similar age, and avoidance of aggregation of pupils with too large of an age of difference.
5. Conducting lessons which involve physical education and school excursions.
6. Organization of different types of summer camps.
7. Supervision of physical development of pupils should have a systematic character, and should constantly concern itself with children who are weak and ill.

The debaters decided that in order to provide well organised hygienic and medical supervision, training of school medical and pedagogical staff was necessary.

An important issue concerning the influence on pupils' health was the cooperation between the school and pupils' families. That is why it was postulated to set up the so called parents' committees working at schools.

The debaters also agreed that it was necessary to conduct scientific research to improve work on development of the school health culture. So, the meeting recommended that steps should be undertaken in the following areas:

1. Setting up a central scientific office, dealing with hygiene and school medicine problems, at the headquarters of TOZ.
2. Setting up doctors' offices of school hygiene which would have a research character.
3. Organising a special meeting of school doctors employed in Jewish schools.
4. Conducting psycho-technical and anthropometric investigations in bigger towns, based on a specially developed standardised procedure. In this field cooperation between TOZ and the Jewish Scientific Institute [*Żydowski Instytut Naukowy*] would be desirable.
5. Introducing standard hygienic supervision in several schools in order to provide: "… an example for hygienic and medical staff in schools and for school authorities [to support them] in their activities to heal school" (Kronika 1928).

It seems that in the aforementioned meeting recommendations of TOZ, established in Warsaw, in 1921 (Beizer 2016a), one can find a continuation of the ideological line of OZE, an organisation existing from 1912, whose aim was also to promote health, hygiene, and support childcare. Founders of OZE believed that Jews had exceptional demographic, biological, and psychological qualities because of their past as a religious group. This past concerned habitation in towns for many centuries where they preserved tradition, endured persecution and isolation. The mission of both organisations was to compensate social and legal discrimination of Jews, to deal with health conditions and inherited diseases which Jews suffered from, and to assist them in cultivating ethnic and religious tradition. Such an attitude was in compliance with assumptions of Jewish liberals, autonomist, and Zionists, but was different from supporters of BUND (General Jewish Labour Bund in Lithuania, Poland and Russia), who believed that medical care should be provided irrespectively of ethnic identity (Beizer 2016b).

7.3 Activity of Jewish Medical Community in the Process of Implementation of Hygiene in Schools in the 1930s

Because of the economic crisis at the beginning of the thirties, work on the health of children and youth based on social initiatives and assistance became, for the Jewish organisations, additionally directed, towards the preparation for emigration. As Ludwik Wulman wrote: "If life in the future does not pan out in this country, the issue of emigration will be a reasonable choice if longer life to be attained. The immigration countries will require people who are physically healthy and adjusted to productive work in these countries" (Wulman 1935).

The centre of gravity of health work provided by the Jewish medical community in social organizations was transferred to the intensification of provision of food to pupils, and to organization and improvement of the summer camps for school children and youth.

At the end of the interwar period, when the World Meeting of Jewish Doctors was summarizing the situation in the Polish Republic, it was emphasised that: "… Jewish doctors in Poland, unlike in other countries, constitute a significant social factor based on a strong organisational bond, and have a significant impact on the outside. This is demonstrated by at least the last meetings of doctors in Warsaw, Lublin, and by widely developed activity of TOZ gathering hundreds of Jewish doctors" (Frydman 1938). TOZ was a massive social organisation, running several hundred medical and hygienic clinics and a network of branches in the territory of Poland. Development of medical and hygienic care in Jewish schools, both secular and religious, and promotion of physical education and hygiene–also via medical and hygienic magazines published in Hebrew and Jewish– were the achievements of this organisation (Wulman 1939).

Representatives of the Jewish medical community got involved in hygiene improvement in school both individually, as doctors-hygienists working in public and private schools, but also on behalf of various Jewish organizations. This was the mission of TOZ; its objectives were defined in the following way: "… to protect the health of the Jewish population, to improve it by all means available to us, to create, in the place of a weak, ill Jew, a Jew that will be fully healthy, powerful and energetic" (Od Wydawnictwa 1939). However, TOZ tasks overlapped partially with the objectives and activities of CENTOS – an organisation taking care of orphans Jewish children. These common areas concerned: organisations of summer camps and summer play centres, supplementary feeding and setting up day rooms and half dormitories (Schaff 1939).

Examination of health condition of pupils in Jewish schools was conducted by doctors, and its results were published in magazines. These results were then used for comparison of the situations in particular centres, and for seeing how one centre looks against the background of pupils' population in public schools in general. Evaluation of pupils' health was, as was mentioned earlier, one of basic obligations of doctors in public schools with permanent medical care, in grammar schools, and teacher colleges. The reports on such health reviews were then sent to MWRiOP and, from the beginning of 1930s, published in official magazines. A characteristic feature of Jewish pupils was noticed: there was a big difference between boys and girls to the disadvantage of the former with respect to weight, height, nutrition and, broadly speaking, physical development and general health condition. The reason for this was discovered in the fact that boys attending 'Talmud-Torah' schools had more school obligations than girls. Such investigations were conducted in: Łódź (Litwak 1932), Warsaw (Lejzerowicz 1935), Suwałki (Rozental 1936), Lvov (Blatt 1939). Earlier investigations conducted by e.g. Leon Wernic in Kalisz or Wiktoryn Kosmowski in Warsaw were compared (Litwak 1932).

As the economic crisis worsened, and significant deterioration of life conditions became more noticeable, the most important task for the doctors was to provide social and medical assistance for children and adolescents (Wulman 1935). The supplementary alimentation action for pupils was commenced because of the growing number of ill and weakened children. This phenomenon was observed especially in the voivodeships of Northern and Eastern Poland that also suffered from natural disasters. At the beginning of the 30s, the central office of TOZ and its local branches organised a campaign directed toward 6000 pupils. Till the end of 1931 the range of assistance was expanded and covered 30,000 pupils. In school, they received a glass of milk, malt coffee, tea and rolls or bread with butter. This campaign was also supported by the American Jewish Joint Distribution Committee and a French organisation – 'Alliance Izraelite' (Wulman 1932).

They emphasised that educational efforts would be ineffective if the basic nutrition needs were not satisfied. In order to justify the food campaign for pupils in Lvov, Lazar Schachner wrote: "Teachers' educational work, coming from the purest regions of their hearts and souls, accompanied by the sincerest of love for children will not bring any effects, if pupils are hungry. Teaching activities, even extremely interesting and beautiful will not make a poor and hungry child focus on them. A hungry child is not open for any sublime thoughts and, when growing without any life-giving juice, the child will suffer from various diseases, especially tuberculosis and sickliness in general. Also, the worst instincts, leading to offences, are born in them and are transform them into festering wounds on the society…. So, the obligations of the school increase with the deepening crisis, and the protective and social work need to become more intense and have wider range". The action of providing food was also connected with distribution of clothes and footwear. Also, it caused pauperization of people, the youngest generation in particular. Because of that some organisations undertook the efforts aimed at providing children and youths with lunches. In Lvov, the Jewish Shelter-House for Homeless [*Żydowskie Schronisko dla Bezdomnych*] and the Jewish Committee of Outpatient Assistance [*Żydowski Komitet Doraźnej Pomocy*], among others, took part in this action (Schachner 1934).

To complete and deepen the school's work for children's health, summer camps and summer play centres were organised. TOZ took the effort to have their own summer camp buildings built. For example, a branch of this organisation in Lvov had a camp centre built in Brzuchowice (Blatt 1939). A pupil's stay at the summer camps had particular significance for his or her health, if we consider poor hygienic and sanitary conditions of both school premises and children's flats (Plenary meeting 1932). Data concerning summer camp action run by TOZ in 1932 indicate that over 8000 pupils were sent to 44 locations (25 – summer camps, 19 – summer play centres). The boys were a bit larger group of participants at the camps (51.5%) than the girls. The pupils of public schools constituted the biggest part of the participants at summer camps (43.8%), the next group were pupils of religious schools – 18.8%, while pupils of Hebrew 'Tarbut' schools accounted for 11.5%. Most participants were between the ages of 8 and 14 (Zestawienie 1932).

J. Landau wrote about the significance of the summer camps in the fight against tuberculosis and in preventing this illness. He described three types of summer camps. The healing summer camp was intended for sick children with active tuberculosis. Holiday settlements were dedicated for healthy, but weakened and frail children. The summer play preventive centres for healthy were the cheapest option (Landau 1932).

Apart from pursuing the main objective of the summer camps, which was defined as "recovery of weakened child", the child's stay at the camp was used in a way in which health education could contribute to deepening knowledge and to strengthening hygienic habits. This information was being conveyed not only via hygienic conditions of the stay, but also via healthy eating habits and organized lectures and conversations about health. The contents of health education were to include: hygiene of flats, hygiene of feeding, hygiene of clothes, work hygiene, physical exercises, epidemiological issues, social diseases, and basic emergency medical services. It was recommended to use such didactic means as: demonstrative boards, a projector, an overhead projector (Biber 1932). The summer camp campaign grew in importance both in quantitative aspect, as well as in the quality of pedagogical work (Wulman 1938; Szwalbe 1938).

Also, in journalistic writing of the thirties one can find signs of concern and care of Jewish doctors about the health of pupils at school. They were especially concerned about private schools, which number grew to exceed 400 in the Second Republic. Izrael Biber, a representative of the Vilnius community expressed very critical remarks about the health culture of these institutions. The buildings were not sufficiently adapted to educational needs. The pedagogical staff of these places did not show proper understanding of pupils' health needs. The conditions at pupils' homes and lack of cooperation with their families contributed to this unfavourable image. Postulates formulated by the aforementioned doctor were convergent with those expressed at the first meeting of TOZ's doctors in Warsaw in the twenties (Biber 1936). The issues of the health culture appeared at the forum of third TOZ meeting in Warsaw, which took place in 1937. Dr. M. Szwalbe presented an outline of the problems in the article entitled: "Experience and Results Obtained in the Area of School Hygiene and Indications for Future Work" (Obwieszczenie 1937).

Next, L. Wulman wrote both from material as well as didactic perspectives about common effects of poverty of many Jewish families and difficult work conditions at school: "… physical development and condition of a child, even the one who develops normally, and who, during the school year did not suffer from any diseases, is disastrous … This concerns particularly children from cheders as they were completely deprived of movement necessary for development of a child's organism, of games necessary for development of the mind, and of filiation." A sure remedy for compensating the lack of health culture was the organisation of summer play centres which were cheaper than summer camps and offered a chance to improve the health of nearly all children covered by TOZ care (Wulman 1938).

The issues of diseases of children and youths at school age, which was raised by Jewish doctors, concerned mainly tuberculosis, recognised as the most dangerous disease at that time. David Nisenson prepared an important work on the statistic of tuberculosis in the school community based on his own research. He indicated different forms of this disease in pupils and showed significant intensification of this disease in children coming from poor communities. He postulated an improvement of the epidemiological proceedings in case of this disease, including access to frequent examination of teachers and other school employees (Nisenson 1925).

In the thirties, Benedykt Glass and Henryk Szpidbaum carried out investigations on the tuberculosis threat to pupils of Warsaw schools. They showed a bigger percentage of children ill of tuberculosis than MWRiOP did, which could result from internal school contaminations. Often it was difficult to indicate any symptoms of this disease. So, it was postulated to conduct screening diagnosis which intended to lead to faster isolation of ill pupils (Glass and Szpidbaum 1933). Other doctors dealing with school hygiene indicated the necessity of X-rays because of divergence in results between physical examination and actual condition of the pupils' health (Hryniewiecki 1934).

The activities of the Jewish medical circle in the field of health promotion in schools can be presented on the example of the activities of the Lvov physicians' school section headed by Maurycy Salpeter and Blanka Jurimowa. They were on board of TOZ (Od Wydawnictwa 1939). It was under their leadership that the school section undertook its activities. In the school year of 1937/38 eight schools were under their care: Ajszel Tora, Talmud Tora, Jesodej Hatora (located on Serbska 4), Jesodej Hatora (located on Łokietka 20), Chorew, Bejt Jakow, Mitet and Hebrew Seminar. In total, there were about 1300 pupils receiving education in those institutions.

Among the tasks undertaken by six doctors of this section, were, first of all, overall screenings of all pupils under their direct care. Such examinations were conducted several times a year. Moreover, in those schools, as well as in all four Lvov Gymnasiums, the doctors of the section held 177 lectures and talks for nearly 9000 people all in the school year of 1937/38. Among the topics discussed were: personal and environmental hygiene, work and fatigue, hygiene of puberty, the importance of vitamins for the body, the importance of sports in maintaining health, first aid in emergencies.

Jewish doctors, acting under the auspices of the school section of TOZ organized campaigns promoting cleanliness. For example, on May 16th to 21st of 1938 there was an action called "The Week of Cleanliness Propaganda". Children from the poorest families were awarded diplomas, toiletries beauty products (soap, toothbrushes, tooth powder, combs, cream) for their keeping of the rules of hygiene (Blatt 1939).

Jewish doctors also referred to the issues of medical pedagogy. Attention was drawn to the overly excessive nervousness of pupils and occurrence of their mental disorders. Examples of solutions applied in this area such as a special centre for mentally disabled children set up by CENTOS in Otwock and the special needs school established in Vilnius on the initiative of Dr. Wita Lewin, were presented (Sznejerson 1934). It was also postulated to organize, apart from already developed psycho-hygienic aid, centres for overwrought children (Rosenblum 1935). The author of this paper was Zofia Rosenblum (1888–1978) – a paediatrician and pioneer of child psychiatry in Poland (Kroszczor 1979; Paciorek 2010).

While treating school as a special place of confrontation of the Jewish culture, of diaspora character, with the Polish national culture. It was recommended by many authors to conduct research on this issue and to look for solutions diminishing tensions resulting from conflict of values between these cultures, and these tensions' influence on Jewish children. F. Schneersohn defined the tasks of the psycho-hygienic work among Jewish children in the following way: "… understanding and thoroughly explaining the extent of the influence of the duality on the sanity of Jewish child, and finding medical and educational means to alleviate trauma created in Jewish child's mind." (Schneersohn 1935).

7.4 Conclusions

The review of selected magazines published in the interwar period allows us to ascertain that the activities of the Jewish medical community for the school health culture concerned all areas of school hygiene. The specificity of Jewish educational centres was taken into consideration by Jewish doctors. Also, the issue of the Jewish doctors' participation in works for the public school health culture was not neglected.

The special character of the press as a historical source did not allowed us to report the past in a fully satisfying way because some ideas were developed only in a shallow, or fragmentary, and even biased way. However, the insight of the presented analyses, noticing complexity and showing nuances of the most important values of the Jewish doctors' journalistic writing. Also encouragement to taking up activities in the areas demanding intervention, account for the value of the Jewish medical community's composes contribution to achievements in the sphere of the school health culture's improvement in the Second Republic of Poland.

References

Beizer, M. (2016a). *TOZ*. Online website: http://www.yivoencyclopedia.org/article.aspx/toz. Accessed 1 Feb 2016.

Beizer, M. (2016b). *OZE*. Online website: http://www.yivoencyclopedia.org/article.aspx/oze. Accessed 2 Feb 2016.

Biber, I. (1932, lipiec-sierpień). Nauczanie higjeny w kolonjach letnich. *Medycyna Społeczna, 7–8*, 96.

Biber, I. (1936, wrzesień-październik). Zagadnienia higieny szkolnej. *Medycyna Społeczna, 9–10*, 2.

Blatt, A. (1939). Działalność TOZ-u. Oddział we Lwowie, rok 1938. In *Almanach Zdrowia* (pp. 9–31). Lwów: TOZ.

Frydman, A. (1938, lipiec-sierpień). Światowy Związek Lekarzy Żydów w r. 1937–38, *Medycyna Społeczna, 7–8*, 1.

Glass, B., & Szpidbaum, H. (1933). Z badań nad zagruźliczeniem dzieci w Warszawie. *Warszawskie Czasopismo Lekarskie, 8*, 360–362.

Hryniewiecki, S. (1934, lipiec-sierpień). Jak lekarz szkolny może skutecznie zwalczać gruźlicę wśród młodzieży szkolnej. *Medycyna Społeczna*, 7– 8, 6.

Instrukcja dla lekarzy szkolnych w zakładach naukowych męskich i żeńskich, ułożona przez komisję międzyministerialną do spraw higjeny szkolnej. (1918). *Dziennik Urzędowy Ministerstwa Wyznań Religijnych i Oświecenia Publicznego Królestwa Polskiego* (Vol. 6, pp. 214–218). Warszawa: MWRiOP.

Kacprzak, M. (1933). Współczesne poglądy na higjenę szkolną. *Praca i Opieka Społeczna, 3*, 246–253.

Kempa, A., & Szukalak, M. (2001). *Żydzi dawnej Łodzi. Słownik biograficzny Żydów łódzkich oraz z Łodzią związanych,* t. I. Łódź; Oficyna Bibliofilów.

Kronika. (1923, lipiec-sierpień). *Opieka nad Dzieckiem* (Vol. 4, pp. 194–195).

Kronika. (1928, wrzesień-październik). *Opieka nad Dzieckiem* (Vol. 5, pp. 379–384).

Kroszczor, H. (1979). *Kartki z historii Żydów w Warszawie XIX–XX w. Sylwetki. Szkice*. Warszawa: Żydowski Instytut Historyczny w Polsce.

Landau, J. (1932, marzec-kwiecień). Kolonje letnie w walce z gruźlicą. *Medycyna Społeczna, 3–4*, 67.

Lejzerowicz, L. (1935, listopad-grudzień). Wynik badania społeczno-higienicznych warunków żydowskiego dziecka w wieku szkolnym. *Medycyna Społeczna, 11–12*, 37.

Litwak, M. (1932, listopad-grudzień), Skala wzrostu i ciężaru dzieci żydowskich. *Medycyna Społeczna, 11–12*, 169.

Mroczka, L. (1997). Dynamika rozwoju i struktura społeczno-zawodowa głównych grup etnicznych w Łodzi w latach 1918–1939. In P. Samuś (Ed.), *Polacy-Niemcy-Żydzi w Łodzi w XIX-XX w. Sąsiedzi dalecy i bliscy* (pp. 99–117). Łódź: Ibidem.

Nisenson, D. (1925). Gruźlica a wiek dziecięcy w świetle liczb. *Warszawskie Czasopismo Lekarskie, 8*, 360–362.

Obwieszczenie. (1937, wrzesień-październik). *Medycyna Społeczna*, 9–10, (cover).

Od Wydawnictwa. (1939). *Almanach Zdrowia* (p. 4). Lwów: TOZ.

Paciorek, M. (2010). Działalność społeczna Żydów w Warszawie pierwszej połowy XX wieku. In Z. Podgórska-Klawe (Ed.), *Polscy Lekarze Żydzi w XIX i XX wieku* (pp. 127–134). Warszawa: Instytut Historii Nauki PAN, Żydowski Instytut Historyczny.

Plenarne posiedzenie Rady "TOZ-u". (1932, marzec-kwiecień). *Medycyna Społeczna, 3–4*, 67–68.

Rosenblum, Z. (1935). Higiena psychiczna w zastosowaniu do dzieci w Europie Zachodniej i w Polsce. *Warszawskie Czasopismo Lekarskie, 18*, 381–384.

Rozental, M. (1936, lipiec-sierpień). Stan sanitarno-higieniczny żydowskich dzieci w Suwałkach. *Medycyna Społeczna*, 7–8, 1.

Schachner, L. (1934, marzec; kwiecień). Opieka społeczna nad dzieckiem szkolnym we Lwowie w roku szkolnym 1932/33. *Przegląd Społeczny, III*, 57–63; *IV*, 94–98.

Schaff, M. (1939). O koordynację pracy społecznej. In *Almanach Zdrowia* (pp. 32–36). Lwów: TOZ.

Schneersohn, F. (1935, wrzesień-październik). Dualizm kulturowy a zadania psychoanalizy. *Medycyna Społeczna*, 9–10, 32.
Sosnowski, K. (1920). *Szkoła a zdrowie*. Kraków: Akademia Handlowa.
Szczepańska, B. (2014). *Higiena szkolna w szkolnictwie ogólnokształcącym w Drugiej Rzeczypospolitej*. Łódź: Wydawnictwo Uniwersytetu Łódzkiego.
Sznejerson, F. (1934, wrzesień-październik). O umysłowo cierpiącym dziecku żydowskim. *Medycyna Społeczna, 9–10*, 10.
Szwalbe, M. (1938, maj-czerwiec). Kolonia a półkolonia. *Medycyna Społeczna*, 5–6, 5.
Więckowska, E. (2004). *Lekarze jako grupa społeczno-zawodowa w II Rzeczypospolitej*. Wrocław: Wydawnictwo Uniwersytetu Wrocławskiego.
Wulman, L. (1932, styczeń-luty). Centrala "TOZ-u" wobec tragicznej nędzy dzieci żydowskich. *Medycyna Społeczna*, 1–2, 35.
Wulman, L. (1935, marzec-kwiecień). Ideologiczne podstawy pracy zdrowotnej wśród Żydów w Polsce. *Medycyna Społeczna*, *3–4*, 17–18.
Wulman, L. (1938, maj-czerwiec). Przekrój głównych zagadnień pracy kolonijnej. *Medycyna Społeczna*, *5–6*, 5.
Wulman, L. (1939). *Dwadzieścia lat pracy zdrowotnej wśród Żydów. Almanach Zdrowia* (pp. 5–8). Lwów: TOZ.
Zestawienie danych statystycznych z akcji TOZ w 1932. (1932, wrzesień-październik). *Medycyna Społeczna*, *9–10*, 149–150.

Chapter 8
A Survey of Jewish Healthcare in Poland After WWII

Ignacy Einhorn and Jakub Einhorn

Abstract The authors of the article analyse the activity of the Society for Safeguarding the Health of the Jewish Population in Poland (TOZ), its functioning in a specific political conditions of 1945–1950 Poland, and evaluate its achievements at the time, while undertaking an attempt to assess the TOZ's autonomy within the structures of the Central Committee of Polish Jews in Poland (CKŻP).

The authors use mainly archival documentation found in the Emanuel Ringelblum's Jewish Historical Institute (archival documentary collection of the Society for Protection of Health of the Jewish People in Poland) and other archives. Information was also found in numerous articles published in Jewish press. Based on the archival documentation and subject literature analysis, the authors conclude that the Society for Protection of Health of the Jewish People in Poland in years 1945–1950 managed to preserve a high level of the autonomy of its functioning within the structures of CKŻP. The phenomenon of this organisation, operating in the difficult post-war period, was based on combination of best Jewish traditions of self-help organisations with the need to give help to the Polish-Jewish Holocaust survivors.

Keywords Polish Jews · Society for Safeguarding the health of the Jewish population · Poland · Jewish medicine · Jewish healthcare · Central Committee of Polish Jews in Poland

I. Einhorn
Medical College in Klodzko, Klodzko, Poland
e-mail: igein@o2.pl

J. Einhorn (✉)
Wroclaw, Social-Cultural Association of Jews in Poland, Wroclaw, Poland
e-mail: nzozkolk@gmail.com

M. Moskalewicz et al. (eds.), *Jewish Medicine and Healthcare in Central Eastern Europe*, Religion, Spirituality and Health: A Social Scientific Approach 3,
https://doi.org/10.1007/978-3-319-92480-9_8

8.1 Introduction

The subject of Jewish healthcare in Poland after WWII has never been properly researched. It was brought up, in the context of the Society for Safeguarding the Health of the Jewish Population (TOZ), by Ignacy Einhorn (2008). More information about activities of Jewish institutions in various regions of Poland can be found in the studies on local Jewish communities.

The specificity of Jewish healthcare in Poland after 1945 relates to the fact that it operated autonomously for merely 5 years – from the end of war until its liquidation and nationalization of last Jewish institutions in 1949. However, in such a short period of time, it managed to create an impressive net of medical establishments which gave assistance in basic and specialist medical care to thousands of Polish Jews saved from the Holocaust. It enabled people not only to recover but also to regain the sense of safety and psychical balance after traumatic war experience.

The basic sources for research on Jewish healthcare in the post-war period are the records of the Society for Safeguarding the Health of the Jewish Population in Poland between 1945 and 1950 assembled in Emanuel Ringelblum's Jewish Historical Institute in Warsaw.

The following text also quotes some articles published in Jewish press of the period and commemorative publications of TOZ.

8.2 Activity of TOZ Until Early 1940s

In the interwar period TOZ was considered the most important Jewish healthcare organization in Poland. In 1936, the organization had 52 branches all over the country, 209 establishments, 11,535 members and 819 employees. Until 1939, it expanded to 368 establishments in 72 places (Epsztein 2001–2011, p. 3). In that period, TOZ mainly propagated hygiene, treated patients, run shelters for mothers with children, and was involved in research on the state of health of the Jewish population. Besides,51 Jewish hospitals were functioning in the 30s with 3500 beds (Kroszczor and Zabłotniak 1978, p. 60).

TOZ also played an important role during WWII bringing help to Jewish people in ghettos in General Government (occupied Polish territories). Its last establishments operated in Warsaw Ghetto until 1942 within the Jewish Social Self-Support scheme (Engelking and Leociak 2013, pp. 258–262).

After the war, the activists of TOZ headed for quick reactivation of the Society. It was possible only in 1946, most of all, due to the establishment of the Central Committee of Polish Jews (CKŻP) which aimed to encompass all of Jewish life in Poland,.

8.3 Health Departments of CKŻP

After the war, before TOZ was reactivated, the health issues of Jewish community were taken care by health departments of Jewish committees. This situation was due to the fact that CKŻP took over those duties of TOZ that before and during the war were within its sphere of activity. Among the main goals of the Society – apart from running social care for Jewish population – was also organization of the medical care for Jews.

Jewish activists were fully aware that the creation of healthcare was crucial necessity for thousands of Jews and the public healthcare was not adequately resourced. The description of the situation can be found in one of the documents: "Due to many years spent in camps, bunkers and forests, malnutrition, physical and spiritual exhaustion and aftermath of camp diseases the people were particularly vulnerable to TB and other infections. Besides, almost all of them leave camps with hunger swelling. People rely on CKŻP help since, in most of the cases, local health care do not provide medical, sanitary or victual aid" (AAN, 5/137, p. 10).

Health department of CKŻP was established on 11th of June 1945 and soon similar departments were established by regional and local Jewish committees.

Until the end of 1945 there were 16 health departments within CKŻP – in Białystok, Bydgoszcz, Częstochowa, Gdańsk, Kalisz, Katowice, Kielce, Lublin, Łódź, Piotrków, Przemyśl, Tarnów, Warszawa, Włocławek and Wrocław-Rychbach (currently Dzierżoniów). Their activities can be traced back based on „The Report of the Health Department of Central Committee of Polish Jews for period between 1st and 31st of December 1945″ (AŻIH, 324/482, pp. 23–24).

10 child care homes were also established – in Zatrzebie near Falenica, Otwock, Helenówek near Łódź, Chorzów, Bielsko, Przemyśl, Częstochowa and Szczyrk, Rychbach, Lower Silesia, and Zakopane, also two other convalescent homes were established: one in Otwock, for 35 children with TB, and one in Rabka, for 100 children. Additionally, in August 1945 the Neuropsychiatric and Educational Out-Patient Centre was established, it cooperated with the Institute of Mental Hygiene in Warszawa and psychologists of Warsaw Municipality Board.

Until December of 1945, there were 14 dispensaries all over Poland including newly opened in Wrocław and Gdańsk. Nevertheless, the biggest problem of Jewish health care was TB. There were in total 1072 patients with TB, registered in the following Jewish committees: Łódź (297), Kraków (250), Wrocław (188) and Częstochowa (127) (AŻIH, 324/482, p. 26).

in 1945 CKŻP had no establishments for patients to recover after TB and sent the people to state owned convalescent homes. The Ministry of Health covered 50% of the cost of treatment of every patient. In Lower Silesia, TB Jewish patients were sent to special convalescent homesin Jar-Sokołowsko (currently Sokołowsko), Krzyżatka (currently Kowary) and Buchwald (currently Bukowiec). The patients were sent with full financial and food support. In December 1945, the organization obtained financial help from two foreign organizations – OSE (*Oeuvre de Secours aux Enfants*), JOINT (*American Jewish Joint Distribution Committee*) as well as from PCK (Polish Red Cross).

8.4 Origins of the Society for Safeguarding the Health of the Jewish Population in Poland After the War

Thanks to the doctors' pressure, prewar activists and foreign organizations financing TOZ (OSE, among others), TOZ was formally reactivated in August 1945. It took place despite serious concern of some of the Jewish activists arguing about potential conflict of interest with CKŻP as both were beneficiaries of foreign financial aids for Jewish community in Poland.

On 12th of August 1945, a special Conciliation Commission passed a resolution on establishing of the Society for Safeguarding the Health of the Jewish Population in Poland by CKŻP. Its purpose was to be an „organization and maintenance of all healthcare for Jewish population in Poland including mother and child care" (Einhorn 2008, p.128). The organization, although still within CKŻP structures maintained highly autonomous; had its own structure, and, most importantly, had independent financing.

The main aim of the Society was: „improvement of health status of Jewish population, prevention of diseases, mother and child healthcare". These were to be achieved by: "establishing hospitals, clinics, medical centers, labs, convalescent homes, preventoria, nurseries, rest homes, TB clinics, VD clinics, pre-marriage centers, X-ray facilities, physiotherapeutic units, but also gathering of medical and material resources as well as sanitary oversight on boarding schools, shelter homes and children homes, soup kitchens and medical protection over them, organization of lectures, discussions and exhibitions" (AŻIH, 324/1, p. 66).

The expansion of TOZ activity took place at the beginning of 1946. In February 1946, TOZ officially took over the health departments of CKŻP including all medical facilities with their staff.

One of the most important responsibilities of TOZ was to take care of several dozen of thousands of Jewish repatriates from the USSR who started to flood into Poland from February 1946. Among them were many diseased and underfed people, in bad physical and psychological shape often demanding specialist medical treatment. Until the end of July 1947 136.579 Jewish repatriates came to Poland and settled mainly in Lower Silesia and also in Szczecin, Łódź, and Upper Silesia (Adelson 1993, p. 397). New TOZ branches were opened in all newly established Jewish settlements in that period, mainly in Lower Silesia.

8.5 TOZ Structure

The headquarters of the Society was located in Łódź, with a special office in Warszawa to coordinate contacts between HQ, the Ministry of Health and JOINT. It also had a consultative role, and role in planning and approval of reorganization plans, as well as development or liquidation of units, offices and medical institutions. Organization's office was directly subjected to it, including its inspectorates – medical, and organizational/budgetary. The first one controlled medical activities of all branches, and the second one controlled all budgets.

The Headquarters had three main offices responsible for medicine, mother and child care, propaganda and statistics. The two first were most important. The medicine office had the following departments: hospital care, dispensaries and convalescent care, medicine distribution control, hygiene and prevention.

Initially, hospital care was based on state facilities since TOZ did not have a hospital on its own. It is worth to mention that the need of having its own hospital was seen as very important from the beginning, and as early as in 1946 the plans to establish a JOINT financed Jewish hospital were made. Meantime, however, there were some attempts to regain hospital facilities owned by Jewish communities or organization in the prewar times, however unsuccessful. The problem of patients' hospitalization was solved by directing them to the municipality, county and medical facilities of Social Insurance system and covering costs of their treatment. Besides, the lack of hospitals was partially dealt by placing patients in the sick bays operating in 6 places – Kraków, Łódź, Warszawa, Białystok, Przemyśl and Tarnów. In the biggest Jewish agglomeration, Lower Silesia, there were no plans to start sick bays since it was natural to start quickly a Jewish hospital there. Yet Lower-Silesian Jews had to wait for its realization until the half of 1948 (Einhorn 2008, pp. 188–194).

Ambulatory treatment was carried out in 47 dispensaries operating in places where TOZ had its branches. Some included laboratories. In smaller Jewish communities (with TOZ units), patients were directed to doctors commissioned by the Society. A similar situation occurred whenever a specialist consultation was necessary.

convalescent homes initially concentrated in few centres – rest house in Krynica and TB convalescent homes in Otwock and Jar (currently Sokołowsko). Unfortunately, the number of places in these facilities was insufficient and patients were also treated in state convalescent homes, mainly in Krzyżatka in Lower Silesia (currently Kowary) and Głuchołazy in Upper Silesia.

TOZ also supervised distribution of medicine by Central Pharmacy Repository. In 1946, the most important problems of this department were shortages of medicine in various sub branches of the Society and inadequate allotment of medicine, incompatible with actual demand. The problem was solved by ordering regional TOZ branches to distribute medicine to their sub branches and units and controlling of actual demand of particular medicine.

Hygiene and prophylaxis were in the main sphere of activities of TOZ in the interwar period. After WWII, for obvious reasons, the sphere became secondary, however, TOZ covered the sanitary and hygienic care in most of the Jewish social institutions. Medical staff of the Society worked in, so called, repatriates' homes, kitchens, canteens of CKŻP and productivity centres.

Mother and child care office of TOZ had 11 special care units in all of Poland. Besides, there were 47 establishments with paediatricians and gynaecologists, although there was a shortage of qualified specialists and trained nurses. Apart from this activity, the Society run complex medical examinations for Jewish children in open, half open and closed TB care establishments. Children qualified for an immediate treatment were directed to convalescent homesand preventoria. In 1946, a TOZ TB convalescent home for children was moved from Otwock to Śródborów. It could treat 80 children at the same time. There was a special ward in Śródborów for children up to the age of two and there were plans for opening a ward for patients with bone tuberculosis and a preventorium. Among medical establishments for children in that period, it is worth to mention convalescent home and preventorium in Gierzcze Puste (currentlyGłuszyca) in Lower Silesia for 80–90 children, Mother and Child home in Soplicowo near Otwock, and Rest Home for Children in Opole.

Important project of Mother and Child Care Department was to create the central record of children with TB. TOZ carried out medical examinations of all Jewish children from children homes and patients of open and half open TOZ facilities to screen for TB (Biuletyn Centrali TOZ 1946, p. 64). The action aimed to disclosed all children with TB or at risk of developing it. The records had also an additional function – it helped to plan building new convalescent homes and preventoria for children and expand activity of open care establishments.

TOZ took particular care of pregnant women and breastfeeding mothers in difficult financial situation. They could count on a special maternity benefits for 5 months (2 months before giving birth and three after). Additionally, TOZ supported underfed children by starting milk drinking units which operated, among other places, in Szczecin (600–700 children), Łódź and in Lower Silesia – Kłodzko and Frydland (currentlyMieroszów).

TOZ expanded its medical and hygienic care to 28 CKŻP schools, 43 kindergartens, 16 child homes, 24 kibbutzes and many other establishments for children. Office of Mother and Child Care had the same problems as adult care units. One of them was a problem with stomatology care which was planned to be solved by the use of mobile dentist ambulances. Thus children without an access to proper dentist's care could be treated.

In 1946 the Society was given the care on mother and child until age the age of two. TOZ became responsible for organization of establishments for small children.

That year, there were 14 nurseries for 400 children only in Lower Silesia and some more were under development. Besides, there were homes for small children – in Szczecin for 70 children and also in Kraków and Łódź for 40 children (Einhorn 2008, pp. 137–138).

Office of Propaganda and Statistics concentrated on health promotion. Any form of statistic research proved to be extremely difficult due to the fluctuation of Jewish population. Since there were no current and detailed data on the number of Jews in Poland it was impossible to estimate morbidity or incidence rate of certain diseases (especially TB). Nevertheless, records of patients were being collected and the relevant offices were constantly informed about an increase of sick.

The office of propaganda activities consisted, among other tasks, concentrated on publishing, including informational leaflets. The materials were prepared in Polish and Yiddish, as it was considered a language of the "Jewish masses". Among other publications there were posters illustrating fighting with TB and treatment and prevention of contagious diseases with information for pregnant and feeding women as well as on child hygiene. Regional TOZ branches also run lectures and popular talks of doctors of various specialties.

One more point stressed by Society activists in 1946 is worth mentioning. Namely, the necessity of taking care of issues of psychological hygiene among Jewish people. Doctors and TOZ activists already knew of the existence of serious mental problems among survivors – both those who had traumatic experiences of ghettos and camps and those who, not having experienced the Holocaust themselves, had lost whole families during the war.

8.6 TOZ Between Years 1946–1949

Until April 1946 the branches of TOZ were opened in the following cities: Białystok, Warszawa, Lublin, Łódź, Olsztyn, Rychbach (Dzierżoniów), Kraków, Katowice and Przemyśl (Biuletyn TOZ-u 1946, p. 11). Elżbieta Ostrowska in her article published in "Krakowski Rocznik Archiwalny" tells a story of the Kraków branch: "There was a sick bay by Długa Street with 11 rooms and 12 doctors' offices where doctors from TOZ were treating Jewish people free of charge" (Ostrowska 1995, p. 69). The Łódź branch was described in „Biuletyn Żydowskiego Instytutu Historycznego" as follows: „At the beginning of 1946 the duties formerly performer by office of healthcare of CKŻP were taken over by the reactivated.

Society for Safeguarding the Health of the Jewish Population in Poland with participation of doctors working in Łódź including Mieczysław Kowalski, Abram Kryński, Leon Szykier, Stefan Chrzanowski. The headquarters of TOZ operated in Łódź between years 1946–1948. The Łodź branch run, among other facilities, a hospital for 60 beds at 30 Stalin Street (currently Piłsudski Alley), an infirmary for adults at 21 Gdańsk Street and also a clinic for mother and child" (Olejnik 1998).

Local branches were opened in smaller towns. In the discussed period, following spas and convalescent homes operated: lung diseases convalescent homein Jar-Sokołowsko, rest home in Krynica, children home in Olin and Soplicowo, and a preventorium for children in Gierzcze Puste (Głuszyca). The most severe problems the TOZ had in that period were lack of adequate space (rooms) and financial problems due to irregular payments from JOINT. The first one was to be solved by regaining Jewish hospitals, convalescent homeand infirmary buildings all over the country.

Among the demands of Medical Office of TOZ formulated in 1946 there was a necessity of opening Jewish hospitals and sick bays. Additionally, in towns with population over 1000 Jews it was called for establishing policlinics (multi-specialist outpatient centres), and in less populated centres it was suggested to hire individual doctors. On the organizational side it is worth to mention the creation of a plan to develop new branches of TOZ. They were to be established in places with population of Jews exceeding 1000 citizens, but also in places with a lesser number of

Jews. Regional branches were asked to start opening therapeutic-prophylactic, mother and child care and repatriation settings.

The number of branches and establishments of TOZ on June 30th 1946 for respective regions was as follows: in Warszawa region there was an all-speciality infirmary for adults, a sick bay, and an infirmary for children, TB children convalescent home in Otwock, Mother and Child Home in Soplicớw near Otwock and a preventorium in Śródborów was in implementation. In Cracow region there was an infirmary for adults, an infirmary for children and a sick bay and clinic for Mother and Child. Additionally, there were offices of TOZ in Tarnów with a sick bay and in Nowy Sącz where an infirmary was established. The structure of the Society in Upper-Silesian region consisted of Regional branch in Katowice and 9 sub branches – Bielsko, Będzin, Bytom, Chorzów, Cieszyn, Gliwice, Opole, Sosnowiec and Zabrze. In 1946, infirmaries were in all aforementioned towns. There was a labour home in Bytom and a rest home for children in Opole. In Pomerania region, with many Jewish repatriates settled in Szczecin and its environs, an infirmary was in an organization stage. In Lublin region TOZ establishments operated only in Lublin itself with an infirmary, a nursery for 20 children and a sick bay. Additionally, there were singular establishments in other regions – infirmaries in Włocławek and in Gdańsk-Wrzeszcz (under organization). The TOZ paid the particular attention to Lower Silesia since there were over 86,000 Jewish repatriates from the USSR. Due to this fact, TOZ transferred 50% of funds, medicine and food to Lower Silesia (AŻIH, 324/7, p. 6).

In 1947, there were 10 regional branches of TOZ – Białystok, Dzierżoniów, Gdańsk, Katowice, Kraków, Lublin, and Łódź, Przemyśl, Szczecin and Warszawa. Local branches operated in Bydgoszcz, Częstochowa and Włocławek and offices in Olsztyn and Poznań. The biggest number of branches – 25 – operated in Lower Silesia. The second region according to the number of TOZ branches was Katowice region (Upper Silesia) where TOZ offices were opened in 7 towns – Bielsko, Bytom, Chorzów, Cieszyn, Opole, Sosnowiec and Zabrze. Next were Kraków and Łódź regions with four offices in the first one – Gorlice, Nowy Sącz, Olkusz and Tarnów, and the second one – in Kalisz, Ozorków, Piotrków Trybunalski and Zduńska Wola. The smallest number of offices operated in Szczecin Region – Golęcin and Żelechów and Gdańsk Region – only one in Sopot. There is no data on sub branches in Białystok, Lublin and Przemyśl regions.

Infirmaries were in all above-mentioned towns, nurseries in 21 of them, prepared for total number of 582 children, however out of 21 nurseries in Poland as many as 16 operated in Lower Silesia.

In 1947, TOZ run 6 sick bays – in Białystok, Kraków, Lublin, Łódź, Przemyśl and Warszawa. Clinics for mother and child operated in 34 places including 16 in Lower Silesia. There were only two labour homes – in Bytom and Łódź and dental surgeries in 9 places – Częstochowa, Jelenia Góra, Kraków, Łódź, Sosnowiec, Świdnica, Świebodzice, Szczecin and Włocławek. TOZ TB establishments were located only in Otwock – for children and adults, and in Jar-Sokołowsko, and preventoria for children endangered with TB in Gierzcze Puste – Głuszyca and Otwock (AŻIH, 324/7, p. 10; 324/584, pp. 13–14).

In 1948, TOZ activities in healthcare were very developed. There was a hospital with internal and maternity wards in Łódź, at 30 Stalin Street. In Upper Silesia TOZ run Labour Home in Bytom, and a sick bay in Kraków. TB convalescent home for adults operated in Jar-Sokołowsko and for children in Śródborów. There was a Health House in Otwock and TOZ preventoria operated in Głuszyca and Otwock. Health centres of the Society operated in 9 places, infirmaries in 44 and nurseries in 19 locations.

Other forms of healthcare of TOZ in 1948 included elderly homes, children homes and less abled people homes. Additionally, there were three training actions: a four-weeks training course for nurses from clinics for mother and child in Lower Silesia, a half-a-year nurses course in Wałbrzych with 40 female participants and conferences for specialist doctors – paediatricians and physiologists (AŻIH, 324/7, pp. 19–23).

In 1949, there is a decrease of TOZ branches. Liquidation of number of establishments was connected with Jewish emigration from Poland or internal migrations (from small Jewish centres to bigger cities) which was especially visible in Lower Silesia. According to "A Survey of sub branches of Regional Lower-Silesian TOZ" in 1949 there were only 15 centres and three medical establishments –convalescent home in Jar-Sokołowsko, a preventorium in Głuszyca and a hospital in Wałbrzych (AŻIH, 324/1169, p. 9). During 1949, the number of branches was reduced by another 7 – branches in Bolków, Kamienna Góra, Lubawka, Ludwików (currently Ludwikowice Kłodzkie), Nowa Ruda, Nowa Sól and Piława Górna were liquidated.

In 1949, another establishment of TOZ was ceremonially opened – „Brijus" convalescent homein Otwock. A Zionist magazine „Mosty-Nasze Słowo" wrote: „One unit of aforementioned convalescent homes for bone tuberculosis for 70 patients is completed and it will be fully equipped in a matter of days. Works in the second unit for lung tuberculosis for 180 patients have entered their final stage and it will be ready and equipped until July 22nd this year." (Otwarcie sanatorium TOZ-u 1949, p. 7). The convalescent homes was opened in a former establishment of the interwar period's Jewish TB Society „Brijus". Its buildings were renovated and adapted for new needs and regulations set for sanatoria.

8.7 TOZ in Lower Silesia

As it was mentioned, the biggest number of TOZ branches was opened in Lower Silesia. It was due to the fact that after the war the biggest Jewish population inhabited the region. According to some researchers, over 100,000 Jews who had survived the Holocaust settled here (Waszkiewicz 1999, p. 18).

The beginnings of TOZ in Lower Silesia were not easy: „Successively, as trains with repatriates were coming to various places, the TOZ committees were being formed; sometimes in cars, at railway stations or in repatriates homes or in synagogues designated as temporary shelters. They were being elected in mass meetings and started to work immediately. Doctors, nurses and hospital attendant were appointed from the newcomers who were forming the new staff of TOZ activists. Such were the circumstances of forming the first branches of TOZ in Świdnica, Piotrolesie, Bielawa, Wrocław, Ludwików and Niemcza" (Bayer 1947, p. 5).

In February 1946, J.H. Bayer MD took charge of the Silesian Health Department by the Regional Jewish Committee. He took the task of organization of medical and sanitary care for Jewish repatriates settling in Lower Silesia. Rafał Ogiński MD and Fryszman MD became his associates. Lower-Silesian TOZ was formally established in June 1946 when the Society's HQ allowed „to open a regional branch of TOZ in Dzierżoniów with the right to open sub branches and offices in Lower Silesia" (AŻIH, 324/554, p. 24).

The first Lower-Silesian TOZ establishments were: a TB convalescent home in Jar-Sokołowsko, a temporary hospital in Głuszyca and a preventorium for children endangered by TB in Gierzcze Puste (currently Głuszyca). The first patients of those establishments were concentration camp prisoners – in Jar the female prisoners from *Stutthoff* camp and prisoners from *Auschwitz-Birkenau* (Jar – sanatorium „TOZ" dla płucnochorych 1946, p. 3). The preventorium in Głuszyca could provide a sty for 100 children and was located in former *Kinderheim* building which, after the war, was part of CKŻP. It started treating patients since March 1946. Among its employees were Tobiasz Cytron MD, Ehrlich MA and Kahanowa MA. There is a description of the beginnings of the establishment: „Głuszyca (Gierzcze Puste) – infamous for location of concentration camps during German occupation, immediately after the war, when Red Army came to the area, the town found itself in very difficult situation. Thousands of diseased Jews released from the camps had to receive an immediate care and treatment which could not have been obtained there. Central Jewish organizations were located in distant Łódź and Warszawa and were not able to provide much help to the Jews liberated from the camps. Among the former prisoners were few energetic people who managed to open 8 hospitals for 3.800 patients including 1.500 with typhoid (…) The hospitals were gradually reduced with decreasing number of patients. The last one was liquidated on the 1st of February 1946. The board of hospitals did not disband and kept on taking care of patients" (AŻIH, 324/587, p. 6).

Until April 1946, 18 local branches and offices of TOZ were opened. Dzierżoniów became the Lower-Silesian centre of the Society. Initially, the centre of the Regional Jewish Committee was also located there. A complex of clinics of all specialities was also opened in the town – a mother and child clinic, a TB clinic, a VD clinic and a dental surgery as well as the Regional Medicine Repository and offices of Regional TOZ board.

TOZ took medical care over children homes, kibbutzim, boarding schools and schools (no matter provenance). Infirmaries, functioning like clinics, operated in Bielawa, Dzierżoniów, Jelenia Góra, Kłodzko, Legnica, Świdnica, Wałbrzych and Wrocław. It is worth to mention that they were also used by non-Jewish population.

On June 30th 1946 TOZ had 23 sub branches in the following places: Biały Kamień, Bielawa, Boża Góra (currently Boguszów-Gorce), Bystrzyca, Duszniki,

Dzierżoniów, Frydberg (currently Mirsk), Jelenia Góra, Kamienna Góra, Kłodzko, Kudowa, Legnica, Lubawa, Nowa Ruda, Piotrolesie (currently Pieszyce), Puszczykowo (currently Polanica-Zdrój), Solice (currently Szczawno-Zdrój), Świdnica, Wałbrzych, Węglowo, Wrocław and Zgorzelec. Additionally, there was a TB sanatorium in Jar-Skołowsko and a preventorium in Gierzcze Puste (Głuszyca) (AŻIH, 324/7, p. 3). Different data can be found in a work of Szyja Bronsztejn who claims the Society had 33 sub branches in Lower Silesia in 1946. The Society was running its own clinics, policlinics, convalescent homes, preventoria and pharmacy units. TOZ branches were located both in large Jewish centres (Dzierżoniów, Wrocław, Wałbrzych), and in small ones (Zgorzelec) with staff of over 100 doctors of many specialities (Bronsztejn 1991, p. 266).

Within the care of small children TOZ organized nurseries and summer play centres for poor working mothers. The first nurseries were opened in Bielawa, Dzierżoniów, Lubawka, Piotrolesie (Pieszyce), Strzegom and Świdnica.

In 1947, Jewish settlement in Lower Silesia entered the phase of some stability. The number of TOZ establishments was biggest in this year. Until January 1st 1947 there were 33 branches in Lower Silesia and three closed establishments – in Jar-Sokołowsko, Głuszyca and Krzyżatka (AŻIH, sygn. 324/554, p. 26). There were 33 infirmaries by the branches, 27 clinics for mother and child, 8 TB clinics, 28 doctors' offices, 30 pharmacies, 16 dentists and one medicine repository in Dzierżoniów. Additionally, TOZ took over the care in 33 half-dormitories, 15 children homes, two hostels, two orphanages, three senior homes, 41 kibbutzes, 17 schools, 9 playschools, and 12 nurseries for 360 children and one rest home.

In 1947, TOZ employed 101 doctors and 52 nurses. Since the Society had no hospital of its own, it financed treatment of their patients in state hospitals – there were about 300 patients monthly.

Dzierżoniów, Wrocław and Wałbrzych became the most important branches in Lower Silesia. Regional Branch in Dzierżoniów was established in March 1946 but it started a dynamic development in May 1946 when its office was moved to a new building where also a multi specialist clinic started to operate. In the words of former activists: in that period, thanks to the efforts of the Circle of Demobilized Jewish Soldiers in Rychbach (Dzierżoniów), TOZ ceased to be homeless. It was given a home where which, after renovation, was to host a clinic of all specialities, clinic for mother and child, a TB clinic, a VD clinic and a dental clinic, the Regional Medicine Repository and offices of the TOZ Regional Board. Thanks to the extraordinary organizational flair and devotion of Doctor Ogiński, a house ruined by bombs was renovated in short time, the doctors' offices equipped with indispensable utilities and instruments" (Towarzystwo Ochrony Zdrowia TOZ 1947, p. 6).

Dzierżoniów TOZ infirmary was also used by other, non-Jewish citizens of the city, among whom the TOZ was very popular. The Society operated also in the days free of work (Saturdays and Sundays) with a doctor and a nurse on duty and an operating pharmacy. It is worth to mention that both TOZ and CKŻP, by which it was affiliated, despite initial pressure from Jewish religious circles, treated Saturday as a normal working day in all of its establishments.

Local branches of the Society attached much importance to the care of children and youth. A TOZ annual summary from Lower Silesia reads: „Because of the particular care for children and youth, TOZ, as an independent organization, spans its medical and hygienic care on all children homes, kibbutzim, boarding schools and schools no matter if run by a Zionist organization, Bund, a religious association or a Jewish Committee – our doctors reach everywhere, our school nurse is at watch everywhere. All children and youth undergo periodic medical examinations. Workers are also examined and these threatened by TB and other contagious diseases are marked for treatment. Patients are placed in applicable establishments or they are treated in surgeries. Medical and hygienic conditions of Jewish establishments improve month by month" (Towarzystwo Ochrony Zdrowia TOZ 1947, p. 7).

According to one of TOZ activists –Norkin MD – Wrocław branch of the Society was established by the initiative of the Society's HQ in March 1946 (Norkin 1947, p. 65). The first Jewish medical establishment in the city was an infirmary operating by local Jewish committee and treated patients for free. There were only two doctors working there – they treated patients on the premises as well as visited them at homes. In time, it gained a proper place which was renovated. The number of medical personnel increased as well. The Society employed an ophthalmologist, a paediatrician, a physiologist, a gynaecologist a neurologist and three dentists. However, due to the arrival of several dozens of thousands Jewish repatriates from the USSR to Lower Silesia, among whom many were in need of specialist treatment, TOZ staff had to face new challenges. One of the most important was the fight with contagious diseases including malaria brought to Wrocław by resettlements from Kazakhstan and Uzbekistan. Thus, TOZ opened a specialist medical office to treat malaria. All patients were registered – those who came to the infirmary as well as those from kibbutzes, children homes and schools. In summer 1947, there were 40 new cases monthly. Thanks to procedures undertaken by TOZ, there was no single new case by July 1947 (Praca TOZ-u na polu zwalczania malarii 1947, p. 7).

Between March 1946 and February 1947 there were 4 out-patient clinics in Wrocław: a TB clinic, a VD clinic, an anti-trachoma clinic and a Care Station for Mother and Child. TOZ had also contacted local hospitals, and X-rays facilities of TB out-patients clinics.

In 1947, „Nowe Życie", a magazine of the Regional Jewish Committee in Lower Silesia, summed up 1 year of TOZ's work in Lower Silesia in the following way: „The protection of Jewish population who had suffered the most, including the matter of health, from atrocities of Hitler's war, is one of the most important issues of Jewish community. TOZ now, as before the war, is strongly committed to the task. The Wrocław branch of TOZ is being managed well despite its limited resources." (Rok pracy TOZ-u we Wrocławiu 1947, p. 7).

Particular role of Wałbrzych, as one of the centres of TOZ, was connected with opening of the only one Jewish hospital in Lower Silesia. The institution was ceremonially opened on July 3rd 1948 in a building adapted for medical needs by a municipal park in Nowe Miasto. The works and equipment was completely financed by JOINT, and Dawid Guzik became a patron of the establishment – the director of the organization in Poland who died tragically. In the speech of TOZ's general secretary on the occasion of the opening we read: „It is not a coincidence but an act of historical justice that the hospital co-funded by JOINT is located in Lower Silesia, the land which used to be a place of concentration camps. Not for the first time Jews build hospitals in Poland. Building of a Jewish hospital on Czyste (in Warszawa), Poznański Family hospital in Łódź, Jewish hospitals in Kraków, Białystok, Lublin and other places – are the Samaritan's acts written down in the history of Polish Jews. (Uroczyste otwarcie szpitala „TOZ-u" 1948, p. 8).

The hospital could contain 150 patients at a time and had three wards: surgical, internal and gynaecological/maternal. It was arranged according to the most recent medical demands and generously equipped with various utensils. There was a policlinic by the hospital as well as nursery for 80 children. In 1948, a half-a-year course for nurses was organized there with 30 female participants (AŻIH, 303/12, pp. 7–8).

Dawid Guzik hospital's openness was its specific feature – it was not exclusively Jewish hospital but served the local population as well. Thus, Polish citizens of Wałbrzych had a chance to benefit from modern medical equipment and medicine financed by JOINT. In „A Survey of TOZ activities" from April 1949 we read as follows: „The statistics of bed occupancy in our hospital in Wałbrzych indicates 28% Jewish and 72% Gentiles. This is a vivid proof that we do not create separatism as it may seem to the people who are not aware of our aims. Our institutions have never been closed or even half-closed. They are wide open for all working people who deserve our help, regardless religion or nationality" (AŻIH, 1193, pp. 111–112).

In the same time (July 3rd 1948), another TOZ establishment was opened – Work Convalescent homein Jar-Sokołowsko. It was opened by a still operating TB covalescent home. The patients of the convalescent home were persons with chronic tuberculosis, closed or open, who did not require any active treatment. According to the character of the establishment, one of the healing factors was work. Number of active hours was estimated by a doctor individually for each patient. Convalescents were employed in the covalescent home until their full recovery and restoration of full working capacity through the gradual increase of working duties. Patients had an opportunity to work in two departments – sartorial and underclothing and leather and goods (AŻIH, 1193, pp. 111–112).

Jewish Dawid Guzik Hospital in Wałbrzych operated as a TOZ unit only until September 1949 when it was taken over by the local government in the process of liquidation of Jewish autonomy in Poland. On September 21st 1949 the Council of Ministers passed a resolution on „taking over, by the local governments, some hospitals run and funded by associations: „the Polish Red Cross" and „the Society for Safeguarding the Health of the Jewish Population" and by congregations, unions and religious foundations". According to the resolution, Municipality of Wałbrzych gratuitously took over movable property and properties hitherto used by the hospital and being properties of TOZ. Additionally, the new owner also took over the authorization and obligations of hospital maintenance.

The building of the hospital of the Society for Protection of Health of the Jewish People in Poland contains a hospital up to date. Rurrently, the Edmund Biernacki Specialist Gynaecologist-Maternity Hospital is located there.

8.8 TOZ Personnel on Lower Silesia

By the example of Lower Silesia – the biggest post-war Jewish concentration in Poland – one can research the structure of employment of TOZ medical personnel. In 1947, in Lower-Silesian branches of the Society, 101 medical doctors, 59 nurses and junior surgeons and 36 persons of auxiliary staff were employed (Towarzystwo Ochrony Zdrowia TOZ 1947, p. 11).

Medical personnel were diverse in both aspects – education and nationality. Most of the doctors were graduates of Polish medical universities in Warszawa, Lwów and Kraków, prestigious universities in Prague and Vienna but also persons who gained crash-course medical qualifications during the war. Among the employees in TOZ Kłodzko branch there were graduates from Lwów University (Maurycy Lauterstein MD, Leon Schneeweiss MD), Prague University (Adolf Friedwald MD), Vienna (Ludwik Landau MD), and also medical universities in the USSR (Maksymilian Kluger MD, Lidia Złotowicz MD) (AŻIH, 324/403, p. 72). Lower medical personnel consisted of junior surgeons and nurses. In Lower Silesia, Jewish Russian females and Russians females belonged to a separate category – they were wives of Polish Jews who were repatriated. The same situation occurred, among others, in Złotoryja (Włodarczyk 2016, p. 104). A local specificity was employment of Germans as lower medical personnel, like, for example, in the hospital for prisoners of concentration camp Gross-Rosen in Głuszyca.

According to TOZ policy, the Society tried to employ Jewish doctors, best, able to communicate in Yiddish enabling communication with patients. However, due to the shortage of manpower, and later to the emigration of Jewish doctors and nurses, TOZ establishments started employing Polish and Russian doctors.

On the basis of an archival document „List of medical personnel and auxiliary medical personnel of TOZ in Poland" the structure of employment of the organization in Lower Silesia in 1949 can be surveyed. In general, in 13 establishments worked 69 persons of medical personnel, including 48 doctors and 21 persons of auxiliary personnel. In Wrocław branch, the personnel consisted of 16 persons, including 11 doctors of various specialities, one pharmacist and 4 trained nurses. The second biggest TOZ centre was Dzierżoniów with 13 specialists including 8 doctors, a junior surgeon, a midwife, two nurses and assistant – pharmacist. In Kłodzko branch, there were 8 persons including 6 doctors of various speciality and two trained nurses. Couple of persons of medical personnel worked in each of the bigger Jewish centres – in Legnica and Żary – 4 (three doctors and a nurse), in Świdnica and Bielawa – three (two doctors and a nurse), in Jelenia Góra – two (a doctor and a junior surgeon) and in Niemcza –two (a doctor and a nurse). Singular doctors worked in Bolków, Strzegom and Ząbkowice. For the overall number of 13 Lower-Silesian centres, the most common speciality among medical personnel was internist, then dentist, paediatrician and gynaecologist (AAN, 25, p. 66).

8.9 End of TOZ

The liquidation of the Society for Safeguarding the Health of the Jewish Population in Poland was one of the elements of limiting Jewish autonomy in Poland, which took place between years 1949–1950. In case of the Society it was preceded by a process of changing of the authorities and internal policy.

A national congress of TOZ was held in April 1949. There was a lot of sharp criticism coming from the group of Communist activists. The Society was accused of bureaucracy in administrative structures, lack of contacts with members of TOZ, and improper attitude towards patients – „in the spirit of unhealthy philanthropy" (AŻIH, 1193, p. 80).

Some of the political threads seemed particularly typical, especially those stressed in speeches of some activists who introduced a new profile of the organization: „TOZ has a duty of treating Jewish population with its often bourgeois way of thinking. One of the forms of depravation of Jewish psychology is a philanthropic way of helping patients in need. It must be fought with. We need to educate a new type of doctor who, together with all working Jewish masses working for TOZ, will have a big range of activities" (AŻIH, 1193, p. 85).

There were some various demands formulated concerning the Society's activities. Among other voices, the necessity to economize within the nation-wide economical action and to end the excessive employment of administration staff, were raised. A new board was elected during the congress and Communist activities hoped: „I believe the next board will lead TOZ towards the ways corresponding with our reality, and new forms of work, coordinated with general principles and directives of general healthcare in the country, will speed up the march towards a Socialist Poland" (AŻIH, 1193, p. 91). There were members of Communist Party in the new board what marked the ultimate end of organization's independence. There were also some changes in the organization's statute (AAN, 476/20, pp. 212–222).

In June 1949, Szymon Zachariasz – a member of the presidium of CKŻP and a member of the Central Commission of Party Control KC PZPR – presented, according to the guidelines of Central Committee, a project of nationalization of Jewish institutions. Hebrew schools, kibbutz, parties and Zionistic organizations were to be liquidated. The project also restricted JOINT activities in Poland (which financed most of the Jewish organization) and Jewish institutions (schools, children homes, theatres, publishing houses) were to be taken over by the state. Zachariasz also envisaged disbandment of CKŻP and of the Jewish Cultural Society, and establishment of a new Jewish organization. The final decision on nationalization of Jewish institutions was made on the meeting of the Secretariat of KC PZPR on August 4th 1949: „In the Fourth point of our daily agenda, comrade Mazur reported the situation of Jewish organization [...] concluding, CKŻP had made a great effort and thanks to it all Jews have normal work conditions and living conditions and do not need any special material aid. Thus. this was decided:

1. All institutions including children homes, boarding schools etc. hitherto run and financed by the Jewish Committee will be taken over by the state from January 1st 1950;
2. terminate all the activities of JOINT from January 1st 1950 r. Inform JOINT about the decision 2–3 months before the deadline" (AAN, 295/VII, p. 72).

The liquidation of TOZ was carried out in two stages. In September 1949, a decision was made to dismiss TOZ employees. A message sent by Main Board to the TOZ's regional branches reads: „Due to the reorganization of operations discharge all employees by 30th of September. The directive does not apply to nurseries personnel. Instruction on further proceedings will be sent by letter" (AŻIH, 324/1193, p. 15). Taking over of all establishments ended in spring 1950. Some institutions were liquidated, others (like Dawid Guzik Hospital in Wałbrzych) was taken by the state.

The reasons of these events can be seen in the complex Polish internal situation and increasing Stalinization. Reasons to preserve the autonomy of Polish Jews ceased to exist since the Polish reality drifted towards Soviet solutions. Polish authorities decided to limit Jewish activities to the spheres of culture and religion (in a very restricted form). It led to the liquidation of the majority of Jewish social and political organizations and dissolution or nationalization of Jewish institutions.

Actually, liquidation of TOZ and other Jewish organizations and parties in Poland marks the end of Jewish autonomy in Poland existing from 1945. It was the end of independent political, social, economic and cultural development of Jewish population in Poland. The process consisted of few factors: nationalization of Jewish education, publishing, separate healthcare (TOZ), withholding of aid form Western organizations (mostly JOINT'S), and liquidation of Jewish cooperatives and the ban of political pluralism.

Bożena Szaynok names TOZ as one of the few Jewish organizations which managed to preserve their independence (Szaynok 2000, p. 185). However, the fact that there were reasonably strong Zionistic influences, since the Society had close contacts with JOINT and other foreign organizations (OSE), had led to focus the communists' attention also on this organization.

8.10 Summary

The history of Jewish healthcare in Poland after WWII is essentially the history of TOZ – an organization extremely merited for Jewish community. Its activity after 1945 was one of the many numerous Jewish social, political, educational and cultural organizations being the elements of Jewish autonomy in Poland.

However, while evaluating TOZ's activity in that period, the Society cannot be perceived as an organization which only gave help to Polish Jews, but it should be treated as a big social movement which encompassed many aspects of Jewish social life. A specific feature of the Society's activity was a particular compilation of traditions of Jewish social and self-help organizations with the modern sense of healthcare and social care. The openness of the organization was also very distinctive, even in the interwar period – its establishments were never closed for Poles and accepted all patients, regardless of religion or nationality.

After the war, TOZ operated in very peculiar socio-political situation. It was not reactivated in its pre-war form making the decision of affiliation by CKŻP. Despite that fact, and thanks to its leaders, TOZ kept its autonomy and independence to the very end, what many times was a source of conflicts in mutual relations of the two organizations. Ability of its activists to adapt to hard, post-war conditions and to fulfil imposed duties (especially to help repatriates) were what made it unique.

TOZ was one of many Jewish organizations which – liquidated in 1949 – did not reactivate in the future. Its liquidation definitely ends a period of the separate Jewish healthcare in Poland.

The Society had a great meaning for the post-war Jewish population – not only as an aid organization but also an institution which helped thousands of Jews saved from the Holocaust to regain dignity after traumatic war experiences. The words of one of leaders of the organization, Leon Gołębiewski, should be the summary of TOZ's activity: „Our efforts might be not sufficient enough compared with what we should do, but in some spheres we were successful and we are proud of it. Our reward is to see all the people we cured and supported in difficult moments. We created an institution essential for Polish Jews. TOZ is not only an organization giving help; it is also a movement concentrating masses of Jewish people" (AŻIH, 1193, pp. 111–112).

References

Adelson, J. (1993). W Polsce zwanej Ludową. W: J. Tomaszewski (red.), *Najnowsze dzieje Żydów w Polsce w zarysie (do 1950 roku)* (s. 387–477). Warszawa: Wydawnictwo Naukowe PWN.

Bayer, J. H. (1947). *Rok TOZ'u na Dolnym Śląsku. Towarzystwo Ochrony Zdrowia „TOZ". Oddział Wojewódzki na Dolnym Śląsku w Dzierżoniowie*. Wrocław: Towarzystwo Ochrony Zdrowia „TOZ".

Biuletyn Centrali Towarzystwa Ochrony Zdrowia Ludności Żydowskiej w Polsce „T.O.Z." przy CKŻP, Nr 4–5, październik-listopad 1946.

Bronsztejn, S. (1991). Ludność żydowska na Dolnym Śląsku po II wojnie światowej (nieudana próba utworzenia skupiska). *Śląski Kwartalnik Historyczny Sobótka, 46*(2), 259–276.

Einhorn, I. (2008). *Towarzystwo Ochrony Zdrowia Ludności Żydowskiej w Polsce w latach 1921–1950*. Toruń: Wydawnictwo Adam Marszałek.

Engelking, B., & Leociak, J. (2013). *Getto warszawskie. Przewodnik po nieistniejącym mieście*. Warszawa: Stowarzyszenie Centrum Badań nad Zagładą Żydów.

Epsztein, T. (2001–2011). *Inwentarz archiwum Towarzystwa Ochrony Zdrowia Ludności Żydowskiej w Polsce przy Centralnym Komitecie Żydów w Polsce (1945) 1946–1949 (1950)*. Warszawa: Archiwum Żydowskiego Instytutu Historycznego im. Emanuela Ringelbluma.

Grabski, A. (2004). *Działalność komunistów wśród Żydów w Polsce 1944–1949*. Warszawa: Wydawnictwo Trio, Żydowski Instytut Historyczny.

Jar – sanatorium „TOZ" dla płucnochorych. (1946). *Nowe Życie, 6*, 3.

Kroszczor, H., & Zabłotniak, R. (1978). Towarzystwo Ochrony Zdrowia Ludności Żydowskiej w latach II Rzeczypospolitej. *Biuletyn Żydowskiego Instytutu Historycznego w Polsce*, 1/105, 53–68.

Norkin, M. (1947). Rok pracy TOZ-u. W: *Towarzystwo Ochrony Zdrowia TOZ. Oddział Wojewódzki na Śląsku w Dzierżoniowie. Rok naszej pracy 1946 marzec 1947*. Wrocław: TOZ.

Olejnik, L. (1998). Wojewódzki Komitet Żydów w Łodzi – powstanie i główne kierunki działalności (1945–1950). *Biuletyn Żydowskiego Instytutu Historycznego w Polsce, 3*(187), 3–22.

Ostrowska, E. (1995). Działalność Komisariatu Wojewódzkiego do Spraw Ludności Żydowskiej w Krakowie 1946-47. *Krakowski Rocznik Archiwalny, 1*, 68–79.

Otwarcie sanatorium TOZ-u odbędzie się 22-go lipca. (1949). *Mosty – Nasze Słowo*, *7*(411), 7.

Praca TOZ-u na polu zwalczania malarii. (1947). *Nowe Życie*, *4*(13), 7.

Rok pracy TOZ-u we Wrocławiu. (1947). *Nowe Życie*, *4*(13), 7.

Szaynok, B. (2000). *Ludność żydowska na Dolnym Śląsku 1945–1950*. Wrocław: Wydawnictwo Uniwersytetu Wrocławskiego.

Szpilman, J. K. (1947). Szpital TOZ'u im. Guzika w Wałbrzychu. W: *Towarzystwo Ochrony Zdrowia „TOZ". Oddział Wojewódzki na Dolnym Śląsku w Dzierżoniowie*. Wrocław.

Towarzystwo Ochrony Zdrowia „TOZ". (1947). *Oddział Wojewódzki na Dolnym Śląsku w Dzierżoniowie. Rok naszej pracy 1946-marzec 1947*. Wrocław: TOZ.

Uroczyste otwarcie szpitala „TOZ-u" w Wałbrzychu. (1948). *Mosty*, *87*(209), 8.

Waszkiewicz, E. (1999). *Kongregacja Wyznania Mojżeszowego na Dolnym Śląsku na tle polityki wyznaniowej Polskiej Rzeczypospolitej Ludowej 1945–1968*. Wrocław: Wydawnictwo Uniwersytetu Wrocławskiego.

Włodarczyk, T. (2016). *Z historii Żydów w regionie legnickim*. Warszawa: Fundacja Rojt.

Archival Sources

Archiwum Akt Nowych (AAN)

KC PZPR, Spuścizna Szymona Zachariasza
476/20: Informacja zespołu PZPR Centralnego Komitetu Żydów w Polsce [25 IV 1949]
Ministerstwo Administracji Publicznej
25: Wykaz personelu lekarskiego i pomocniczo-lekarskiego TOZ-u w Polsce
PZPR, Sekretariat
295/VII: Narada Sekretariatu KC PZPR [4 VIII 1948]
Urząd Rady Ministrów, Prezydium Rady Ministrów.
5/137: Jedenaste sprawozdanie z działalności Referatu dla Spraw Pomocy Ludności Żydowskiej za maj 1945 r.

Archiwum Żydowskiego Instytutu Historycznego Im. Emanuela Ringelbluma (AŻIH)

Prezydium CKŻP
303/12: Protokół posiedzenia Prezydium CKŻP z przedstawicielami TOZ [7 X 1948]
Towarzystwo Ochrony Zdrowia Ludności Żydowskiej w Polsce przy Centralnym Komitecie Żydów w Polsce
324/1: Odpis Statutu Towarzystwa Ochrony Zdrowia Ludności Żydowskiej w Polsce przy Centralnym Komitecie Żydów Polskich

324/7: Spis oddziałów i zakładów TOZ-u na Dolnym Śląsku [30 VI 1946]
324/7: Spis oddziałów i zakładów „TOZ-u" na terenie województw warszawskiego, krakowskiego, górnośląskiego, pomorskiego i lubelskiego [30 VI 1946 r.]
324/7: Pismo do American Joint Distribution Committee [28 V 1947]
324/7: Sprawozdanie z działalności TOZ-u w 1948 r.
324/403: Spis lekarzy zatrudnionych w Kłodzku
324/482: Sprawozdanie Wydziału Zdrowia Centralnego Komitetu Żydów Polskich za okres od 1 grudnia do 31 grudnia 1945 r.
324/482: Lista chorych podopiecznych Wydziału Zdrowia przy komitetach żydowskich
324/554: Pismo Centrali TOZ-u do Urzędu Wojewódzkiego we Wrocławiu [18 VI 1946]
324/554: Pismo Towarzystwa Ochrony Zdrowia Oddziału Wojewódzkiego w Dzierżoniowie do *American Joint Distribution Committee* [17 II 1947]
324/584: Pismo z 3.06.1947 r. do American Joint Distribution Committee [3 VI 1947]
324/587: Pismo Towarzystwa Ochrony Zdrowia Oddziału Wojewódzkiego w Dzierżoniowie do Centrali TOZ w Łodzi [23 XI 1946]
324/1169: Spis pododdziałów Wojewódzkiego Oddziału TOZ-u na Dolnym Śląsku
324/1193: Depesza Zarządu Głównego „TOZ" do Wojewódzkiego Oddziału TOZ-u we Wrocławiu [30 IX 1949]
324/1193: Protokół Obrad I-go Ogólnokrajowego Zjazdu Delegatów TOZ-u odbytego w Warszawie w salach Domu Kultury Żydowskiej przy ul. Jagiellońskiej 28 [22-23 IV 1949]
324/1193: Sprawozdanie dra Chaina-Gołębiowskiego z działalności TOZ-u wygłoszone na zjeździe delegatów w Warszawie [23–24 IV 1949]

Part III
Shared Identities

Chapter 9
German-Jewish Doctors as Members of the Colonial Health Service in the Dutch East Indies in the First Half of the Nineteenth Century

Philipp Teichfischer

Abstract In the first half of the nineteenth century, a large number of German doctors entered the colonial health service in the Dutch East Indies. Among them were also a significant number of German-Jews, some of whom came from modern-day Poland. In this article, I will demonstrate that the professional outlook in the Dutch colonial service for Jewish doctors, who were often discriminated in their German home countries, was just as good as for their non-Jewish German counterparts. To this end, I will compare the time for promotion of both groups of doctors. To embed this topic into a broader context, I will start by presenting some general background information on the employment of German doctors in the Dutch military service. Then, I will focus on the Jewish doctors therein by giving some biographical details of each individual – information about their family background, education, career, decoration, and scientific work. Subsequently, I shall make some suggestions about the reasons why German-Jewish doctors entered the Dutch colonial health service, as well as about their "Jewish identity".

Keywords Colonial history · Colonial medicine · Dutch East Indies · Dutch-German history · German-Jewish doctors · German-Jewish history · German-Polish history · Jewish military doctors

9.1 Introduction

Since the beginning of the seventeenth century, large parts of modern-day Indonesia were the property of the so called "Dutch East India Company" (VOC: Vereenigde Oostindische Compagnie), one of the first multinational corporations and one of the

P. Teichfischer (✉)
Otto-von-Guericke-Universität Magdeburg, Magdeburg, Germany
e-mail: philipp.teichfischer@med.ovgu.de

M. Moskalewicz et al. (eds.), *Jewish Medicine and Healthcare in Central Eastern Europe*, Religion, Spirituality and Health: A Social Scientific Approach 3,
https://doi.org/10.1007/978-3-319-92480-9_9

biggest trading enterprises in the seventeenth and eighteenth centuries. VOC's wealth and power came mainly from its control over the so called "Spice Route" from South-East India to Europe. At the end of the eighteenth century, as a result of the 4th Anglo-Dutch War (1780–1784), as well as of fraud and corruption in their own ranks, the company got into serious financial difficulty (Gelder 2004, pp. 30–31). This, in addition to contemporary political events, especially Napoleon's march into the Netherlands and the proclamation of the Batavian Republic (1795), resulted in the liquidation of the VOC on 1 January 1800. Subsequently, the Dutch East Indies were administered by the government of the Batavian Republic. After the end of Napoleon's satellite state in 1806, administration was assumed by the Ministry of Colonies of the newly created Kingdom of Holland (1806–1810). As a result of the Congress of Vienna (1815) – after being for a short time a British Crown colony (1811–1816) – the country, which was usually called "Dutch East Indies" or "Netherlands East Indies", became officially a colony of the newly founded United Kingdom of the Netherlands (Hack and Rettig 2006; Ricklefs 2008).

The employment of Germans in the Dutch colonial health service in the nineteenth century forms part of a longer tradition of German-Dutch labour migration. Already during the VOC-period a significant part of the ship's doctors and surgeons working in the few VOC-hospitals came from German-speaking parts of the Holy Roman Empire (Bruijn 2009). Besides, Germans can also be found as medical actors in the health care services in the Netherlands itself. So far, I could identify 45 Germans who were hired between 1774 and 1815 as military doctors for the so called "Landmacht" in the Netherlands.[1] Some of these doctors later switched to the colonial health service in the Dutch East Indies.

This early employment of German doctors suggests that the Netherlands, as a result of their colonial expansion, already since the beginning of the seventeenth century had problems in meeting their needs for qualified medical staff from among their own ranks. Foreign, especially German surgeons and physicians became more and more necessary to fill this gap, thereby ensuring that the ambitious Dutch colonial health policy succeeded. This situation did not change until the middle of the nineteenth century when the absolute number of graduates from the so called *'s-Rijkskweekschool voor militaire geneeskundigen* in Utrecht increased significantly (Knecht-Van Eekelen 1988). This school was founded in 1818 in Leiden for the training of Dutch military doctors. In 1822 it moved to Utrecht and, finally, in 1865/66 to Amsterdam (Lieburg 1988). As the amount of the graduates from the *'s Rijkskweekschool* entering the colonial health services rose, the number of German doctors in the colonial troops decreased considerably (see below Fig. 9.1). This increase of Dutch doctors in the colonial health service was due in no small part to financial incentives, introduced by the Dutch government, for students of the *'s-Rijkskweekschool* in Utrecht, as well as for graduates determined to enter the colonial services (Kerkhoff 1989, pp. 14–15).

---

[1]Dutch National Archives, The Hague, Inspecteur-Generaal van de Geneeskundige Dienst der Land- en Zeemacht 1774–1868, catalog reference number 2.13.62.06, inventory number 237.

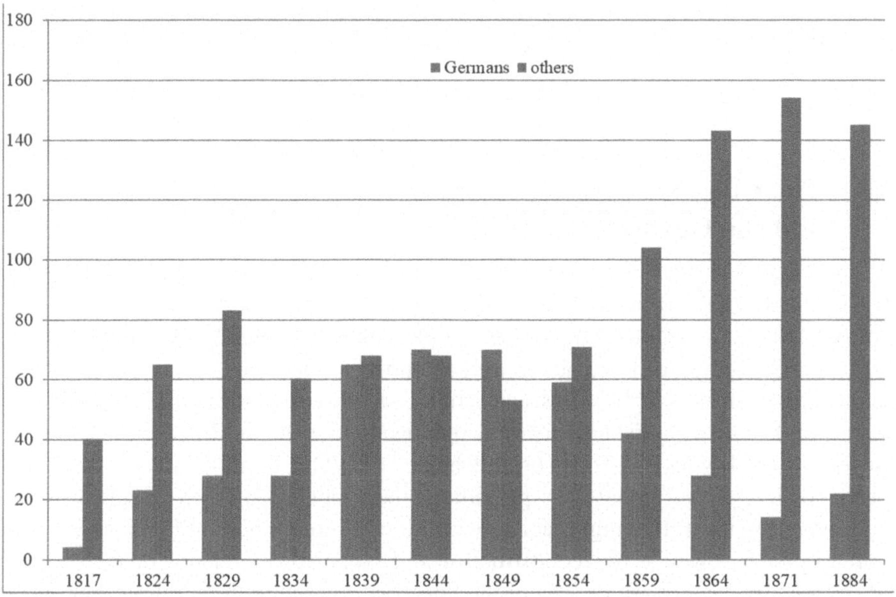

Fig. 9.1 Proportion of German doctors among the military health staff in the Dutch East Indies (1817–1884)

9.2 State of Research

Although a large proportion of the medical personnel in the Dutch East Indies in the nineteenth century consisted of German doctors and pharmacists, this historical fact is almost unknown in the contemporary historiography (Teichfischer 2016). It is therefore hardly surprising that there is no specific research literature about the (German-) Jewish doctors in the Dutch East Indies. Moreover, there is almost no literature about Jews employed as colonial officials or soldiers, or about Jews working in the Dutch East Indies as traders or in companies (Christiaans 1992b). This is also a consequence of the fact that, apparently, only since the beginning of the nineteenth century was the colonial administration open to Jews (Blom and Cahen 2002, pp. 287–288; Hadler 2004; Franke 2013). Another reason is that during the whole nineteenth century there was only a relatively small migration of citizens to the Dutch East Indies – this colony was considered more as a "colony for exploitation" than a "colony for settlement" (Bosma 2010, p. 18). In contrast, there existed a bigger Jewish community in the West Indies (Netherlands Antilles, Suriname), whose settlement was, even at an early stage, supported by the Dutch government. About this Jewish community and about their doctors and pharmacists some research has been done (Emmanuel & Emmanuel 1970; Vink 2010; Davis 2016). Also, some research has been performed on German-Jewish doctors in the German states, as well as about the Jewish communities from which the doctors, targeted in this work, originated (Horovitz 1886; Landau 1895; Lewin 1909; Kracauer 1925–1927;

Felsenthal 1930; Richarz 1974; Rosenthal 1981; Schlich 1990; Stude 2007; Richter-Hallgarten 2013). Some of these research results serve as a basis for the following analysis.

9.3 The Colonial Health System in the Dutch East Indies and the German Participation

The medical health service, which had been established by the Ministry of Colonies since 1815/16 in the Dutch East Indies, was based on the model of health system in the United Kingdom of Holland, as well as on the foundations of the former VOC' health system (Kerkhoff 1989). Because of the steady rise of the colonial population and the ambitious colonial health policy pursued by the Dutch Ministry of Colonies, the few hospitals which had been established by the VOC (Bruijn 2009, pp. 98–116), were no longer sufficient for providing the population with medical help. Consequently, the following years saw building of many new military and a few civil hospitals, as well as the reconstruction and expansion of existing ones (Schoute 1937, pp. 116–117). Most of the colonial doctors worked in these military or civilian hospitals, which were located all over the Dutch colonial residencies. Based on the daily average number of patients, military hospitals were divided into three categories: (1) military hospitals, (2) garrison hospitals (of 1st, 2nd, and 3rd class), and (3) infirmaries (of 1st, 2nd, 3rd, and 4th class). In 1867 we find a total number of 3 hospitals in the first category, 34 in the second, and 43 in the third category (Schoute 1937, p. 159). Of course, in the event of war military doctors had to work in mobile field hospitals as well.

At every military base a specific number of military doctors were available depending on the size of the military unit. At the head of the whole health services was the so called "chief, first health officer" ("Chef, Eerste Officier van Gezondheid") in the rank of a colonel. Then came a small number of "commanding health officers of 1st and 2nd class" ("Dirigerende Officieren van Gezondheid") in the ranks of lieutenant colonels and majors, who were in charge of the bigger military units. In the lower positions we find the ordinary "health officers of 1st, 2nd and 3rd class" ("Officieren van Gezondheid") in the ranks of captains and lieutenants (1st and 2nd class). Until about 1845 the ordinary health officers of the 1st class were mostly called "Surgeon-Majors" ("Chirurgijn-Majoors"), the health officers of the 2nd class "surgeons of the second class" ("Chirurgijn van de tweede Klasse"), and the health officers of the 3rd class "surgeons of the third class" ("Chirurgijn van de derde Klasse"). The term "surgeon" was most common for these officers, regardless of their specific (academic or non-academic) qualification or profession. Around 1845, a general shift to the designation "health officer" ("officier van gezondheit") can be observed in the Dutch sources. Doctors employed for the civil sector were usually called "(civile) geneesheer", best translated as "civil doctor".

A particularly high proportion of Germans among the military health staff in the Dutch East Indies can be found from the middle of the 1830s to the middle of the 1850s: In 1839 around 48.9% of all health officers were Germans (65 Germans: 68

others), in 1849 it increased to 56.9% (70: 53), in 1854 it fell to 45.4% (59: 71), in 1864 to 16.4% (28: 143), in 1871 it dropped down to 8.3% (14: 154), and finally in 1884 it rose again to 13.2% (22: 145).[2] The changing proportion of German doctors in the colonial health service is shown in the graph below (Fig. 9.1).

9.4 List of the German-Jewish Doctors Who Worked for the Colonial Services in the Dutch East Indies

So far, I have found only four German-Jewish doctors who entered the colonial health service in the Dutch East Indies outside our period of investigation (1815–1850/60): Julius Schülein from Eschenau in Bavaria – employed from 1877 to 1890,[3] Victor Lehmann from Berlin – employed from 1887 to 1891,[4] Albert Paradies from Wreschen in the Prussian province of Posen (today's Września in Poland) – employed from 1898 until his death in 1899[5]; and last but not least Salomon Hurwitz from Königsberg (today's Kaliningrad in Russia) – employed from 1923 until his death in 1946.[6,7] There is only little evidence of the presence of German-Jewish

[2] These numbers I calculated by comparing my database about German doctors in the Dutch colonial health service with the information about colonial health staff given in the *Regerings-Almanak van Nederlandsch-Indië* (1817 ff.) where all military doctors are yearly listed under the category "Militaire geneeskundige dienst".

[3] Schülein was born in 1850 and died 1915 in The Hague. For his service time in the Dutch army see: Dutch National Archives, The Hague, Dienstaten en stamboeken der officieren van de Koninklijke Landmacht en van de koloniale troepen in Nederland 1814–1940, catalog reference number 2.13.04, inventory number 405 (fol. 87).

[4] Lehmann was born 1861, his death date is unknown. For his service time in the Dutch army see: Dutch National Archives, The Hague, Dienstaten en stamboeken der officieren van de Koninklijke Landmacht en van de koloniale troepen in Nederland 1814–1940, catalog reference number 2.13.04, inventory number 408 (fol. 91).

[5] Paradies was born in 1872 and died shortly after his arrival in Indonesia in 1899. For his service time in the Dutch army see: Dutch National Archives, The Hague, Dienstaten en stamboeken der officieren van de Koninklijke Landmacht en van de koloniale troepen in Nederland 1814–1940, catalog reference number 2.13.04, inventory number 659 (fol. 1).

[6] Hurwitz was born in 1888. According to the Netherlands War Graves Foundation he died in 1946 in Melbourne as a war victim (https://oorlogsgravenstichting.nl/persoon/67979/salomon-hurwitz). Japan had occupied Indonesia from 1942 to 1945 – during this time a lot of Dutch people got imprisoned in camps. Since Hurwitz got naturalized in 1938 he was considered to be Dutch – additionally he was Jewish, so even his German roots couldn't prevent him from being persecuted and imprisoned: Dutch National Archives, The Hague, Ministerie van Binnenlandse Zaken: Stichting Administratie Indische Pensioenen (SAIP), Stamboekgegevens KNIL-militairen, met Japanse Interneringskaarten 1942–1996, (Japan), catalog reference number 2.10.50.03, inventory number 433. For his service time in the Dutch army see the same archive, Dienstaten en stamboeken der officieren van de Koninklijke Landmacht en van de koloniale troepen in Nederland 1814–1940, catalog reference number 2.13.04, inventory number 663 (fol. 17).

[7] Another German-Jewish doctor, Hermann Aschheim (1821–1850), born in Birnbaum (today's Międzychód in Poland) as the son of the Jewish towel manufacturer Benjamin A., entered the Dutch colonial health services on 7 April 1847. Because he was already dismissed on 9 July 1847,

doctors in the Netherlands Antilles or any other of the former Dutch colonies. But I can provide evidence of some German-Jewish doctors working for the other two big military branches – the Dutch Navy[8] and the Dutch Army in the Netherlands itself. A discussion of these doctors however, does not fall within the scope of my current contribution.

The so far identified 19 German-Jewish doctors, who entered the colonial health services in the Dutch East Indies between 1824 and 1844, will be presented in more details in this chapter. For this purpose, I will provide some brief information concerning their birth and death dates, family backgrounds, education, scientific achievements, and career paths. Most of this information can be found in the so called "stamboeken" (a kind of military registration system) which are stored in the National Archives in The Hague. As far as possible, I tried to verify this information by cross-referencing with research done in German archives.

[no 1] **Bensbach**, Heinrich (Hayum) Leopold, born on 6 December 1806 [1801][9] in Mannheim (Grand Duchy Baden) as the first son of the rabbi Simon (Simche) Bensbach (1774–1845) and Mirle Isaak (~ 1779–1821)[10]; studied in Würzburg (1830–1832) and Heidelberg (1832–1837)[11]; obtained his doctorate in Gießen on 5 Juny 1837 (Dr. med.)[12]; entered the military health service on 29 August 1837 as an appointed health officer 3rd class; arrived on 26 April 1838 in Batavia; retired on 17 February 1858 as health officer 3rd; died on 2 April 1859 in Dutch Celebes (Sulawesi).[13]

---

he never went to the Dutch East Indies. Most probably he stayed in the Kingdom of the Netherlands where he died on 23 September 1850 in Utrecht. See Dutch National Archives, The Hague, Dienstaten en stamboeken der officieren van de Koninklijke Landmacht en van de koloniale troepen in Nederland 1814–1940, catalog reference number 2.13.04, inventory numbers 255 (fol. 101). For Aschheim's Jewish background see Genest and Marquardt 2003, p. 366, no. 5150.

[8] E.g. Abraham Lilienfeld, born on 24 September 1821 in Marburg – cf. Dutch National Archives, The Hague, Stamboeken Marinepersoneel 1813–1940, catalog reference number 2.12.14, inventory number 20 (fol. 1044).

[9] According to the registers of the family Simon (Simche) Bensheim (later Bensbach), which are stored in the Municipal Archive of Mannheim, in 1807 Heinrich was already 6 years old! Either the Dutch sources are incorrect in this point, or Bensbach made himself younger on purpose – in a recruiting appeal from the Dutch government in the Bavarian newspaper *Bayerischer Volksfreund* from 22 Mai 1837 it can be read that doctors, who wanted to apply for a job as a military doctor in the Dutch East Indies, should not have been older than 30 years. See also Teichfischer 2016.

[10] Municipal Archive of Mannheim, family registers (1807–1900), family of Simon (Simche) Bensheim, later Bensbach.

[11] Verzeichnis des Personals und der Studierenden/Würzburg 1830/31, p. 11; 1832, p. 12. Toepke 1904, p. 496; Adressbuch der Ruprecht-Karls-Universität/Heidelberg,1836/37, p. 13.

[12] Kössler 1970, p. 6. Apparently, Bensbach was not enrolled as a student of the university in Gießen, see Kössler 1976. A written doctoral thesis also doesn't exist. In the so called "Dekanatsbuch" of the medical faculty of Gießen (University Archive of Gießen, Med C 1, vol. 4, p. 38) it is written that he graduated on 5 June 1837 and intended to enter the Dutch services on Java.

[13] Dutch National Archives, The Hague, Dienstaten en stamboeken der officieren van de Koninklijke Landmacht en van de koloniale troepen in Nederland 1814–1940, catalog reference number 2.13.04, inventory numbers 253 (fol. 74), 390 (fol. 106), 650 (fol. 44), 706 (fol. 360).

[no 2] **Bensinger**, Michael, born on 17 June 1811 in Mannheim (Grand Duchy Baden) as the first son of a butcher Löb Mayer Bensinger (~ 1768–1854) and Adelheid Bensheim (~ 1789–1846)[14]; studied in Heidelberg (1837–1841)[15]; entered the military health service on 22 May 1841 as an appointed health officer 3rd class; arrived on 6 May 1842 in Batavia; died on 21 September 1848 in the rank of health officer 3rd class in Fort Willem I (Java).[16]

[no 3] **Berliner**, Philipp, born on 6 August 1820 in Breslau (Province Silesia, Kingdom of Prussia; today's Wrocław in Poland) as a son of a health officer Jacob Ludwig Berliner (1796–1853)[17] and Henriette Heymann; entered the military health service on 13 May 1843 as an appointed health officer 3rd class[18]; arrived on 14 February 1844 in Batavia; decorated in 1858 with the "Insignia for long lasting service as officer"[19]; retired on 9 January 1863 as health officer 2nd class; died on 13 October 1888 in Breslau.[20]

[no 4] **Borck**, Daniel Adolf, born on 12 September 1812 in Zirke (Province Posen, Kingdom of Prussia; today's Sieraków in Poland) as the son of Abraham Borck and Amalia Sachs; entered the military health service of the Dutch colonial navy on 12 August 1836 as an appointed surgeon 3rd class[21]; arrived on 12 April 1837 in Batavia; after the liquidation of the Dutch colonial navy[22] in 1841, he changed to the Dutch colonial army in the rank of surgeon 3rd class; decorated

---

[14] Municipal Archive of Mannheim, family registers (1807–1900), family of Löw/Löb Mayer Benzinger, former Bensheim.

[15] Toepke 1904, p. 591; Adressbuch der Ruprecht-Karls-Universität/Heidelberg 1840/41, p. 10.

[16] Dutch National Archives, The Hague, Dienstaten en stamboeken der officieren van de Koninklijke Landmacht en van de koloniale troepen in Nederland 1814–1940, catalog reference number 2.13.04, inventory numbers 254 (fol. 117), 392 (fol. 110), 651 (fol. 26), 665 (fol. 44), 706 (fol. 417).

[17] J. L. Berliner entered the military health service of the Dutch Navy on 15 April 1826 as an appointed health officer 3rd class; see Dutch National Archives, The Hague, Stamboeken Marinepersoneel 1813–1940, catalog reference number 2.12.14, inventory number 12 (fol. 227). In 1820, the year of Philipp's birth, he was already divorced from his mother Henriette, as one can take from Breslau's Jewish register of births – State archives of Saxony, StA-L, inventory 22,310 Familiengeschichtliche Sammlungen des Reichssippenamtes, Jüdische Personenstandsunterlagen, Sig. AS 0824 (Breslau 1812–1820), p. 240, no. 110.

[18] So far it is unclear if Berliner studied medicine at a university.

[19] The Dutch designation is "Onderscheidingsteken voor Langdurige Dienst als officier" – it was awarded after at least 15 years of service time.

[20] Dutch National Archives, The Hague, Dienstaten en stamboeken der officieren van de Koninklijke Landmacht en van de koloniale troepen in Nederland 1814–1940, catalog reference number 2.13.04, inventory numbers 255 (fol. 12), 393 (fol. 29), 651 (fol. 49), 665 (fol. 94). Stamboeken en pensioenregisters van militairen KNIL in Oost- en West-Indië 1815–1949, catalog reference number 2.10.50, inventory number 625 (fol. 487). At the end of 1859, Berliner got a 2-year sick leave in Europe from which he apparently didn't return to the Dutch East Indies.

[21] As in the case of Berliner, it is unclear if Borck studied medicine at a university.

[22] The Dutch *colonial* navy was not identical to the Dutch navy at this time – it was a special branch of the colonial army and existed only from 1815 to 1841, see also Hanefeld, Royen 2001, p. 254.

in 1846 with the Order of the Netherlands Lion ("Orde van de Nederlandse Leeuw")[23] for his participation in military conflicts on Borneo in 1845, and in 1851 with the "Insignia for long lasting service as officer"; retired on 8 April 1858 as health officer 2nd class; returned in 1859 to Germany and died there on 5 August 1874 in Berlin.[24]

[no 5] **Feist**, Benjamin, born on 5 March 1812 in Bad Kreutznach (Kingdom of Prussia) as the second son of a merchant (of clover seeds) Carl Feist (1775–1817) and Philippina Clara Wolf (1782–?) (Fink 2001, p. 30); studied in Bonn (1831–1832) and Heidelberg (1832–1833)[25]; entered the military health service on 8 June 1835 as an appointed surgeon 3rd class; arrived on 29 January 1836 in Batavia; decorated in 1850 with the "Insignia for long lasting service as officer"; retired on 10 October 1855 as health officer 2nd class; worked after his official retirement as a civil doctor in Cirebon (Java) and as a sugar manufacturer (Christiaans 1992a, b, p. 13); wrote a critical treatise on the so called Cultivation System which was implemented in the 1830s for increasing the exploitation of the natural resources primarily in Java (Feist 1865); returned in 1864 to the Netherlands (Feist 1865, p. 3) and died there on 22 September 1871 in Amsterdam.[26]

[no 6] **Gobée**, August Heinrich Adolf, born on 25 June 1812 in Bruchsal (Grand Duchy Baden) as the second son of a soaper Moses (Moritz) Gumbel (surname variants: Gumpil, Gobe, ~ 1778–1830) and Schanette Sundheimer (~ 1782–1817)[27]; studied in Heidelberg (1830–1832)[28]; converted on 3 May 1832 to

[23] The Dutch name is "Orde van de Nederlandse Leeuw", for more information: https://en.wikipedia.org/wiki/Order_of_the_Netherlands_Lion

[24] Dutch National Archives, The Hague, Dienstaten en stamboeken der officieren van de Koninklijke Landmacht en van de koloniale troepen in Nederland 1814–1940, catalog reference number 2.13.04, inventory number 390 (fol. 12), 650 (fol. 20); Stamboeken en pensioenregisters van Militairen KNIL in Oost- en West-Indië 1815–1949 (1954), catalog reference number 2.10.50, inventory number 624 (fol. 359). Borck's death date is documented here: Prussian Privy State Archives (GStA PK), VIII. HA J 1 registers of Jews and dissidents (Berlin), no. 61 deceased, 1872–1874.

[25] Verzeichnis der Studirenden/Bonn SS 1831, n. p.; Toepke 1904, p. 503; Adressbuch der Ruprecht-Karls-Universität/Heidelberg 1832/33, p. 18; 1833, p. 14.

[26] Dutch National Archives, The Hague, Dienstaten en stamboeken der officieren van de Koninklijke Landmacht en van de koloniale troepen in Nederland 1814–1940, catalog reference number 2.13.04, inventory numbers 252 (fol. 108), 389 (fol. 35), 393 (fol. 37), 649 (fol. 64), 706 (fol. 325); Ministerie van Koloniën: Stamboeken Burgerlijke Ambtenaren 1836–1936, catalog reference number 2.10.36.22, inventory number 903 (fol. 161, Register: K); Stamboeken en pensioenregisters van militairen KNIL in Oost- en West-Indië 1815–1949, catalog reference number 2.10.50, inventory number 625 (fol. 551).

[27] The original name of A. H. A. Gobée was Abraham Moses Gumbel (Gumpil), see Schlitz 2015, p. 2971. His older brother – Carl (Salomon Moses) Gobée (1804–1875) – worked also for the Dutch health services, not in the colonial services, but in the service of the Dutch Army in the Kingdom of the Netherlands itself. This member of the Gobée-family even got an entry in the biographical lexicon of Lindeboom, see Lindeboom 1984, pp. 682–683. For Moses Gumbel see also Stude 2007, p. 78.

[28] Toepke 1904, p. 438; Verzeichniß der sämmtlichen Studirenden/Heidelberg 1830/31, p. 28; 1831, p. 29; Adressbuch der Ruprecht-Karls-Universität/Heidelberg 1831/32, p. 21.

Protestantism (Schlitz 2015, p. 2734); entered the Dutch military health service on 16 August 1832 as an appointed surgeon 3rd class; from 1832 to 1834, he participated in several military actions during the Belgian revolution; afterwards he worked for the Dutch army in Utrecht until June 1836; changed on 16 June 1836 to the Dutch colonial army as an appointed surgeon 2nd class; arrived on 26 January 1837 in Batavia; decorated in 1847 with the "Insignia for long lasting service as officer"; retired on 4 February 1853 as health officer 1st class; naturalized on 1 May 1856; died on 30 April 1874 in Batavia (Java).[29]

[no 7] **Hahn**, Philipp, born on 1 December 1800 in Frankfurt a. M. as the fourth son of a merchant Isaak Abraham Hahn zum weißen Hirsch (1773–1847) and Bella Gumpertz (Dietz 1907, p. 139); studied in Würzburg (1819–1823) (Merkle 1922, p. 924); obtained his doctorate (Dr. med.) in Würzburg in 1823 (Hahn 1823; Mälzer 1994, p. 88); entered the military health service on 15 April 1823 as an appointed surgeon 3rd class; worked until July 1826 for the Dutch army in the Netherlands; changed on 13 July 1826 to the Dutch colonial army as an appointed surgeon 3rd class; arrived on 29 January 1827 in Batavia; decorated in 1831 with the "Java war medal",[30] and in 1838 with the "Order of the Netherlands Lion" for his participation in the war on Sumatra's west coast from May 1836 to August 1837, and in 1844 with the "Insignia for long lasting service as officer"; retired on 27 September 1845 as health officer 1st class; died on 25 October 1881 in Frankfurt a. M.[31]

[no 8] **Hahn**, Salomon Isaak, born on 25 February 1799 in Frankfurt a. M.; older brother of [no 7] – the third son of his parents; studied in Würzburg (1819–1822) and Heidelberg (1824–1825)[32]; obtained his doctorate (Dr. med., chir. et obst.) in Würzburg in 1821 (Hahn 1822; Mälzer 1994, p. 85); entered the military health service of the Dutch navy on 17 May 1826 as an appointed surgeon 3rd class; arrived around 1826/27 in Batavia, where he first worked as a so called "civil

[29] Dutch National Archives, The Hague, Dienstaten en stamboeken der officieren van de Koninklijke Landmacht en van de koloniale troepen in Nederland 1814–1940, catalog reference number 2.13.04, inventory numbers 253 (fol. 23), 390 (fol. 3), 650 (fol. 16), 706 (fol. 348); Inspecteur-Generaal van de Geneeskundige Dienst der Land- en Zeemacht 1814–1940, catalog reference number 2.13.62.06, inventory number 238 (fol. 461); Ministerie van Justitie 1813–1876, catalog reference number 2.09.01, inventory number 4862; Gobée's death date is documented here: Centre for Genealogy (CBG/The Hague), VIBDNI006448. See also Regerings-Almanak voor Nederlandsch-Indië, 1875, p. 276.

[30] The Dutch designation is "Medaille van den Oorlog op Java 1825–1830" – it was awarded for participation in the Java war.

[31] Dutch National Archives, The Hague, Dienstaten en stamboeken der officieren van de Koninklijke Landmacht en van de koloniale troepen in Nederland 1814–1940, catalog reference number 2.13.04, inventory numbers 194 (fol. 54), 387 (fol. 31), 649 (fol. 19), 705 (fol. 177); Inspecteur-Generaal van de Geneeskundige Dienst der Land- en Zeemacht 1814–1940, catalog reference number 2.13.62.06, inventory number 238 (fol. 259). For Hahn's death date see Institute for City History/Frankfurt a. M., register of deaths 1881, registry office I, vol. 4, p. 569, no. 236.

[32] Merkle 1922, pp. 931, 949. Toepke 1904, p. 286; Verzeichniß der sämmtlichen Studirenden/Heidelberg 1824/25, p. 24.

geneesheer", i.e. a kind of civil doctor; during 1827 he switched again to the military service as an appointed surgeon 2nd class; died on 4 September 1829 in the rank of health officer 2nd class.[33]

[no 9] **Hartzfeld**, Joseph, born on 15 August 1814 in Heidingsfeld (Kingdom of Bavaria) as the third son of the merchant Jacob Hartzfeld (1764–after 1851) and Babette Schwarzschild (Rosenstock 2008, p. 258); studied in Würzburg (1833/34–1836; 1838/39–1839)[34]; obtained his doctorate (Dr. med., chir. et obst.) in Würzburg in 1839 (Hartzfeld 1839; Mälzer 1994, p. 156); entered the military health service on 19 April 1841 as an appointed officer of health 3rd class; arrived on 22 November 1841 in Batavia; naturalized on 22 December 1861; decorated in 1856 with the "Insignia for long lasting service as officer"; retired on 15 January 1869 as commanding health officer 1st class; died on 2 January 1885 in The Hague.[35]

[no 10] **Heymann**, Salomon (Samuel) Löb, born on 6 May 1804 in Diez (Duchy Nassau) as the second son of a wine merchant Löb Heymann and Bella Mayer[36]; studied in Gießen (1821–1824), Bonn (1824–1825/26) and Göttingen (1826–1827)[37]; entered the military health service on 20 December 1827 as an appointed surgeon 2nd class; arrived on 8 October 1828 in Batavia; retired on 11 June 1849 as commanding health officer 2nd class; published a book about tropical diseases and medicine in 1855 (Heymann 1855); decorated in 1846 with the "Insignia for long lasting service as officer"; died on 7 February 1859 in Würzburg.[38]

[no 11] **Jolberg**, Albert/Anton, born on 22 March 1800 in Kassel (Electorate of Hesse) as the son of a merchant Joel Itzig Jolberg (~ 1743–1833) and Alwine

[33] Dutch National Archives, The Hague, Ministerie van Koloniën 1814–1849, catalog reference number 2.10.01, inventory number 3096 (fol. 1); Stamboeken Marinepersoneel 1813–1940, catalog reference number 2.12.14, inventory number 17 (fol. 772). It is unknown when exactly S. I. Hahn came to the Dutch East Indies. Maybe he arrived together with his brother in the beginning of 1827. His death date can be found here: Institute for City History/Frankfurt a. M., register of deaths 1829, vol. 53, p. 566. According to this record Hahn died in "Komeet/Indonesia" – "Komeet" is not an existing place name, but the name of a Dutch war ship from this time, so maybe he died on board of this ship in Indonesia.

[34] Verzeichnis des Personals und der Studierenden/Würzburg 1833/34, p. 14.

[35] Dutch National Archives, The Hague, Dienstaten en stamboeken der officieren van de Koninklijke Landmacht en van de koloniale troepen in Nederland 1814–1940, catalog reference number 2.13.04, inventory numbers 254 (fol. 113), 392 (fol. 81), 651 (fol. 11), 665 (fol. 40), 706 (fol. 413); Ministerie van Justitie 1813–1876, catalog reference number 2.09.01, inventory number 4864; Stamboeken en pensioenregisters van militairen KNIL in Oost- en West-Indië 1815–1949, catalog reference number 2.10.50, inventory number 626 (fol. 699).

[36] Central State Archive of Hesse/Wiesbaden, HHStAW Abt. 221 no. 1174.

[37] Kössler 1976, p. 78; Verzeichnis der dermaligen Lehrer und Studirenden/Giessen 1823, p. 7; Verzeichnis der Studirenden/Bonn 1824/25, n. p.; 1825, n. p.; Selle 1980 (1937), p. 765.

[38] Dutch National Archives, The Hague, Dienstaten en stamboeken der officieren van de Koninklijke Landmacht en van de koloniale troepen in Nederland 1814–1940, catalog reference number 2.13.04, inventory numbers 251 (fol. 128), 387 (fol. 45), 649 (fol. 35); Stamboeken en pensioenregisters van militairen KNIL in Oost- en West-Indië 1815–1949, catalog reference number 2.10.50, inventory number 623 (fol. 97).

Fleersheim[39]; apprentice to the municipal surgeon C. G. Altmueller in Kassel (1817–1820); studied in Marburg (chirurg. cand., 1820–1821/22)[40]; entered the military service in 1822 as a fusilier (kind of soldier); arrived around 1822/23 in Batavia[41]; entered the military health service in 1824 as an appointed surgeon 3rd class ("Tabellarisch overzigt" 1859, pp. 264–265); died on 2 June 1824 with the rank of a military surgeon 3nd class in Weltevreden (Java).[42]

[no 12] **Lenheim**, Lazard, born on 25 September 1800 in Fulda (Electorate of Hesse) as the son of a merchant Martin (Moses Simon) Lenheim (~ 1765–1847) and Johanna Wolf (Schlich 1990, p. 116); studied in Würzburg (1820–1824) and Marburg (1824–1825)[43]; obtained his doctorate (Dr. med., chir. et obst.) in Marburg in 1825 (Lenheim 1825); entered the military health service on 31 May 1827 as an appointed surgeon 3rd class; arrived on 29 January 1828 in Batavia; decorated in 1831 with the "Java war medal", and in 1846 with the "Insignia for long lasting service as officer"; retired on 17 May 1848 as health officer 2nd class; died on 16 July 1865 in Frankfurt/Main.[44]

[no 13] **Lindmann**, Lazarus, born on 17 March 1814 in Mannheim (Grand Duchy Baden) as the third son of the teacher Jacob Lindmann (1773–1852) and Hendle Werner (1780–1851)[45]; studied in Heidelberg (1835–1839)[46]; entered the military health service on 5 May 1841 as an appointed officer of health 3rd class; arrived on 24 November 1841 in Batavia; decorated in 1852 with the "Military

[39] Jolberg's birthdate can be found in an archival record that deals with his legacy: HStAM inventory 9 a, no. 2395, vol. 5 (1841–1847). In this archival record his forename is written as "Albert"; the name of his parents can be found there as well. Schlich writes "Anton" and gives only the name of the father, see Schlich 1990, p. 105.

[40] Birt 1913, p. 534. In 1825, at least according to Schlich, Jolberg was also enrolled as a "chirur. cand." at the University in Kassel, which is impossible because Jolberg already died in 1824, see Schlich 1990, p. 105.

[41] His ship set sail to the Dutch East Indies on 15 June 1822 – at that time it took between 4 and 6 months to sail from the Netherlands to the Dutch East Indies.

[42] Thus far, I could not find any information about Jolberg's employment in the Dutch National Archives in The Hague. That Jolberg was employed as a military surgeon in the Dutch East Indies I found in this list: "Tabellarisch overzigt", 1859, pp. 264–265. This list says that he was engaged in 1824 and died in the same year. According to the investigations of Schlich, in 1824 Jolberg was still in Hesse. He had not emigrated before 1826 to the Netherlands, Schlich 1990, p. 105f. For now, I decided to rely on the Dutch lists because their content is also confirmed by the information given here: State Archive of Hesse/Marburg, HStAM inventory 9 a no. 2395, vol. 5 (1841–1847). The date and place of his death can also be found here: Regerings-Almanak voor Nederlandsch-Indië, 1825, p. 172.

[43] Merkle 1922, p. 941; Birt 1913, p. 552.

[44] Dutch National Archives, The Hague, Dienstaten en stamboeken der officieren van de Koninklijke Landmacht en van de koloniale troepen in Nederland 1814–1940, catalog reference number 2.13.04, inventory numbers 387 (fol. 40), 649 (fol. 32), 666 (fol. 89); Stamboeken en pensioenregisters van militairen KNIL in Oost- en West-Indië 1815–1949, catalog reference number 2.10.50, inventory number 623 (fol. 56).

[45] Municipal Archive of Mannheim, family registers (1807–1900), family of Jacob Lindmann.

[46] Toepke 1904, p. 559; Adressbuch der Ruprecht-Karls-Universität/Heidelberg 1835/36, p. 16; 1839, p. 18.

Order of William"[47] 4th class for his engagement in the Palembang Highlands Expeditions (1851–1859), and in 1856 with the "Insignia for long lasting service as officer"; naturalized on 22 June 1862; retired on 4 November 1869 as commanding health officer 1st class; died on 14 April 1877 in Amsterdam.[48]

[no 14] **Morawski**, Gustav Adolph, born on 4 September 1804 in Posen (Province Posen, Kingdom of Prussia; today's Poznań in Poland) as the son of the scrivener Samuel Morawski and his wife Bertha (Witkowski 2009, p. 37); entered the military health service on 1 March 1836 as an appointed officer of health 3rd class[49]; arrived on 4 September 1836 in Batavia; died on 8 March 1849 with the rank of health officer 3nd class in the district of Lampung (Sumatra).[50]

[no 15] **Pariser**, Bernhard Nathan, born on 19 July 1806 in Fraustadt (Province Posen, Kingdom of Prussia; today's Wschowa in Poland) as a son of the merchant Jacob Thomas Nathan (Pariser)[51] and Bela Goldfaenger (Pariser 1830, *curriculum vitae*, n. p.); studied in Berlin (1824–1828;1835)[52]; obtained his doctorate (Dr. med. et. chir.) in Berlin in 1830 (Pariser 1830; Erman 1973 (1899), p. 133); worked after his doctorate as a hospital and district doctor in Prussia (until November 1831); entered the military health service on 16 March 1836 as an appointed officer of health 3rd class; arrived on 3 December 1836 in Batavia; honourably discharged and retired on 29 December 1842 as health officer 3rd class; died on 3 March 1845 in Utrecht.[53]

---

[47] The Dutch term is "Militaire Willems-Orde", for more information: https://en.wikipedia.org/wiki/Military_Order_of_William

[48] Dutch National Archives, The Hague, Dienstaten en stamboeken der officieren van de Koninklijke Landmacht en van de koloniale troepen in Nederland 1814–1940, catalog reference number 2.13.04, inventory numbers 254 (fol. 115), 392 (fol. 82), 651 (fol. 15), 665 (fol. 43), 706 (fol. 415); Ministerie van Justitie 1813–1876, catalog reference number 2.09.01, inventory number 4865; Stamboeken en pensioenregisters van militairen KNIL in Oost- en West-Indië 1815–1949, catalog reference number 2.10.50, inventory number 626 (fol. 762).

[49] So far, I couldn't find any information about Morawski's time as a student. Maybe he studied in Breslau or Königsberg.

[50] Dutch National Archives, The Hague, Dienstaten en stamboeken der officieren van de Koninklijke Landmacht en van de koloniale troepen in Nederland 1814–1940, catalog reference number 2.13.04, inventory numbers 389 (fol. 71), 650 (fol. 3). Morawski's death date and place are also documented here: Regerings-Almanak voor Nederlandsch-Indië, 1850, p. 451.

[51] Apparently, the family assumed the surname "Pariser" later – in the Jewish birth register of Fraustadt the name of Bernhard Nathan's father is simply "Jacob Thomas Nathan" (see State archives of Saxony, StA-L, inventory 22,310 Familiengeschichtliche Sammlungen des Reichssippenamtes, Jüdische Personenstandsunterlagen, Sig. AS 3556–3 (Fraustadt 1802–1833), p. 16). Even in the matriculation register of the university in Berlin his name is given only as "Bernhard Nathan", without the surname "Pariser" (Matrikel-Nr. 302).

[52] Pariser enrolled at the Friedrich-Wilhelms-University in Berlin on 17 April 1824 under the surname "Nathan" – Bahl and Ribbe 2010a, p. 252, no. 302. On 6 May 1835 he enrolled for the second time, now under the surname "Pariser". Finally, he was exmatriculated on 10 August 1835.

[53] Dutch National Archives, The Hague, Dienstaten en stamboeken der officieren van de Koninklijke Landmacht en van de koloniale troepen in Nederland 1814–1940, catalog reference number 2.13.04, inventory numbers 253 (fol. 9), 389 (fol. 90), 650 (fol. 10), 706 (fol. 343); Stamboeken en pensioenregisters van militairen KNIL in Oost- en West-Indië 1815–1949, catalog reference num-

[no 16] Reiß, Emanuel, born on 1 September 1806 in Stadtlengsfeld (Grand Duchy Saxe-Weimar-Eisenach) as the second son of a merchant Wolf Reiß (~ 1770–?) and Henriette Reiß; studied in Heidelberg (1832–1836)[54]; obtained his doctorate in Heidelberg in 1836 (Dr. med., chir. et obst.)[55]; entered the military health service on 28 September 1836 as an appointed officer of health 3rd class; arrived on 10 August 1837 in Batavia; died on 13 May 1846 with the rank of health officer 2nd class in Ponorogo (Java).[56]

[no 17] Robinow, Marcus Octavius, born on 4 October 1811 in Hamburg as the third son of a merchant Ruben Marcus Robinow (1770–1840) and Emma Beit (1784–1830) (Steckmest 2013, pp. 6–15); studied in Heidelberg (April 1831–1835)[57]; obtained his doctorate (Dr. med., chir. et obst.) in Heidelberg in 1835 (Robinow 1836)[58]; entered the military health service on 4 June 1839 as an appointed officer of health 3rd class; arrived on 3 January 1840 in Batavia; died on 22 November 1849 in the rank of health officer 2nd class in Gombong (Java).[59]

[no 18] Schwarz, Julius, born on 19 December 1804 in Schwerin an der Warthe (Province Posen, Kingdom of Prussia; today's Skwierzyna in Poland) as a son of Jacob Abraham Schwarz and Maria Nathan; studied in Königsberg (1824) and Berlin (1824–1828; 1829–1833)[60]; worked from May 1831 to November 1831 as a doctor in Warsaw during the Polish-Russian War (1830–1831) (Schwarz 1833, *curriculum vitae*, n. p.); obtained his doctorate (Dr. med. et. chir.) in Berlin in 1833 (Schwarz 1833; Erman 1973 (1899), p. 152); entered the military health service on 16 March 1836 as an appointed officer of health 3rd class; arrived on 3 November 1836 in Batavia; decorated in 1851 with the "Insignia for long lasting service as officer"; retired on 30 October 1856 as health officer 3rd class; died on 9 October 1880 in Szczecin.[61]

---

ber 2.10.50, inventory number 621 (fol. 1326). Pariser's death certificate is documented in Het Utrechts Archief – Burgerlijke Stand Utrecht, Toegang 481, inv. nr. 1189–03, aktenr. 322. Here one can read that he died in a hospital for mentally ill people (the so called "Krankzinnigengesticht").

[54] Toepke 1904, p. 499; Adressbuch der Ruprecht-Karls-Universität/Heidelberg 1832/33, p. 28; 1836, p. 21.

[55] University Archive of Heidelberg, inventory H-III, Sign. 111/038, fol. 352–383.

[56] Dutch National Archives, The Hague, Dienstaten en stamboeken der officieren van de Koninklijke Landmacht en van de koloniale troepen in Nederland 1814–1940, catalog reference number 2.13.04, inventory numbers 253 (fol. 39), 390 (fol. 47), 650 (fol. 28), 706 (fol. 350).

[57] Toepke 1904, p. 449; Adressbuch der Ruprecht-Karls-Universität/Heidelberg 1831/32, p. 33; 1835, p. 19.

[58] University Archive of Heidelberg, inventory H-III, Sign. 111/036, fol. 63–70, 74–100; H-III, Sign. 111/038, no. 11; fol. 41.

[59] Dutch National Archives, The Hague, Dienstaten en stamboeken der officieren van de Koninklijke Landmacht en van de koloniale troepen in Nederland 1814–1940, catalog reference number 2.13.04, inventory numbers 391 (fol. 86), 650 (fol. 78), 706 (fol. 386).

[60] Erler 1911, p. 747; Bahl and Ribbe 2010a, p. 263, no. 596, b, p. 418, no. 363.

[61] From 1854 to 1856, Schwarz got a 2-year home leave after which he didn't return to the Dutch East Indies. Dutch National Archives, The Hague, Dienstaten en stamboeken der officieren van de Koninklijke Landmacht en van de koloniale troepen in Nederland 1814–1940, catalog reference number 2.13.04, inventory numbers 253 (fol. 8), 389 (fol. 82), 650 (fol. 7), 706 (fol. 342);

[no 19] Straus(s), Heinrich, born on 13 December 1800 in Frankfurt a. M. as the third son of a merchant Isaak Hirsch Straus(s) (1767–1853) and Rosa Schiff (1776–1853) (Straus und Ullmann 1880); prepared himself for his university studies at the Collegium Carolinum in Braunschweig (1819/20) (Düsterdieck 1983, p. 40, no. 1853); studied in Berlin (1821–1822), Halle (1822–1823), Göttingen (1823–1824), and Würzburg (1824–1825)[62]; obtained his doctorate (Dr. med. et chir.) in Göttingen in 1824 (Strauß 1825)[63]; entered the colonial civil health service on 27 July 1826 as a civil doctor; arrived around 1826/1827 in Batavia; entered the colonial military health service on 25 May 1827 as an appointed officer of health 2nd class; left the Dutch East Indies in February 1834 for a 2-year home leave; came back around 1836/37 and worked then until 1855 as a civil doctor in Batavia; died on 10 June 1858 in Frankfurt a. M.[64]

9.5 Religious Affiliation and Identity

Religious affiliation cannot be taken from the "stamboeken". This information I collected from German parish registers and municipal archives. In our period of investigation (1824–1844) a total of 124 German doctors entered the colonial health service in the Dutch East Indies. According to the current state of knowledge, 66 (= 53.2%) of them were Protestants, 31 were Catholics (= 23.1%), and 19 (= 15.3%) were Jews. The religious affiliation of the remaining 8 persons (= 6.5%). The names and origins of three of them indicate a Jewish origin as well: Josephus Franciscus Xaverius Henricus Maria Hubertus Falk from Rawitsch (today's Rawicz in Poland),[65] Bernhard Nathan Danziger from Fraustadt (today's Wschowa in Poland),[66]

Stamboeken en pensioenregisters van militairen KNIL in Oost- en West-Indië, 1815–1949, catalog reference number 2.10.50, inventory number 624 (fol. 331).

[62] Bahl and Ribbe 2010a, p. 179, no. 301; University Archive Halle-Wittenberg, UAHW, Rep. 46, Nr. 9 (1820–1826); Selle 1980 (1937), p. 702; Merkle 1922, p. 972.

[63] University Archive of Göttingen, med. Prom. 1824, Heinrich Straus.

[64] Dutch National Archives, The Hague, Dienstaten en stamboeken der officieren van de Koninklijke Landmacht en van de koloniale troepen in Nederland 1814–1940, catalog reference number 2.13.04, inventory numbers 387 (fol. 34); Ministerie van Koloniën 1814–1849, catalog reference number 2.10.01, inventory number 3096 (fol. 2). Strauss' death date is documented here: Kallmorgen 1936, p. 426.

[65] Falk was born in 1806 and died in 1849 in Palembang (Indonesia). The name of his father, who was a merchant, was "Hermann Judas F."; his mother's name was "Rebecca Gints". I could not find any birth certificate that documents the birth of Falk – neither in the catholic nor in the protestant parish registers of Rawitsch. Falk's "stamboeken" can be found here: Dutch National Archives, The Hague, Dienstaten en stamboeken der officieren van de Koninklijke Landmacht en van de koloniale troepen in Nederland 1814–1940, catalog reference number 2.13.04, inventory numbers 255 (fol. 64, 79), 393 (fol. 125), 651 (fol. 78), 668 (fol. 31, 41).

[66] Danziger was born in 1814 – his death date is unknown (after 1843). Danziger's parents were Eduard Danziger and Caroline Jakobowska – both surnames can be found in the Jewish birth register of Fraustadt - State archives of Saxony, StA-L, inventory 22,310 Familiengeschichtliche

and Heinrich Bloch from Meseritz (today's Międzyrzecz in Poland)[67] – all three birthplaces were located in what was then the Prussian province of Posen. If Danziger and Bloch were born as Jews, both must have been converted before or during the time of their study. The majority of the remaining doctors were most probably Protestant, thus leading to an overall Protestants' share of 53–56%, while the percentage of Catholics is estimated to be 25–27%. Given that Falk was also a Jew, their share would rise to 16.1%.

Figure 9.2 shows these results by combining the religious affiliation with the origin of the German doctors:

According to our investigation, the share of Jews (40%) was particularly high among the doctors coming from the Grand Duchy of Baden. In the Free Cities of Frankfurt and Hamburg their proportion was significant as well (33%).

It is hard to say anything about "Jewish identity" in the Dutch East Indies or about the influence of the employment as a colonial doctor on the religious belief. Maybe the lack of a Jewish community life and the limited possibilities to practice their religion lead in some cases to the renunciation of religious belief. On the other hand, the apparent Jewish intermarriage may be seen as a sign of the upholding of Jewish identity. Philipp Berliner [see no. 3], for example, married Amalia Catharina Strauss (1830–1859), one of the daughters from Heinrich Strauss [see no. 19], in 1848 in Batavia (Straus und Ullmann 1880, pp. 41–42).[68] Strauss himself got married to Bertha Speyer (1807–?), a daughter of a Jewish merchant Meyer Gottschalk Speyer, during a 2-year home leave. The same applies to Joseph Hartzfeld [see no. 9], who married Ida Ascherberg (1837–1906),[69] who was Ludwig Ascherberg's

Sammlungen des Reichssippenamtes, Jüdische Personenstandsunterlagen, Sig. AS 3556–3 (Fraustadt 1802–1833). In the matriculation register of the University of Leipzig where he studied one can find the information that he was catholic. Maybe his family got converted to Catholic before his birth, as I couldn't find him or his parents in the birth register. His "stamboeken" can be found here: Dutch National Archives, The Hague, Dienstaten en stamboeken der officieren van de Koninklijke Landmacht en van de koloniale troepen in Nederland 1814–1940, catalog reference number 2.13.04, inventory numbers 253 (fol. 6), 389 (fol. 77), 650 (fol. 4), 706 (fol. 338).

[67] Bloch was born in 1808 and died in 1860 in Breda (NL). The name of his father, who was a merchant, was Is(aac) Behrend Bloch (especially the forename of his father is a typical Jewish one, but also the surname was very common among Jews); his mother's name was Johanna Broddi. In the *curriculum vitae* of his dissertation Bloch wrote that he was Protestant. Since I couldn't find him in the Protestant parish register of Meseritz, his family was maybe Jewish and he or his family got converted later. Unfortunately, the Jewish birth register of Meseritz starts only in 1817, so I couldn't check it for Bloch's birth. His "stamboeken" can be found here: Dutch National Archives, The Hague, Dienstaten en stamboeken der officieren van de Koninklijke Landmacht en van de koloniale troepen in Nederland 1814–1940, catalog reference number 2.13.04, inventory numbers 391 (fol. 81), 650 (fol. 79), 706 (fol. 388).

[68] Institute for City History Frankfurt/Main, Senatssupplikationen 205/770.

[69] Ida Ascherberg was born on 10 August in 1837 in Köthen as the daughter of Itzig Ascherberg, who was the brother of Ludwig – this information I got from the municipal archive in Köthen. Her death certificate is stored in The Hague's municipal archive: Haags Gemeentearchief, 's-Gravenhage Overlijden 1906 serie, Akte 3617.

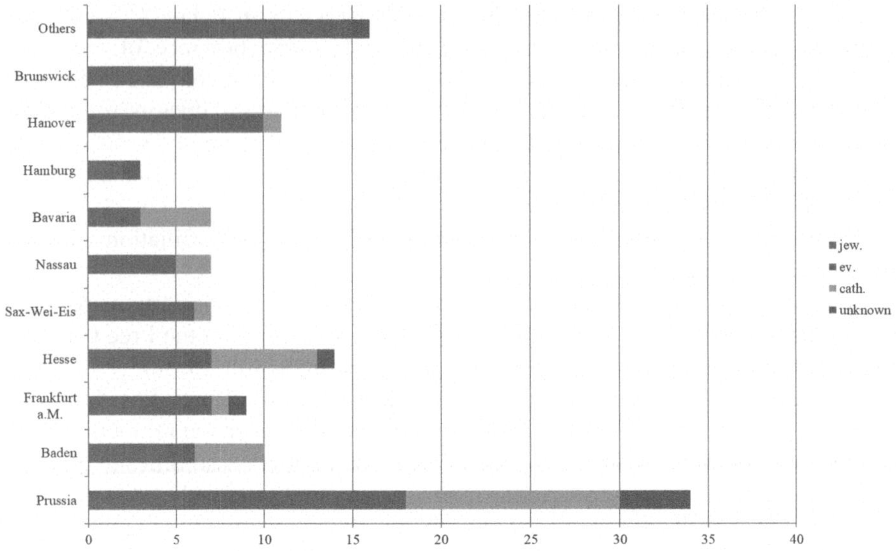

Fig. 9.2 Origin and religious affiliation (total number: 124)

(1810–1866) niece, another German-Jewish doctor originated from Köthen (Duchy of Anhalt-Köthen). Ascherberg had entered the military health service as a ship's surgeon of the Dutch navy in 1832 and sailed in 1843 to the Dutch East Indies.[70]

However, there is at least one German-Jewish doctor, August Heinrich Gobée [see no. 6], who had converted to Protestantism 3 months before he entered the Dutch military service. Whether or not his conversion was inspired by the hope of better career opportunities in the Dutch army is hard to say. Against the background of the liberal attitude of the Dutch military administration, as will be shown later in this article, it is not very likely that German-Jewish doctors converted for career reasons.

9.6 Motives to Enter the Colonial Service

In general, bad professional outlook and bad living conditions in the former German states were a strong motive to enter foreign services for both Jewish and non-Jewish doctors. Because of manifold discrimination, the situation of the Jewish doctors was even worse than that of their non-Jewish counterparts. For example, during the first half of the nineteenth century it was almost impossible for Jewish doctors to build

[70] Dutch National Archives, The Hague, Stamboeken Marinepersoneel 1813–1940, catalog reference number 2.12.14, inventory number 10 (fol. 47).

up a university or military career. Furthermore, the quickly growing number of Jewish medical students did not result in an equally growing number of positions for Jewish doctors (Huerkamp & Spree 1982, pp. 82–86; Richarz 1974, pp. 83–131; 172–178).

In the Free City of Frankfurt, where at the beginning of the nineteenth century around 3000 "local" and 1000 "foreign" Jews lived (Richter-Hallgarten 2013, pp. 76, 80, 114), only three Jewish doctors were allowed to practice medicine until 1850 (Kallmorgen 1936, p. 350). Additionally, they were limited, at least officially, to treating only Jewish patients (Kallmorgen 1936, p. 454). Many of the civil privileges that were awarded to the Jews by Elector Karl Theodor von Dalberg (1744–1817) were taken from them after the Congress of Vienna (1815) and the so called Hep-Hep riots (1819). This was accompanied by many restrictions for Jews, e.g. concerning their professions and wedding licenses (Richter-Hallgarten 2013, pp. 66–81).

In the case of two Frankfurt doctors S. I. Hahn and H. Strauss some interesting archival records concerning their application for a practice license are preserved in the Institute for City History in Frankfurt/Main. According to these records, between 1823 and 1826 S. I. Hahn [see no. 8] tried to get a license to practice medicine three times without success.[71] His first application was rejected without any reason by the municipal government of Frankfurt. One year later, after his second attempt, he was permitted to participate in the compulsory exams, which were carried out by the Sanitary Bureau of Frankfurt. He failed these exams in all three subjects. Whether he failed because of a lack of knowledge or for other reasons, remains to be seen. The description of a similar case given by Richter-Hallgarten makes it at least doubtful that Hahn failed only because of poor knowledge (Richter-Hallgarten 2013, pp. 82–93). Subsequently, it was suggested to Hahn to enrol once more at a university to improve his medical knowledge. In 1826, after complying with this request and studying another year at the University of Heidelberg, Hahn applied for the third time. Again, his application was dismissed – this time due to a supposed offence Hahn had committed during his studies and for which he had already been punished in Heidelberg. The commission decided Hahn's application could not be taken into account as long as he could not give a good explanation of his behaviour in this case. Apparently, this was Hahn's last try before he entered the Dutch military health services on 17 May 1826, for which his younger brother Philipp [see no. 7] had already been working since 1823. He did not respond to those accusations of the Sanitary Bureau, and thus it is unclear what kind of "offence" Hahn committed.

However, bad professional outlook and bad living conditions cannot be considered as the only motives for working migration or emigration. This can be shown by the case of the doctor H. Strauss [see no. 19] who, in 1824, also applied for a license to practise in his home town of Frankfurt. After he passed the exams in 1825, he received a license, as well as citizenship of Frankfurt.[72] Nevertheless, he preferred

[71] Institute for City History Frankfurt/Main, Senatssupplikationen 146/10.

[72] Institute for City History Frankfurt/Main, Senatssupplikationen 205/770.

to enter the Dutch medical services on 27 July 1826. Unfortunately, we have no further clues as to his reasons. Even when he was allowed in 1835 to marry the daughter of a Frankfurt Jewish merchant, which was a kind of "honour" or "luck" considering the fact, that in Frankfurt only 15 Jewish couples a year were allowed to marry (Richter-Hallgarten 2013, p. 79), Strauss and his new wife went back to the Dutch East Indies.

Overall, it must be stated that the often precarious situation of Jews in the German states in the first half of the nineteenth century was surely a strong motive for some of them to seek their fortune abroad (Schlich 1990, pp. 206–207). The (United) Kingdom of the Netherlands as a comparatively liberal and Jewish-friendly state seemed to be quite an attractive destination until, especially in the second half of the nineteenth century, other countries like the United States became more attractive and the career prospects for Jewish doctors in Germany improved.

9.7 Military Careers: Positions and Promotion Times

To answer the initial question of whether German-Jewish doctors had the same career opportunities in the Dutch military health service as their non-Jewish German colleagues, we can have a look at the ranks they achieved and the time they needed to ascend in rank.[73] We can thereby assume that discrimination of German-Jewish doctors should have led to relatively lower ranks and longer promotion times.

In 1827, a German doctor Georg Joseph Peitsch (1788–1838), who was appointed "chief, first health officer" of the whole colonial health service in the Dutch East Indies (1827–1837), introduced a new promotion system called "concours" (French: "competition"). Under this system, promotions were no longer based on the number of completed service years, but on the results of special exams the health officers could apply for. Because of the fact that these exams took place in a few of the bigger military hospitals, located in the main Dutch residencies, the new system had major disadvantages for those doctors who were placed on remote posts (Schoute 1937, pp. 127–128; Kerkhoff 1987, p. 52; Kerkhoff 1989, pp. 14–15).

A total of 50 persons from the 124 German doctors investigated here, hold at the end of their career, the lowest position of officer 3rd class in the health service. Most of them also had started their military career with this rank (and thus had not been promoted at all during their time of service). Six of them started as so called "élèven" (French: "students/pupils") and were promoted to the rank of health officer 3rd class after passing an exam.[74] We must add that 25 of those 50 persons already died

[73] Information about their promotion I took from the "stamboeken" of the health officers.

[74] There is also one German doctor – Ernst Wilhelm Rauch (1802–1839), born in Hessen/Osterwieck (Duchy of Brunswick) – who entered the colonial service as a soldier and died as an appointed student of surgery ("élève van chirurgie"), see also Fig. 9.3.

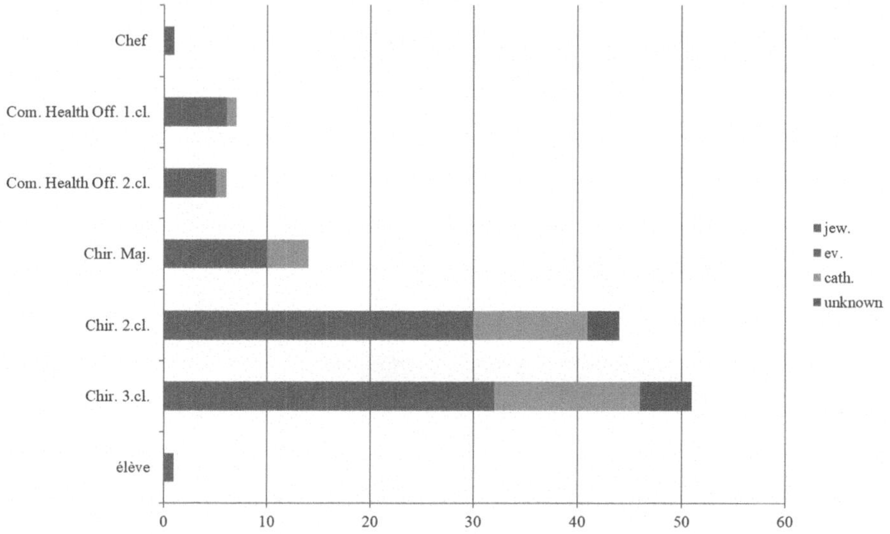

Fig. 9.3 Positions at the end of the career (total number: 124)

during the first 3 years after their arrival in the Dutch East Indies – taking into consideration the average promotion time (see below Fig. 9.4) these persons had hardly a chance to get promoted. A further 44 persons had risen by the end of their career to the position of health officer 2nd class – four of them had already entered the military health service in this position. Another 14 doctors we find in the rank of health officer 1st class. Three of them had already entered the service in this position. Another six doctors made their way up to the rank of commanding health officer 2nd class, and even seven rose to the rank of commanding health officer 1st class. One of the German doctors who entered the services in our period of investigation (1824–1844) even reached the position of "chief, first health officer" with the military rank of a colonel: Carl Wilhelm Reinhard Voigt (1812–1886), born in Neustaedt/Werra (which is nowadays a part of Gerstungen located in the former Grand Duchy of Saxe-Weimar-Eisenach). He entered the service in 1836 as a health officer 3rd class and headed the Dutch colonial health service from 1856 to 1859.[75] Fig. 9.3 shows these results:

Considering the average promotion time for the different ranks, we receive the following results: on average, the 59 doctors who were promoted from the position of health officer 3rd class to that of health officer 2nd class required approx. 5 years and 7 months for their promotion. Among them were also 9 German-Jewish doctors whose average promotion time was around 5 years and 4 months,

[75] Dutch National Archives, The Hague, Dienstaten en stamboeken der officieren van de Koninklijke Landmacht en van de koloniale troepen in Nederland 1814–1940, catalog reference number 2.13.04, inventory numbers 253 (fol. 14), 389 (fol. 90), 706 (fol. 346).

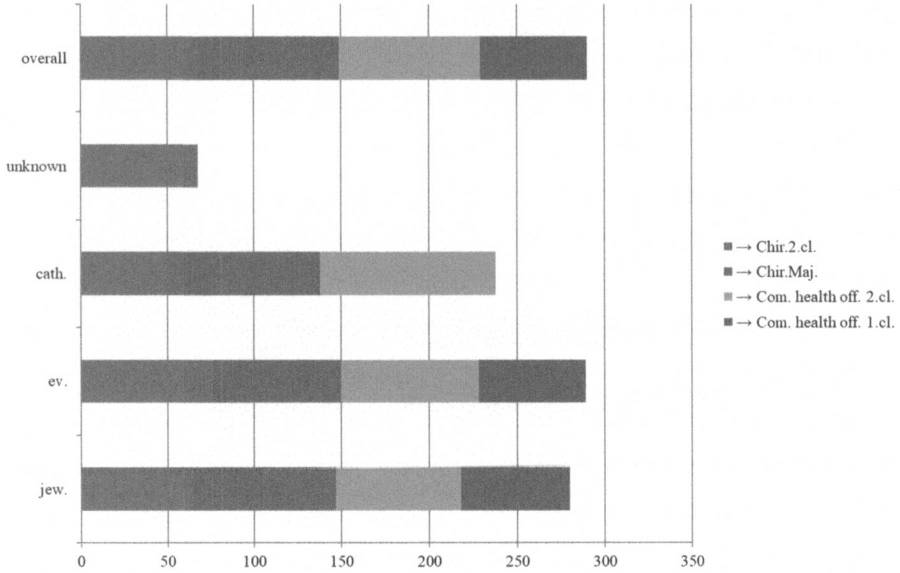

Fig. 9.4 Average time for promotions to higher positions (in months)

and thus a bit lower than the overall average. In case of the 23 health officers 1st class the average promotion time took around 6 years and 6 months, while the average time of the herein included 5 German-Jewish doctors was a bit higher with 6 years and 9 months. Promotion to the rank of commanding health officer 2nd class required an average time of approx. 6 years and 6 months for the 14 doctors – the 3 German-Jewish among them needed, on average, 9 months less. Last but not least, to reach the rank of commanding health officer 1st class an average time of 5 years and 1 month for both groups can be measured. Fig. 9.4 shows these numbers at a glance:

On the hypothesis that promotion times correlate with career opportunities, it can be concluded that German-Jewish doctors were clearly not discriminated against. This result aligns with a conclusion Kerkhoff drew on the promotion times of German doctors in the colonial health service compared with their Dutch counterparts: also in this case the time needed for promotion was not higher for German doctors than for the Dutch doctors. Rather Kerkhoff found out that German doctors were promoted a bit faster than their Dutch colleagues (Kerkhoff 1987). Based on these results, it can be concluded that German-Jewish doctors were not discriminated against at all: neither relative to their (non-Jewish) German nor relative to their Dutch colleagues. Comparing the total number of German-Jewish doctors (= 2) with the total number of non-Jewish German doctors (= 5) who reached the position of commanding health officer 1st class, and taking into account that only 15.3% of the here 124 investigated German doctors were Jews, it could even be concluded that they were "overrepresented" in these ranks.

9.8 Conclusions

In summary, it must be stated that the specific religious affiliation apparently did not play a big role in the career prospects of health officers in the Dutch colonial army. According to our investigations neither Jews nor Catholics (see Fig. 9.4) were disadvantaged because of their religion, even though the Kingdom of the Netherlands was a mainly Protestant country at that time. Provided that there was at least some contact between Jews in the Dutch East Indies and their families and friends in their homelands (which we do not know for sure), we can also assume that the prospect of a better professional outlook, was one of the main reasons for Jewish doctors to enter the Dutch colonial service.

References

Adressbuch der Ruprecht-Karls-Universität zu Heidelberg. (1831–1845). Heidelberg.

Bahl, P., & Ribbe, W. (Eds.). (2010a). *Die Matrikel der Friedrich-Wilhelms-Universität zu Berlin 1810–1850: Teil 1: Die Matrikel für das 1. bis 23. Rektoratsjahr (1810–1833)*. Berlin [u. a.]: de Gruyter.

Bahl, P., & Ribbe, W. (Eds.). (2010b). *Die Matrikel der Friedrich-Wilhelms-Universität zu Berlin 1810–1850: Teil 2: Die Matrikel für das 24. bis 40. Rektoratsjahr (1833–1850)*. Berlin [u. a.]: de Gruyter.

Birt, T. (1913). *Catalogus studiosorum scholae Marpurgensis. 1811–1822*. Marburg: Friedrich.

Blom, J. C. H., & Cahen, J. J. (2002). Jewish Netherlanders, Netherlands Jews, and Jews in the Netherlands, 1870–1940. In J. C. H. Blom & R. G. Fuks-Mansfeld (Eds.), *The history of the Jews in the Netherlands* (pp. 230–295). Oxford: The Littman Library of Jewish Civilization.

Bosma, U. (2010). *Indiëgangers. Verhalen van Nederlanders die naar Indië trokken*. Amsterdam: Bert Bakker.

Bruijn, I. (2009). *Ship's surgeons of the Dutch East India company. Commerce and the progress of medicine in the eighteenth century*. Leiden: Leiden University Press.

Christiaans, P. A. (1992a). *Genealogische bronnen betreffende Europeanen te Cheribon*. 's-Gravenhage: Indische Genealogische Vereniging.

Christiaans, P. A. (1992b). Joden in Nederlands-Indië. *Misjpoge, 1*, 9–11.

Davis, N. Z. (2016). Physicians, healers, and their remedies in colonial Suriname. *Canadian Bulletin of Medical History, 33*(1), 3–34.

de Knecht-Van Eekelen, A. (1988). De invloed van 's-rijksweekschool voor militaire geneeskundigen op de geneeskunde in Nederlandsch Oost-Indie. In D. d. Moulin (Ed.), *'s- Rijkskweekschool voor Militaire Geneeskundigen te Utrecht. 1822–1865* (pp. 95–109). Amsterdam: Rodopi.

Dietz, A. (1907). *Stammbuch der Frankfurter Juden. Geschichtliche Mitteilungen über die Frankfurter jüdischen Familien von 1349–1849, nebst einem Plane der Judengasse*. Frankfurt a. M: Goar.

Düsterdieck, P. (1983). *Die Matrikel des Collegium Carolinum und der Technischen Hochschule Carolo-Wilhelmina zu Braunschweig, 1745–1900*. Hildesheim: Lax.

Emmanuel, I. S., & Emmanuel, S. A. (1970). *History of the Jews of the Netherlands Antilles*. Cincinnati: American Jewish Archives.

Erler, G. (1911). *Die Matrikel der Albertus-Universität zu Königsberg i. Pr.: II. Band: Die Immatrikulationen von 1657–1829*. Leipzig: Duncker & Humblot.

Erman, W. (1973 (1899)). *Verzeichnis der Berliner Universitätsschriften 1810–1885. Nebst einem Anhang enthaltend die ausserordentlichen und Ehren-Promotionen*. Hildesheim: Olms.

Feist, B. (1865). *De werking van het cultuurstelsel op Java en de nieuwe cultuurwet*. Amsterdam: Schlömann.

Felsenthal, S. (1930). Jüdische Ärzte in Alt-Mannheim. *Sudhoffs Archiv für Geschichte der Medizin, 23*, 184–196.

Fink, A. (2001). *Jüdische Familien in Kreuznach. Vom 18. Jahrhundert bis zuem Ersten Weltkrieg; eine Dokumentation*. Bad Kreuznach.

Franke, E. (2013). Searching for traces of Judaism in Indonesia. In M. Hutter (Ed.), *Between Mumbai and Manila. Judaism in Asia since the founding of the state of Israel* (pp. 39–51). Göttingen: V & R Unipress.

Gelder, R. V. (2004). *Das ostindische Abenteuer. Deutsche in Diensten der Vereinigten Ostindischen Kompanie der Niederlande (VOC) 1600–1800*. Convent: Hamburg.

Genest, A., & Marquardt, S. (2003). *Quellen zur Geschichte der Juden in polnischen Archiven. Bd. 1: Ehemalige preußische Provinzen Pommern, Westpreußen, Ostpreußen, Preußen, Posen, Grenzmark Posen-Westpreußen, Süd- und Neuostpreußen*. München: Saur.

Hack, K., & Rettig, T. (Eds.). (2006). *Colonial armies in Southeast Asia*. London: Routledge.

Hadler, J. (2004). Translations of antisemitism: Jews, the Chinese, and violence in colonial and post-colonial Indonesia. *Indonesia and the Malay World, 32*(94), 291–313.

Hahn, S. I. (1822). *Inaugural-Abhandlung über Nachgeburtszögerungen*. Würzburg: Becker.

Hahn, P. (1823). *Betrachtungen über die Theorien der Epigenese und Evolution*. Würzburg: Becker.

Haneveld, G. T., & Royen, P. C. (2001). *Vrij van zichtbare gebreken: de medische zorg bij de Nederlandse zeemacht in de negentiende eeuw*. Amsterdam: Bataafsche Leeuw.

Hartzfeld, J. (1839). *Der Wasserbruch*. Würzburg: Eine Inaugural-Abhandlung.

Heymann, S. L. (1855). *Versuch einer pathologisch-therapeutischen Darstellung der Krankheiten in den Tropenländern*. Würzburg: Stahel.

Horovitz, M. (1886). *Jüdische Aerzte in Frankfurt a. M.* Kauffmann: Frankfurt am Main.

Huerkamp, C., & Spree, R. (1982). Arbeitsmarktstrategien der deutschen Ärzteschaft im späten 19. und frühen 20. Jahrhundert. In T. Pierenkemper & R. Tilly (Eds.), *Historische Arbeitsmarktforschung. Entstehung, Entwicklung und Probleme der Vermarktung von Arbeitskraft* (pp. 77–116). Vandenhoeck und Ruprecht: Göttingen.

Kallmorgen, W. (1936). *Siebenhundert Jahre Heilkunde in Frankfurt am Main*. Diesterweg: Frankfurt a. M.

Kerkhoff, A. H. M. (1987). German physicians and pharmacists in the Dutch East Indies. *Medical Corps international: Forum for Disaster & Military Medicine, 2*, 52–54.

Kerkhoff, A. H. M. (1989). The organization of the military and civil medical service in nineteenth century. In A. M. Luyendijk-Elshout, G. M. van Heteren, A. de Knecht-van Eekelen, & M. J. D. Poulissen (Eds.), *Dutch medicine in the Malay Archipelago. 1816–1942* (pp. 9–24). Amsterdam: Rodopi.

Kössler, F. (1970). *Verzeichnis der Doktorpromotionen an der Universität Giessen von 1801–1884*. Giessen: Univ.-Bibliothek.

Kössler, F. (1976). *Register zu den Matrikeln und Inscriptionsbüchern der Universität Gießen. WS 1807/08–WS 1850*. Giessen: Univ.-Bibliothek.

Kracauer, I. (1925–1927). *Geschichte der Juden in Frankfurt/Main, 1150–1824*. 2 vol. Frankfurt am Main: Kauffmann.

Landau, R. (1895). *Geschichte der jüdischen Ärzte*. Berlin: Karger.

Lenheim, L. (1825). *De morborum generatione. Dissertatio inauguralis*. Marburg: Krieger.

Lewin, A. (1909). *Geschichte der badischen Juden seit der Regierung Karl Friedrichs (1738–1909)*. Karlsruhe: Braun.

Lindeboom, G. A. (1984). *Dutch medical biography. A biographical dictionary of Dutch physicians and surgeons 1475–1975*. Amsterdam: Rodopi.

Mälzer, G. (Ed.). (1994). *Würzburger Hochschulschriften 1804–1885. Bestandsverzeichnis*. Würzburg: Universitätsbibliothek Würzburg.

Merkle, S. (1922). *Die Matrikel der Universität Würzburg. 2. Hälfte*. München [et al.]: Duncker & Humblot.

Pariser, B. N. (1830). *Historia opinionum, quae de sanguinis circulatione ante Harvaeum viguerint*. Berlin: Brüschckianis.
Regerings-Almanak van Nederlandsch-Indië. (1815–1942). Batavia.
Richarz, M. (1974). *Der Eintritt der Juden in die akademischen Berufe. Jüdische Studenten und Akademiker in Deutschland 1678–1848*. Tübingen: Mohr.
Richter-Hallgarten, H. G. (2013). *Die Beteiligung jüdischer Ärzte an der Entwicklung der Dermatologie zu einem eigenständigen Fach in Frankfurt am Main*. München. Retrieved from https://edoc.ub.uni-muenchen.de/16176/
Ricklefs, M. C. (2008). *A history of modern Indonesia since c. 1200*. Basingstoke [et al.]: Palgrave Macmillan.
Robinow, M. O. (1836). *Quaedam de dysomphalotocia sive de funiculi umbilicalis circumvolutionibus, contorsionibus et prolapse*. Hamburg.
Rosenstock, D. (2008). *Die unterfränkischen Judenmatrikeln von 1817. Eine namenkundliche und sozialgeschichtliche Quelle*. Würzburg: Schöningh.
Rosenthal, B. (1981, reprint of the edition from 1927). *Heimatgeschichte der badischen Juden. seit ihrem geschichtl. Auftreten bis zur Gegenwart*. Magstadt bei Stuttgart: Bissinger.
Schlich, T. (1990). *Marburger jüdische Medizin- und Chirurgiestudenten. 1800–1832. Herkunft – Berufsweg – Stellung in der Gesellschaft*. Marburg: Elwert.
Schlitz, M. M. (2015). *Ortsfamilienbuch Bruchsal* (Vol. 3). Weißenthurm: Cardamina-Verl.
Schoute, D. (1937). *Occidental Therapeutics in the Netherlands East Indies during three centuries of Netherlands settlement. (1600–1900)*. Batavia.
Schwarz, J. (1833). *De febris infarctus effectu*. Berlin: Nietackianis.
Steckmest, S. (2013). *Die Robinows: eine Hamburger jüdische Kaufmannsfamilie*. Hamburg: Ina Lorenz.
Straus, H. J. & Ullmann, E. (1880). *Verzeichniß der von Hirsch Herz Straus Frankfurt a. M. stammenden Familien*. Frankfurt a. M.
Strauß, H. (1825). *Ueber den Rhythmus in den Lebenserscheinungen*. Göttingen.
Stude, J. (2007). *Geschichte der Juden in Bruchsal*. Ubstadt-Weiher: verlag regionalkultur.
Tabellarisch overzigt van het personeel der geneeskundige dienst. (1859). *Geneeskundig tijdschrift voor Nederlandsch-Indië, 7*, 237–317.
Teichfischer, P. (2016). Transnational entanglements in colonial medicine. German medical practitioners as members of the health service in the Dutch East Indies (1816–1884). *Histoire, médecine et santé. (Special issue: Guerre, maladie et empire), 10*, 63–78.
Toepke, G. (1904). *Die Matrikel der Universität Heidelberg. Teil 5: Von 1807 bis 1846*. Heidelberg: Winter.
Verzeichnis der dermaligen Lehrer und Studirenden auf der Großherzoglich-Hessischen Universität zu Giessen. (1823–1829). Gießen: Georg Friedrich Heyer.
Verzeichnis des Personals und der Studierenden an der Julius-Maximilians-Universität zu Würzburg. (1830–1836). Würzburg: Carl Wilhelm Becker.
Verzeichniß der sämmtlichen Studirenden auf der Universität Heidelberg. (1818–1831). Heidelberg.
Verzeichniß der Studirenden auf der Rheinischen Friedrich-Wilhelms Universität zu Bonn. (1825–1867). Bonn.
Vink, W. (2010). *Creole Jews: Negotiating community in colonial Suriname*. Leiden: KITLV Press.
von Selle, G. (1980, reprint of the edition from 1937). *Die Matrikel der Georg-August-Universitat zu Göttingen (1734–1837)*. Hildesheim/Leipzig: Lax.
Witkowski, R. (2009). Spis dorosłych Zydów mieszkających w Poznaniu sporządzony 26 marca 1832 r. *Kronika Miasta Poznania (KMP), 1*, 25–41.

Chapter 10
Jewish Students from Silesia Studying at the Medical Faculty of Vienna University in the Years 1850–1938 According to the Records Regarding University Promotion and Requirements

Joanna Lusek and Horst Doležal

Abstract In the second half of nineteenth century a large number of people studying at the medical faculty of Vienna University were Jewish, including Jews from Silesia. In the years 1850–1938 medicine at the University of Vienna was studied by 202 people of Jewish origin from Silesia. They came mainly from the territory of Austrian Silesia, less often from Prussian Silesia. In 1938, after the seizure of Austria, the university became an arena of racist and political persecution, which resulted in irreversible losses of the medical faculty's intellectual potential – both lecturers and students. Approximately one third of Jewish students were forced to stop their studies.

The topic raised in this study is a reference to the seldom discussed issues regarding medical education of Jews on the territory of Silesian borderland in the context of supracultural assimilation, i.e. coexistence of Jews in the academic circles of Vienna as well as the development of professional careers of Jewish people in the German speaking area. This work has been based on archival materials: promotion records prepared for the needs of the rector's office of Vienna University and requirements records kept for the needs of the medical faculty for the years 1818–1938, which are available at the Archives of Vienna University. They allow establishing the exact number of Jewish students from Silesia who were studying at the medical faculty in the years 1850–1938. Apart from personal data, which include the dates of obligatory and promotion examinations, the archival materials allow analysing such details as the place and date of birth, the place of taking secondary school final examinations, father's property or job.

J. Lusek (✉)
Upper Silesia Museum, Bytom, Poland
e-mail: j.lusek@muzeum.bytom.pl

H. Doležal
Heraldisch-Genealogische Gesellschaft "Adler", Wien, Austria
e-mail: office@adler-wien.at

M. Moskalewicz et al. (eds.), *Jewish Medicine and Healthcare in Central Eastern Europe*, Religion, Spirituality and Health: A Social Scientific Approach 3,
https://doi.org/10.1007/978-3-319-92480-9_10

Keywords Vienna University · Medical faculty · Prussian Silesia students · Jewish students

10.1 The Development of the Medical Faculty of the University of Vienna Until the Second World War

The Rudolf IV foundation deed for the oldest university in the German-language area was issued on 12. March 1365. Three months later, on 18. June 1365, Pope Urban V gave his consent to the foundation, but with the exclusion of the theological faculty. This was almost imperative for a mediaeval university and was finally agreed to by Pope Urban VI on 20. February 1384. Albrecht III's – in some instances supplementary – charter of confirmation (known as the Second Foundation Deed) confirmed the Rudolfine foundation and enlarged it in some points in 1384. Faculty records dating back to 1399 show that the university was called upon to act as a mediator in disputes between barber surgeons, midwives and landlords (Tuisl 2014, pp. 20–22; Rexroth 2015, pp. 14–27; Jordak 1965; Gall 1965; Uiblein 1965).

Following the Great Papal Schism a number of scholars, in particular the theologian, Heinrich von Langenstein, left the university in Paris. Many of them came to Vienna and university life there began to blossom, a period which lasted until the first 25 years of the sixteenth century. During this time the University of Vienna became the most important German university. After years of plague and the death of Maximilian I. (1519) urgent attempts to reform the university failed to establish themselves. During the Counter-Reformation the Societas Jesu was called to Vienna to take up two theological chairs after 1558. Subsequently the Jesuits strengthened their influence by means of the 1623 "Sanctio pragmatic", which gave them the right to hold the chairs in the philosophy and theology faculties. This change, and the reform introduced by Ferdinand I in 1554, meant that efforts to re-catholicise the university proved successful, but the law and medical faculties became utterly insignificant. The main point now was to pass on knowledge and scarcely any scholars were involved in research and publishing their results.[1]

After several ineffective attempts at reform, radical changes finally came into effect under Maria Theresia in the middle of the eighteenth century, when the foundation stone for the first medical school in Vienna was laid. Amongst the teachers and research workers in Vienna at the time were Anton de Haen, Maximilian Stoll, Johann Lorenz Gasser, Anton von Störck and the man who discovered percussion, Leopold Auenbrugger. The state extended its influence even as far as being able to appoint professors. The old rights and freedoms were practically wiped out and the

[1]Institut für Österreichische Geschichtsforschung (Ed.), *Die Matrikel der Universität Wien. Quellen zur Geschichte der Universität Wien.* Bd. 1. 1377–1450, Wien 1954; Bd. 2. 1451–1518, Wien 1967; Bd. 3. 1518/II–1579/I, Wien 1971; Bd. 4. 1579/II–1658/59, Wien 1974; Bd. 5. 1659/60–1688/89, Wien 1975; Bd. 6. 1689/90–1714/15, Wien 1993; Bd. 7. 1715/16–1745/46, Wien 2011; Bd. 8. 1746/47–1777/78, Wien 2014.

influence of the church reduced to a minimum by the annulment of the Societas Jesu in 1773. The well-known medical doctor, Gerard van Swieten, was entrusted with reforming the University in 1749. He was the driving force behind the changes that were made in the spirit of the Enlightenment. Up until then, students had been forced to make a public profession of their Catholic faith but this was abolished by the Patents of Toleration issued by Josef II in 1781. Now students of any religion were allowed to attend lectures. Two years later university jurisdiction was abolished and the state took over the University completely along with its property and wealth. Now the emphasis was on knowledge that would be useful to the state and this hindered the development of academic research with the exception of the legal and, even more so, the medical faculty, both of which had been hitherto shabbily neglected. The creation of the General Hospital of the City of Vienna in 1784 provided an impulse for a positive development. Thus in the pre-March years, the medical faculty became globally famous thanks to names like Carl Frh. v. Rokitansky, Joseph L. Skoda, Ferdinand v. Hebra and Josef Hyrtl. These were also the precursors of the Second Vienna Medical School (Beck von Mannagetta and C. von Kelle 1906).

The university reforms introduced under the Minister for Culture and Teaching, Count Leo *Thun*-Hohenstein (1849–1860), following the year of revolution in 1848 gave a hitherto unknown boost to all universities in Austria. Modelled on Prussian universities, the universities of the Monarchy (Vienna, Graz, Innsbruck, Padua, Pavia, Prague, Krakau, Lviv and Pest) were revamped into places of teaching and research (Drimmel 1959). New student regulations came into force in the University of Vienna in 1850. Following the German model, new systems of teaching and research were introduced and several departments were re-structured. Those of chemistry, mineralogy, botany and zoology that had previously been a part of the medical faculty, were now moved to the faculty of philosophy. The restrictions on teaching posts that had up to then had only been open to Roman Catholics were abolished in 1873. New university buildings designed by Heinrich von Ferstel were opened between 1873 and 1884. Women were finally officially admitted as students in 1897 (faculty of philosophy), 1900 (medicine) and 1919 (law) (von Bruch 1998, pp. 133–156; Höflechner 1988).

The social and political caesura caused by the First World War, the reduced boundaries of the state and the corresponding material difficulties were countered to some extent by the fact that scholarly activities in the University continued and it was able to maintain its academic reputation. In 1938, following the annexation of Austria into the German Reich, the University of Vienna was forced to conform to Nazi policies and a huge number of teachers and students were driven out on racist and political grounds. The university in Vienna lost almost half its teachers, including the medical faculty. This loss included the following Silesians: Albert Herz – internal medicine, Bruno Klein – dentistry, Eugen Pollak – anatomy and pathology of the nervous system, Erich Ruttin – otology, Friedrich Silberstein – pathology, and Richard Balthasar Wasiscky – pharmacognosy. The loss was felt for many decades (Mühlberger 1993, p. 7; Bauer-Merinsky 1980).[2] Despite heavy war

[2] *Gedenkbuch für die Opfer des Nationalsozialismus an der Universität Wien 1938*. http://gedenkbuch.univie.ac.at. Accessed 24 May 2015.

damage to buildings and chaotic staff and administrative structures, teaching activities were taken up once more at the end of 1945. The Austrian regulations, valid before the 13. March 1938, came into force once more and a student self-administration was set up.

10.2 Sources on the Theme

The list of Jewish students from Silesia is based on the register of doctoral students kept by the Vice Chancellor's office of the University of Vienna, and the minutes of the doctoral thesis disputation (Rigorosum) kept by the faculty between 1818 and 1938. These are held in the archive of the University of Vienna.[3] On average, the students are named twice, once in the Rigorosum minutes and once in the minutes of the doctorate title: there are roughly 52,000 records on around 26,000 graduates. Both sets of minutes were taken in order to enhance the safety of the records and, more so, to raise the data width. Whereas the doctorate minutes as a rule contain the first and family names, place and land of origin, alongside the handwritten signature of the doctoral student, the Rigorosum minutes provide us additionally with the date of birth, religious confession, information on previous schooling as well as data and the grades given for the Rigorosum, and often the day on which the students received their doctorate; albeit in a number of different variations.

Another stock of evaluated documents consists of 50 (originally there were 58) Rigorosum volumes covering the years 1822–1873, with information on the examinations for additional studies not contained in the basic medical training. The field of study (surgeon, dentist, midwife etc.) is set down alongside the data on the Rigorosum. As a rule the documents reproduce the data of the first Rigorosum, since the majority of the succeeding second Rigorosum mostly consist of the grade awarded and seldom the day of the examination. Since surgeons had to undergo practical training and were organised in guilds, the successful completion of the first Rigorosum was a condition for the receipt of a diploma that gave them the right to practice surgery in the whole area under the Austrian-Hungarian monarchy. The second Rigorosum led to a guarantee (Sponsion: Magister Chirurgum – Mag. chir.).

[3]Archiv der Universität Wien/Medizinische Fakultät (AU/MF), 170, Rigorosenprotokoll 1821–1871, Sign. Med. 12.1; 177, Rigorosenprotokoll 1872–1894, Sign. Med. 12.2; 195, Rigorosenprotokoll 1894–1910, Sign. Med. 12.3; 196, Rigorosenprotokoll 1903–1930, Sign. Med. 12.4; 197, Rigorosenprotokoll 1922–1934, Sign. Med. 12.5; 175, Promotionsprotokoll 1818–1840, Sign. Med. 33.1; 176, Promotionsprotokoll 1840–1854, Sign. Med. 33.2; 181, Promotionsprotokoll 1854–1865, Sign. M. 33.3; 182, Promotionsprotokoll 1865–1873, Sign. M. 33.4; 186, Promotionsprotokoll 1874–1890, Sign. M. 33.5; 187, Promotionsprotokoll 1890–1894, Sign. M. 33.6; 188, Promotionsprotokoll 1894–1898, Sign. M. 33.7; 189, Promotionsprotokoll 1898–1904, Sign. M. 33.8; 190, Promotionsprotokoll 1904–1912, Sign. M. 33.9; 191, Promotionsprotokoll 1912–1919, Sign. M. 33.10; 192, Promotionsprotokoll 1919–1923, Sign. M. 33.11; 193, Promotionsprotokoll 1923–1929, Sign. M 33.12; 194, Promotionsprotokoll 1929–1941, Sign. M. 33.13.

Occasionally courses ended with a Doktor Chirurgum – Dr. chir. This differentiation disappeared in 1873 when the title of Doctor of Medicine was introduced – Doctor of General Medicine, thus also of surgery. Until 1873 surgeons were trained by practical surgeons, and frequently by medical students – including those who had already passed their doctorate exams. In such cases the Rigorosum is as a rule noted in the register books I. and II.

Candidates studying medicine as a main course had to study for 10 semesters and were subject to theoretical and practical tests during Rigorosums I.–III. The first Rigorosum examined their knowledge of the following subjects: general biology, medical physics, medical chemistry, anatomy, histology and physiology. The second Rigorosum comprised the following obligatory tests: pathological anatomy and histology, general and experimental pathology, pharmacology and knowledge of prescriptions, internal medicine, paediatrics, psychiatry and neuropathology. The third Rigorosum tested students' knowledge of surgery, gynaecology and midwifery, ophthalmology, dermatology, hygiene and forensic medicine. New subjects were added at the beginning of the twentieth century. These included laryngology, otology, dentistry and immunology, plus one course in each of the following: operation, percussion and auscultation. Each test was taken by the appropriate examiner under the supervision of superiors. During the second and third Rigorosum the tests were also supervised by a government commission. The theoretical tests lasted roughly a quarter of an hour. The practical tests lasted longer because of the demands involved. The grades given by the examiners were either "outstanding", "satisfactory" or "unsatisfactory". After they had completed their Rigorosum successful candidates were given a report: "Herr...... born in (place and country)... after completing ... semesters of medical studies at the ... fulfilled the... Rigorosum with satisfactory/outstanding success".[4]

In addition 18,000 records were taken from catalogues of studies undertaken by medical students at the University Vienna. These records were taken every eighth semester and referred exclusively to Jewish students between 1862 and 1938, regardless of the particular semester. As a rule the records also contain the students' address in Vienna,[5] as well as the names and professions of their fathers.[6]

There are gaps in some of the early Rigorosum minutes. Thus, in the first 2 years there is no information on the religious confession of almost 50% of the students.

---

[4] "Nr. 672. Erlaß des Ministeriums für K. u. U. vom 20. Februar 1887 (an das medizinische Dekanat in Wien, betr. der Abhaltung von „Kursen" and "Nr. 674. Erlaß des Ministeriums für K. u. U. vom 25. März 1896 (an alle medizinischen Dekanate, betr. die Behandlung von Gesuchen der Studierenden um Abkürzung von Reprobationsfristen und um andere Begünstigungen bei Ablegung von Rigorosum)" and "Nr. 679. Erlaß des Ministeriums für K. u. U. vom 8. Mai 1903 (an alle Dekanate der medizinischen Fakultäten, womit eine Instruktion zu der mit Verordnung vom 14. April 1903 /RGBNr. 102/ erlassenen medizinischen Rigorosenordnung kundgemacht wird)" (Beck von Mannagetta, von Kelle, 1906, pp. 879–907).

[5] Corresponding to the official way of writing e.g. "IX., Glasergasse" for Vienna, 9. District, Glasergasse.

[6] AU/MF, 134, Nationalien/Studienkataloge 1862–1938, Sign. WS 1862/63, WS 1866/67, WS 1870/71.

During the 1930s there was a strong rise in the lack of information on a student's religion due to the political situation at the time. Independent of this, it is no surprise after such a long time that the entries in all the volumes worked on, differ greatly in form and informative value. For this reason it might be helpful to make targeted research in other archive stocks when connecting factors are found.

Where place names occur on multiple occasions, in case of doubt it has been decided to add (Austrian/Prussian) "Silesia" almost without exception. When only one birth year is given, this is mostly given by working back from the information on the given age. An aptitude certificate (Maturazeugnis) is also listed even when such a certificate was not expressly issued to show that the student had visited a teaching institute immediately before beginning his or her term at the university. At the end of the entry, the last known date is given (say, the academic year, Rigorosum, the thesis defence, or the date of the doctorate). Where the data do not refer to a general medical practitioner this is expressly mentioned (for example, surgeon or midwife).

The "Josephinische Academie" set up by Josef II in 1781 existed – with breaks – until 1873. This is where military surgeons, and also "Doktores chirurgum" received their training. Some of the minutes of the "Academie" still exist in the university archive. Evaluations were made of the doctorate minutes for the years 1854–1869,[7] and the catalogue of students between 1781 and 1785.[8]

The following list of persons contains the most important biographical data from all these sources, occasionally supplemented with the help of non-university, partly unpublished sources, like for example the "Medicinal-Schematismus der österreichischen Monarchie 1848 and Oesterreichischer Medizinal-Schematismus 1859" by Josef Nader[9]; register books drawn up by the Israelite community in Vienna,[10] a chronicle of doctors from Lower Austria by Berthold Weinrich (Weinrich 1990) and specialist articles about medical students from Silesia by Horst Doležal (2007, pp. 154–195; Idem 2010, pp. 127–144; Idem 2012, pp. 121–150) created from a private data base.[11]

[7] AU/MF, 260/01, Promotionsprokoll Josephs Academie 1854–1869, Sign. Jo 34.1.

[8] AU/MF, 260/02, Catalog der Zöglinge der Josephs-Academie 1781–1785, Mikrofilm Sign. 1411 (Abteilung Kriegsarchiv).

[9] AU/MF, 253/1, Josef Nader, Medicinal-Schematismus der österreichischen Monarchie 1848, Wien o. J.; 253/2, Josef Nader, Oesterreichischer Medizinal-Schematismus, Wien 1859.

[10] Wiener Stadt- und Landesarchiv, Matriken der Israelitischen Kultusgemeinde Wien auf Mikrofilmen (WSuL/MIKW), Sign.100, Geburten; Sign. 200, Trauungen.

[11] H. Doležal, A database of medical students at the University of Vienna 1818–1938, Mödling 1998 (not published). The database comprises ca. 175.000 data sets. They include rare sources, but above all excerpts from church register books, from archives of the city of Vienna and the region of Vienna, also from the archive of the University of Vienna dealing with medical students, and guild documents on medical professions.

10.3 Silesia as a Research Area – Jewish Medical Students at the University of Vienna

One of the consequences of the first Silesian War in 1742 was that the overwhelming majority of Silesia was handed over to Prussia. Roughly a seventh of the area of the whole of Silesia remained with Austrian-Hungary as Austrian-Silesia. In 1920 a part of Teschen Silesia and in 1922 also Eastern Upper Silesia were incorporated into Poland as the autonomous County of Silesia. From 1815 onwards the Prussian part of the province of Silesia with the governing district of Liegnitz, Breslau and Opole was integrated into Silesia. From 1919 to 1938 and from 1941 to 1945 this province of Silesia was split up into the provinces of Lower and Upper Silesia. Upper Silesia made up the south-eastern part of the historical region of Silesia that now lies for the most part in Poland in the counties of Opole and Silesia. The town of Opole was generally regarded as the historical capital of Upper Silesia. Lower Silesia was the northern part of the region of Silesia, just as most of Upper Silesia now lies in Poland – in the County of Lower Silesia. It stretches around the central course of the River Oder with its capital, Breslau. By contrast the Western part of Silesia – Austrian-Silesia remained with Austria until 1918 – belongs to the Czech Republic (Žáček 2004, pp. 56–58; Conrads 1994; ed. P. Haslinger and Kreft 2010).

Austrian-Silesia was the official term for the Duchy of Upper and Lower Silesia. It incorporated part of the territory belonging to the Bohemian crown of the Austrian monarchy. Between 1850 and 1918 it was the Crown land of the Empire of Austria, respectively of Austria-Hungary. In the 1867 Dual Monarchy Austria-Hungary, Silesia was a part of Cisleithania in the Austrian part. It later sent (resp. elected) members of parliament to the Reich Council in Vienna. After the collapse of Austria-Hungary at the end of the First World War, German occupied areas were not incorporated into German Austria. From 22 November until 3 December 1918, decisions made in Troppau by the government of Sudetenland were published in the official Gazette of the Duchy. The definitive establishment of the province of Sudetenland, which had taken in the German occupied parts of Austria-Silesia, was hindered by the occupation of the area by Czechoslovakian troops at the turn of 1918/1919. The land was incorporated into Czechoslovakia. The Eastern part of the area around Bielitz was given to Poland and this made up the basis for the County of Silesia. The Czechoslovakian part remained an independent administrative unit – země Slezsko – until 1928 when it was united with Moravia to become Moravian-Silesia – země Moravskoslezská (Alexander 2008; Conrads 1994; ed. Weczerka 1977).

The individual periods in the statistics within the period dealt with are demarcated by the inception of the Thun higher education reform on the one hand and, on the other hand, by the outbreak of the First World War in 1914 with its consequences and effects on the planning of curricula and the training of the next generation of doctors.

Eight hundred and forty one Silesians were amongst the total number of medical students in Vienna in the years between 1818 and 1938, including 534 from Upper Silesia and 337 from Lower Silesia. These were registered according to their gender,

Table 10.1 Medical students from Upper and Lower Silesia – Statistics (1818–1938)

Period	Gender			Confession			
	Male	Female	Together	Cath.	Mos.	Evang.	Unknown
1. Students from Upper Silesia, including students of Jewish confession							
1818–1849	72	–	72	27	6	4	35
1850–1913	297	1	298	154	84	13	47
1914–1938	142	22	164	72	59	14	19
Totality	**511**	**23**	**534**	**253**	**149**	**31**	**101**
2. Students from Lower Silesia, including students of Jewish confession							
1818–1849	83	–	83	34	2	4	44
1850–1913	203	3	206	95	34	6	70
1914–1938	44	4	48	23	17	2	6
Totality	**330**	**7**	**337**	**152**	**53**	**12**	**120**
Totality (1. And 2.)	**841**	**30**	**871**	**405**	**202**	**43**	**221**

The table also contains surgeons and doctoral students of surgery in the Josephinum
Sources: AU/MF, 170, Sign. Med. 12.1; 177, Sign. Med. 12.2; 195, Sign. Med. 12.3; 196, Sign. Med. 12.4; 197, Sign. Med. 12.5; 175, Sign. Med. 33.1; 176, Sign. Med. 33.2; 181, Sign. M. 33.3; 182, Sign. M. 33.4; 186, Sign. M. 33.5; 187, Sign. M. 33.6; 188, Sign. M. 33.7; 189, Sign. M. 33.8; 190, Sign. M. 33.9; 191, Sign. M. 33.10; 192, Sign. M. 33.11; 193, Sign. M 33.12; 194, Sign. M. 33.13; 256/1, Sign. Med. 9,1; 256/2, Sign. Med. 9,2; 256/5, Sign. Med 9,5; 134, Sign. WS 1862/63, WS 1866/67, WS 1870/71; 260/01, Sign. Jo 34.1; 260/02, Sign. 1411; 253/1; 253/2; WSuL/MIKW, Sign.100; Sign. 200

religious confession and the time when they began their study (see Appendix to the text). Two hundred and two of the 871 medical students were of Jewish confession, of whom 149 (27.9%) came from Upper Silesia and 53 (15.7%) from Lower Silesia (Table 10.1). The students who came from places in Lower Silesia preferred to study at the medical faculties, above all at the universities in Breslau and Berlin, less commonly in Vienna. The greatest number of Jewish students from Silesia came to Vienna at the turn of the twentieth century. More than 58% of all medical students were of Jewish faith in the period under research (Table 10.1).

Seventy birthplaces of medical students are named in the sources (Table 10.2), including 12 (17.1%) places in Lower Silesia and 58 (82%) in Upper Silesia. This number contains 50 from Teschen Silesia (71.4%). The majority of Jewish students from Silesia who studied medicine in Vienna came from large towns in Teschen Silesia, above all from Bielitz (21.8%), Teschen (13.8%) and Troppau (10.9%). Twenty nine towns were mentioned as places where A-level examinations occurred: Bielitz, Breslau, Brünn, Freudenthal, Friedek, Güstrow, Komotau, Krems, Kremsier, Lemberg, Letschau, Lieblitz, Moravian Ostrava, Moravian Weisskirchen, Nikolsburg, Oberhollenbrunn, Olomouc, Poznan, Prague, Prerau, Przemysl, Reschen, Teschen, Troppau, Uhersky Brod, Uhersky Hradiště, Weidenau and Vienna (Tables 3–5). The greatest number came from Bielitz (45 persons), Teschen (25 persons), Troppau (25 persons) Vienna (24 persons) and Moravian Ostrava (11 persons).

Table 10.2 Birthplaces of medical students

Alphabetical list of birthplaces
Alt Lublinitz
Benisch, Bielitz, Bludowitz, Breslau, Brieg, Brosdorf, Bunzlau
Czechowitz
Deutsch Wartenburg, Deutschleuthen, Dombrau, Dzieditz
Eidlitz, Ernsdorf
Freiberg, Freiheitsau, Freistadt, Freiwaldau, Freudenthal, Friedek
Golleschau, Gross Hernsdorf, Groß Knezendorf, Groß-Olbersdorf, Grossglogau, Grudek
Hirschberg, Holeschau, Hotzenplotz, Hruschau
Jablonkau, Jauernig, Jägerndorf
Kattowitz, Königsberg
Landsberg, Lautsch, Liegnitz
Marklowitz, Martinau, Myslowitz
Niedek, Niezlachowitz
Oderberg, Odrau, Olbersdorf, Orlau, Ostrava
Pless, Polanka, Polish Ostrava
Raycza, Reichwaldau, Roppitz
Schibitz, Schumbarg, Skotschau, Stauding, Strzebowitz
Terlitzko, Teschen, Troppau, Tworog
Ustron
Waldenburg, Weidenau, Wendrin, Wiglstadtl
Zülz

Sources: AU/MF, 170, Sign. Med. 12.1; 177, Sign. Med. 12.2; 195, Sign. Med. 12.3; 196, Sign. Med. 12.4; 197, Sign. Med. 12.5; 175, Sign. Med. 33.1; 176, Sign. Med. 33.2; 181, Sign. M. 33.3; 182, Sign. M. 33.4; 186, Sign. M. 33.5; 187, Sign. M. 33.6; 188, Sign. M. 33.7; 189, Sign. M. 33.8; 190, Sign. M. 33.9; 191, Sign. M. 33.10; 192, Sign. M. 33.11; 193, Sign. M 33.12; 194, Sign. M. 33.13; 256/1, Sign. Med. 9,1; 256/2, Sign. Med. 9,2; 256/5, Sign. Med 9,5; 134, Sign. WS 1862/63, WS 1866/67, WS 1870/71; 260/01, Sign. Jo 34.1; 260/02, Sign. 1411; 253/1; 253/2; WSuL/MIKW, Sign.100; Sign. 200

The medical students came from different social classes. More than 30 paternal professions are mentioned: arable farmer, doctor or veterinary surgeon, railway director, architect, baker, civil servant, accountant, manufacturer, forester, innkeeper, grocer, police constable, house owner, timber merchant, hotelier, merchant, teacher, tenant farmer, private person, rabbi, butcher, tailor, blacksmith, shoe maker, soldier, haulage contractor and upholsterer. The fathers of future doctors were mostly merchants (65 persons), civil servants (21 persons) and doctors (19 persons), in this case the children followed in the tradition of their fathers (Appendix to the text). The students from Silesia primarily lived in districts II, VIII and IX (Appendix to the text). This can be explained by the fact that the II district in Vienna has always traditionally contained the highest proportion of Jewish citizens. The VIII district and particularly the IX district were traditional favourites with students because of their proximity to the university.

10.4 Summary

New conditions of study were introduced into the University of Vienna in 1850. Following the German pattern, new departments and research subjects were introduced into the curriculum alongside restructuring measures like the moving of the departments of chemistry, mineralogy, botany and zoology from the medical faculty to the philosophical faculty. The development of the university, i.e. the new courses and the outstanding professorial staff, attracted students from all over Europe, including Silesia. In the years between 1850 and 1938, 871 candidates, including 202 persons of Jewish confession from Upper and Lower Silesia, studied at the medical faculties of the University of Vienna. The racist and political persecutions following the seizure of power by the National Socialists were an enormous intellectual loss for the university and their effects were felt for many years afterwards.

Appendix

List of Jewish students from Silesia at the medical faculty of the University of Vienna.

1. Jewish students from Silesia at the medical faculty of the University of Vienna before 1850

Name, first name	Birthplace, date	Study Rigorosum
Jägerndorf, David	Jägerndorf/Wagstadt, 4. May 1802	Prom. 6. August 1834
Kohn, Joseph Bernhard	Hotzenplotz,27. September 1820	Prom. 30. September 1847
Lazar, Wilhelm	Troppau, 13. January 1817	Rigor. 31. January 1842 (chir.)
Loewy, Isac	Weidenau, is now baptized: Lehnhard Joseph	Rigor. 1836 (chir.), Olomouc
		3. August 1839 (dent.)
Mayer, Ignatius	Schlackenwerth/Troppau, 19. July 1813	Prom. 9. June 1840 (med.)
		Prom. 13. January 1843 (chir.)
Schlesinger, Samuel	Gross Glogau, 2. January 1776	Prom. 7. August 1801
Weissenberg, Jacobus	Teschen, 9. January 1813	Prom. 5. August 1843

2. Jewish students from Silesia at the medical faculty of the University of Vienna between 1850 and 1913

Name, first name	Birthplace, date	Father's name, Father's profession	Study Rigorosum A-Level	Student's address in Vienna
Amster, Rudolf	Bielitz, 9. May 1868	Ignatz Amster, Hotelier in Bielitz	Prom. 16. March 1895 Me. Bielitz, 24. September 1886	II.[a], Czerninplatz 7
Aufricht, Emil	Ernsdorf, 13. May 1865	Julius Aufricht, Private owner in Bielitz	Prom. 18. May 1889 Me. Bielitz, 25. June 1883	IX., Höfergasse 7/I
Aufricht Friedrich	Ernsdorf, 2. November 1866	–	Rigor. 1. March 1904 Me. Bielitz, 14. July 1884	–
Baum, Gustav	Bielitz, 10. April 1879	Father died	Prom. 24.July 1905 Me. Bielitz, 6. July 1899	IX., Mariannengasse 27/I/9
Berger, Simon	Freiberg, 11. February 1863	Jakob Berger, Innkeeper in Freiberg	Prom. 24. July 1888	XV., Pelzgasse 16/I/7, II., Pazmanitengasse15/4/20
Berl, Heinrich	Freudenthal, 26. December 1874	Max Berl, Wood trader in Freudenthal	Prom. 27. May 1898 Me. Olomouc, 7. July 1892	VI., Rahlgasse 3
Berl, Victor	Freudenthal, 13. November 1873	Max Berl, Wood trader in Freudenthal	Prom. 19. March 1898 Me. hung. Hradisch, 28. September 1892	VI., Rahlgasse 3/II/I/40
Berliner, Schefftel	Breslau, 16. May 1874	Abraham Berliner, Merchant in Vienna II	Prom. 22. March 1902 Me. Vienna, 21. September 1896	II.,Rauscherstrasse 2
Blumenthal, Emil	Teschen, 30. August 1846	Michael Blumenthal, Tradesmann	Prom. 3.April 1879 Me. Teschen, 29. July 1867	IX., Schwarzspanierstrasse 3 Hof, III/VII/I
Blumenthal, Emil	Teschen, 1848	–	1. June 1879 as a senior physician	–
Blumenthal, Emil	Teschen, 1849	–	Prom. 1882	I., Salzgries 10
Blumenthal, Emil	Teschen, 1894	–	–	–

(continued)

Name, first name	Birthplace, date	Father's name, Father's profession	Study Rigorosum A-Level	Student's address in Vienna
Brandeis, Simon	Freistadt, 15. September 1869	Bernhard Brandeis, Father died	Prom. 17. July 1896 Me. 4. July 1888	–
Buber, Oskar	Jägerndorf, 12. Juni 1870	A.M. Buber, Merchant in Jägerndorf	Prom. 1. December 1894 Me. Troppau, 5. July 1888	IX.,Ackergasse 7
Buchbinder, Walter	Troppau, 10. July 1889	Dr. L. Buchbinder, Physician in Troppau	Prom. 22. October 1913 Me. 9. July 1907	VIII., Laudongasse 25/ III/19
Burstein, Moritz	Jablunkau, 12. May 1849	Emanuel Burstein, Tradesmann in Vienna	Prom. 14. July 1880 Me. Vienna, 20. July 1870	II., Afrikanergasse 9 Augartenalleestrasse 27/I
Deutsch, Karl	Bielitz, 25. June 1879	Father died, Teacher	Prom. 4. July 1905 (Lipnik) Me. Vienna, 10. July 1899	II., Castellezgasse 12/ II/I/22
Deutsch, Richard	Troppau, 29. August 1867	Bernhard Deutsch, Teacher in Nikolsburg	Prom. 29. July 1892 Me. Nikolsburg, 7. July 1886	II., Gr. Sperlgasse 39/4/6, IX., Rote Löwengasse 17/II/37
Ehrmann, Ignatz	Teschen, 19. February 1832	–	Prom. 3. August 1860 (med.) Prom. 19. March 1861 (chir.) Me. Teschen, 19. September 1854	–
Eichenwald, Leopold	Orlau	–	Prom. 1906	–
Eichenwald, Paul	Orlau, 4. May 1885	Leopold Eichenwald, Doctor of Medical Sciences	Prom. 23. January 1909 Me. Kremsier, 7. July 1903	IX., Lacknergasse 5/7
Ernst, Moritz	Hotzenplotz, 27. Februar 1845	Joachim Ernst, Troppau, not working due to illness	Prom. 29. November 1870 (med.) Prom. 18. February 1874 (chir.) Me. Troppau, 1865	VII., Mariahilfer Hauptstrasse 46

(continued)

Name, first name	Birthplace, date	Father's name, Father's profession	Study Rigorosum A-Level	Student's address in Vienna
Fasal, Hugo	Freiheitsau, 10. November 1873	Ferdinand Fasal	Prom. 27. May 1898 Me. Vienna, 9. July 1892	III., Pfefferhofgasse 5/III
Feiner, Adolf	Wendrin, 28. December 1866	Markus Feiner, Merchant in Teschen	Prom. 18. March 1893 Me. Teschen, 16. July 1886	IX., Roßauergasse 4, VIII. Laudongasse 16/3/1/15
Fränkel, Oskar	Jablunkau, 6. Oktober 1884	Vorm. Adolf Löwy, Merchant in Jablunkau	Prom. 2. July 1913 Me. Moravian Ostrava, 16. July 1904	I., Judenplatz 7/2/6, IX., Porzellangasse 22/11, IX., Harmoniegasse 4/3/15
Fränkel, Wilhelm	Bielitz, 22. Juny 1862	Simon Fränkel, Cloth salesman in Bielitz	Prom. 24. July 1899 Me. Bielitz, 25. Juny 1883	VIII., Stolzenthalergasse 15/8, VI., Windmühlgasse 39
Fiedetzki, Arthur	Troppau, 13. December 1884	Jakob Friedetzki, Merchant in Troppau	Prom.10. May 1909 Me. Troppau, 11. June 1903	II., Pragerstrasse 14/III/16
Friedmann, Conrad (Konrad)	Teschen, 16. May 1868	S. Friedmann, Circlerabbiner in Teschen	Prom. 14. May 1892 Me. Troppau, 11. July 1885	IX., Pelikangasse 13, IX., Lazarettgasse 5
Goldeberger, Josef	Olbersdorf, 25. August 1874	Leopold Goldberger, Merchant in Oberndorf	Prom. 23. December 1899 Me. Troppau, 5. July 1893	II., Leopoldsgasse 6/14
Gross, Siegmund	Bielitz, 17. March 1873	Abraham Gross, Merchant in Bielitz	Prom. 27. July 1897 Me. Bielitz, 14. July 1891	IX., Brünnlbadgasse
Hatschek, Rudolf	Troppau, 26. May 1865	Bernhard Hatschek, Regimental doctor in Troppau	Prom. 10. November 1888 Me. Troppau, 4. July 1882	VIII., Josefstädterstrasse 20/I
Hecht, Robert	Bielitz, 1866	Lazarus Hecht, Official in the department of culture in Bielitz	Prom. 1886	XIV., Feldgasse 6, VII., Stiftsgasse 33
Hechter, Ignaz	Golleschau, 27. June 1870	Baruch Hechter, Merchant and Innkeeper in Ustron	Prom. 29. September 1896 Me. Bielitz, 16. July 1890	VIII., Alser Strasse 27/II, dann 7/16

(continued)

Name, first name	Birthplace, date	Father's name, Father's profession	Study Rigorosum A-Level	Student's address in Vienna
Heller, Samuel	Biala/ Glischowitz, 26. August 1839	Wilhelm Heller, Private owner	Rigor. 24. February 1868 Me. Vienna, 3. October 1860	VI., Am Glacis 23
Herz, Albert	Freudenthal, 2. December 1876	Dr.med. J. Herz, Staff physician in Bisenz, Moravia	Prom. 31. January 1901 Me. Nikolsburg, 19. July 1894	II., Donaustrasse 63, IX., Wasagasse 24
Huppert, Ludwig	Friedek, 29. June 1872	Leopold Huppert, Merchant in Friedek	Prom. 18. July 1896 Me. Troppau, 27. September 1890	IX., Schlagergasse 10/II, Sechsschimmelgasse 8/I/21, Brünndlbadgasse 6/I/5
Karplus, Johann Paul	Troppau, 25. October 1866	Gottlieb Karplus, Commercial Council in Vienna II.,	Prom. 30. December 1890 Me. Vienna, 1. June 1884	II., Taborstrasse 27
Karplus, Oskar/Oscar	Troppau, 28. December 1877	Berthold Karplus, Merchant in Troppau	Prom. 23. December 1905 Me. Bielitz, 17. July 1897	IX., Hahngasse 12, Lenaugasse 11,Garnisongasse 12
Katz, Willy	Brieg, 17. December 1878	Max Katz, Freight forwarder in Vienna II	Rigor. 14. February 1902 Me. Vienna, 8. July 1898	II., Jägerstrasse 26
Kellermann, Max	Bielitz, 26. January 1868	–	Rigor. 26. January 1892 Me. Bielitz, 12. July 1886	–
Klauber, Arnold	Troppau, 26. March 1867	D. Klauber, Merchant in Troppau	Prom. 24. February 1894 Me. Brünn, 14. July 1887	IX., Dietrichsteingasse 6/ III/16
Klein, Bruno	Raycza/Bielitz, 1879	–	–	–
Kohn, Alois	Teschen, 7. May 1832	–	Prom. 9. July 1858 Student of philosophy, Olomouc and Vienna	–
Kohn, Arthur	Teschen, 9. February 1867	Alois Kohn, Doctor of Medical Sciences in Teschen	Prom. 25. November 1893 Me. Teschen, 11. July 1884	VIII., Florianigasse 64/1
Kohn, Heinrich	Tierlitzko, 5. December 1834		Prom. 19. May 1865	IX., Alservorstadt 123

(continued)

Name, first name	Birthplace, date	Father's name, Father's profession	Study Rigorosum A-Level	Student's address in Vienna
Kohn, Josef	Strzebowitz, 21. February 1843	Jacob Kohn, Merchant in Strzebowitz	Prom. 22. March 1869 (med.) Prom. 14. May 1877 (chir.)	VIII., Josefstadt, Lammgasse 1
Kohn, Julius	Teschen, 20. March 1870	Alois Kohn, Doctor of Medical Sciences in Teschen	Prom. 1890 Me. Teschen	VIII., Schlösselgasse 16/ II/14
Kohn (nunc Kaulbach), Richard	Benisch, 19. April 1887	Bernhard Kohn, Farm owner in Benisch	Prom. 23. December 1910 Me. Troppau, 13. July 1905	II., Förstergasse 10, III., Gärtnergasse 2
Kohn, Rudolf	Hotzenplotz, 1861	Joachim Kohn, Merchant in Troppau	Scholar, 1878	I., Maximilianstrasse 7/22
Kohn, Siegmund	Friedek, 25. April 1830	–	Rigor. 12. October 1865 (chir.) Rigor. 12. January 1866 (dent.)	–
Königer, Hugo	Einsiedel, 22. June 1874	Leopold Königer, Merchant in Einsiedel	Prom. 2. April 1900 Me. Weidenau, 7. July 1894	IX., Glasergasse 19, Wasagasse 24
Kuffler, Moritz	Friedek, 14. November 1870	Leopold Kuffler, Merchant in Vienna	Rigor. 6. July 1896 Me. Vienna, 12. Jule 1888	II., Ferdinandstrasse 14
Kupfermann, Wilhelm	Schibitz, 23. November 1882	Daniel Kupfermann, Merchant in Teschen	Prom. 23. December 1905 Me. Teschen, 3. July 1900	IX., Dietrichsteingasse 3/19
Kuznitzky, Albert	Myslowitz, 1862	Isaac Kuznitzky, Merchant in Myslowitz	Scholar, 1886	VIII., Lederergasse 32/ III
Kuznitzky, Simon	Tworog, 25. February 1870	Eugen Kuznitzky, Director of the local railway in Brünn	Prom.31. January 1896 Me. Brünn, 18. June 1889	IX., Harmoniegasse 1/6, VIII., Langegasse 58/ II/17, Lederergasse 20/ II/17
Landsberger, Otto	Friedek, 4. April 1888	Leopold Hitschmann, Industrialist in Nachod	Prom. 26. January 1912 Me. Friedek, 10. July 1906	I., Eberndorferstrasse 3, IX., Lazarettgasse 3

(continued)

Name, first name	Birthplace, date	Father's name, Father's profession	Study Rigorosum A-Level	Student's address in Vienna
Lanzer, Hermann	Deutschleuten, 28. January 1855	Ludwig Lanzer, House owner in Deutschleuten	Prom. 4. July 1885 Me. hung. Hradisch, 21. June 1878	VIII., Piaristengasse 31
Lanzer, Otto	Ustron, 16. September 1851	Ignaz Lanzer, Innkeeper and homeowner in Ustron	Prom. 14. July 1876 Me. Teschen, 11. August 1870	IX., Schwarzspanierstrasse 3/II/XII/45,VIII., Wickenburggasse 21
Lederer, Ernst	Hotzenplotz, 6. February 1876	David Lederer, Gendarmerie Sergeant in Troppau	Prom. 27. March 1901 Me. Troppau, 13. July 1895	–
Lichtwitz, Leopold	Troppau, 19. December 1858	Emanuel Lichtwitz, Manufacturer in Troppau	Prom. 14. March 1883 Me. Troppau, 7. July 1876	IX., Rotes Haus I/179 [Schwarzspanierstrasse 3]
Lichtwitz, Robert	Hotzenplotz, 15. November 1871	Adolf Lichtwitz, Tradesman in Hotzenplotz	Prom. 18. July 1896 Me. Troppau, 27. September 1890	IX., Porzellangasse 49a/II/I
Lindner, Ignatz	Skotschau, 15. May 1842	Samson Lindner, Merchant in Skotschau	Prom. 26. July 1872 (med.) Prom. 2. July 1873 (chir.)	–
Lindner, Leopold	Teschen, 11. December 1843	Carl Lindner, Grocer in Hruschan	Prom. 10. December 1869 (med.) Prom. 3. August 1870 (chir.)	IX., Ludwigsgasse 6/II/I/6
Lindner, Sigismund	Teschen, 11. July 1840	Carl Lindner, Homeowner zu Hruschan	Prom. 24. June 1870 (med.) Prom. 3. August 1870 (chir.) Me. Reschen, 27. July 1861	IX., Hauptstrasse 138
Loewe, Richard	Zülz, 1860	Mayer Loewe, Merchant in Zülz	Scholar. 1886	IX., Lazarettgasse 29/12
Mahrer, Victor/Viktor	Freistadt, 27. May 1880	Ferdinand Mahrer, Doctor of Medical Sciences in Freistadt	Prom. 26. July 1906 Me. Bielitz, 27. September 1900	II., Taborstrasse 65/8

(continued)

Name, first name	Birthplace, date	Father's name, Father's profession	Study Rigorosum A-Level	Student's address in Vienna
Manasse, Paul	Breslau, 1868	Vorm. Albert Kauffmann, Merchant in Breslau	1866 in the second semester	I., Elisabethstrasse 8
Markus, Maximilian	Groß-Olbersdorf, 1864	Vorm. S. Fried, House owner in Odrau	Scholar, 1886	IX., Alserbachstrasse 22
Mendel, Gustav	Bunzlau, 1856		1886, practical Doctor	VIII., Wickenburggasse 22/I/5
Munk, Alfred	Friedek, 22. December 1859 (or 1860)	Vorm. Emanuel Munk, Manufacturer in Friedek	Scholar. 1882	IX., Schwarzspanier Strasse 15
Neumann, Alfred	Bielitz, 11. January 1872	Albert Neumann, Merchant in Bielitz	Prom. 16. July 1897 Me. Bielitz, 11. July 1891	IX., Brünnlbadgasse 6/3/8
Neumann, Rudolf	Radwanitz/ Freistadt, 13. October 1871	Leopold Neumann	Prom. 18. July 1896 Me. Vienna, 7. July 1890	II., Praterstrasse 14/II
Perl, Julius	Skotschau, 8. September 1870	Samuel Perl, Doctor of Medical Sciences in Neutitschein	Prom. 25. July 1894 Me. Moravia Weisskirchen, 30. September 1888	IX., Lazarettgasse 6/II/14
Perls, Wilhelm		Doctor of Medical Sciences, Father of Else Perls	Scholar, 1926	–
Philipp, Johann Paul	Kattowitz, 16. May 1870	–	Prom.20. October 1894 (Haan, Bohemia) Me. Przemysl, 25. July 1888	–
Pollak, Hugo	Troppau, 12. July 1874	–	Prom. 21. July 1902 Me. Krems, 23. September 1895	VIII., Strozzigasse 26
Pollak, Salomon	Nezlachowitz, 27. Juny 1836	–	Prom. 29. January 1861 (med.), 31. May 1861 (chir.) Me. Troppau, 9. August 1853	–

(continued)

Name, first name	Birthplace, date	Father's name, Father's profession	Study Rigorosum A-Level	Student's address in Vienna
Popper, Adolf	–	Doctor of Medical Sciences, Father of Hans Popper	Scholar, 1918	–
Poppe(r), Josef	Niklasdorf, 24. February 1880	Josef Popper, Blacksmith in Niklasdorf	Prom. 28. November 1908, Me. Weidenau, 20. July 1901	IX., Georg-Sigl-Gasse 3/III/3
Porges, Dr., Max	Jägerndorf		Scholar, 1886	–
Presser, Daniel	Teschen, 1. July 1863	Moriz Presser, Merchant and landowner in Teschen	Prom. 20. July 1901 Me. Teschen, 1. February 1883	XIX., Oberdöbling, Hauptstrasse 39
Rauchamann, Leopold	Ustron, 13. February 1876	Samuel Rauchmann, Butcher in Ustron	Prom. 24. December 1903 Me. Bielitz, 19. September 1896	IX., Bleichergasse 13
Rhoden, Edgar	Oderberg, 21. September 1886	Maximilian Rhoden, Physician in Oderberg	Prom. 13. December 1909 Me. Troppau, 7. July 1904	IX., Fluchtgasse 6/5
Roth, Ignaz	Bielitz, 5. April 1834		Prom. 24. April 1863 Me. Teschen, 7. August 1857	–
Ruttin, Erich	Bielitz, 13. November 1880	Isaak Ruttin, Oficial in Bielitz	Prom. 23. December 1905 Me. Bielitz, 27. June 1900	IX., Mariannengasse 27/III/10
Ruttin, Moriz	Bielitz, 17. March 1872	Isaak Ruttin, Merchant in Bielitz	Prom.15. July 1907 Me. Bielitz, 11. July 1891	IX., Dietrichsteingasse
Sachs, Otto	Bielitz, 9. January 1870	Louis Sachs, Shoemaker in Bielitz	Prom. 31. March 1898 Me. Bielitz, 17. July 1890	IX., Lazarettgasse 8/III/21
Sachs, Richard	Bielitz, 4. January 1871	Louis Sachs, Shoemaker in Bielitz	Prom. 19. Juny 1897 Me. Bielitz, 18. July 1890	IX., Berggasse 20/III/18, Lazarettgasse 8/III/21
Sachs, Theodor	Troppau, 14. February 1855	Abraham Sachs, Senior Physician in Vienna III and Pressburg	Rigor. 1879 Me. Vienna, 19. July 1873	III., Ungargasse 42/I/9, VIII., Schlösselgasse 23

(continued)

Name, first name	Birthplace, date	Father's name, Father's profession	Study Rigorosum A-Level	Student's address in Vienna
Schmid, Julius	Teschen, 1880	Heinrich Schmid, Cutter in Teschen	Scholar, 1902	IX., Wasagasse 28, Fluchtgasse 9
Schmied, Julius	Teschen, 6. Januar 1881	–	Prom. 24. Juli 1905 Me. Teschen, 4. Juli 1899	–
Schorr, Ernst	Bielitz, 13. June 1874 19. November 1894 evangelically baptized Vienna-Währing	Moritz Schorr, Doctor of Medical Sciences	Prom. 23. March 1899 Me. Bielitz, 22. June 1893	VIII., Alser Strasse 27/3
Schwarz, Otto Josef	Olbersdorf, 24. September 1882	Moriz Schwarz, Builder in Schibitz/Teschen	Prom. 29. November 1907 Me. Teschen, 3. July 1902	VIII., Skodagasse 19/4
Siebenschein, Eugen	Oderberg, 30. November 1867	Abraham Siebenschein, Customs official in Troppau	Rigor. 1891 Me. Troppau, 7. July 1886	IX., Berggasse 43/II/II/40, Berggasse 37/II/IV/13
Silberstein, Friedrich	Teschen, 20. November 1888	Jakob Silberstein, Merchant in Teschen	Prom. 26. January 1912 Me. Teschen, 4. July 1906	I., Biberstrasse 15, XVIII., Anastasius-Grün-Gasse 100
Singer, Berthold	Alt Lublitz, 21. November 1872	Salomon Singer, Guest house owner in Jaktar/Troppau	Prom. 24. March 1896 Me. Troppau, 4. July 1890	I., Esslinggasse 1/III
Sofer, Leo	Friedek, 12. December 1872	Protector: Herrmann Löw, Economist in Colloredow/Moravia	Prom. 6. March 1897 Me. Troppau, 27. September 1890	VIII., Langegasse 19/3, VII., Neubaugasse, Neustiftgasse 28/32
Sonnenschein, Gustav	Königsberg, 17. December 1868	Markus Sonnenschein, Merchant in Troppau	Prom. 20. May 1893 Me. Troppau, 7. July 1886	IX., Mariannengasse 12/II
Sperber, Wolf/Wilhelm	Holeschau/Altstadt, 1. June 1841	Leopold Sperber	Prom. 6. März 1867 (med.), Prom. 17. März 1874 (chir.) Me. Teschen, 12. August 1858	–

(continued)

Name, first name	Birthplace, date	Father's name, Father's profession	Study Rigorosum A-Level	Student's address in Vienna
Spieler, Friedrich/Fritz	Teschen, 8. October 1875	Vorm. Jakob Spieler, Merchant in Koritschan/ Moravia	Prom. 7. July1899 Me. Vienna, 8. July 1893	IV., Prinz-Eugen-Strasse 19
Spitzer, Albert	Friedek, 25. May 1836	–	Prom. 8. November 1867 Me. Teschen, 9. August 1856	–
Spitzer, Jakob	Grudek, 20. July 1849	–	Prom. 31. May 1880 Me. Teschen, 27. July 1868	–
Spitzer, Moriz/ Mauritius	Polanka, 14. March 1835 (or 1838)	Ignatz Spitzer in Teschen	Prom. 1. March 1864 (med.), 31. December 1864 (chir.)	IX., Alser Strasse 212
Strauss, Josef	Bielitz, 16. December 1868	Philipp Strauss, Merchant in Bielitz	Prom. 14. May 1892 Me. Bielitz, 13. July 1886	II., Herminengasse 6/III, IX., Höfergasse 2, Lazarettgasse 17
Teschner, Max	Troppau, 6. September 1865	–	Prom. 27. July 1896 Me. Troppau, 7. July 1884	–
Thumim, Leopold	Breslau, 1. December 1870	David Thumin, Official in Vienna, IX. St.-Bez.	Prom. 16. March 1894 Me. Vienna, 12. October 1888	IX., Liechtensteinstrasse 8/III
Tobias, Emanuel	Deutschleuten, 24. February 1869	Moritz Tobias, Merchant in Deutschleuten	Prom. 23. February 1895 Me. Bielitz, 8. July 1887	IX., Spitalgasse 9/II/ III/26
Trebitsch, Erwin	Troppau, 25. July 1890	Jakob Trebitsch, Accountant in Vienna	Rigor. 14. December 1911 Me. Vienna, 13. July 1909	II., Wolfgang-Schmälzl-Gasse 3
Treu, Emil	Jauernig	Moritz Treu, Merchant in Jauernig	Scholar, 1890 in the fourth semester	IX., Garnisongasse 1/I
Treu, Viktor	Jauernig, 31. August 1861	Moritz Treu, Merchant in Jauernig	Prom. 27. February 1885 Me. Troppau, 15. July 1879	IX., Allgemeines Krankenhaus [Alser Strasse]

(continued)

Name, first name	Birthplace, date	Father's name, Father's profession	Study Rigorosum A-Level	Student's address in Vienna
Tugendhat, Julius	Teschen, 7. May 1844	Daniel Tugendhat, Manufacturer in Teschen	Prom. 31. December 1869 (med.), 31. December 1870 (chir.) Me. Teschen, 9. August 1863	IX., Lammgasse 1
Tugendhat, Max	Bielitz, 21. December 1870	Moritz Tugendhat, Merchant in Bielitz	Prom. 10. February 1897 Me. Bielitz, 18. July 1890	II., Lilienbrunngasse 9, IX., Schwarzspanierstrasse 6/11
Waldmann, Robert	Landsberg, 1. February 1846	Abraham Waldmann, Gastroenter in Landsberg	Rigor. 1873 Prom. 23. March 1875 Me. Güstrow, 21. September 1869	IV., Schleifmühlgasse 8 Margaretenstrasse 6 IX., Garnisongasse 4
Wechsberg, Friedrich	Orlau, 22. November 1873	Ferdinand Wechsberg, Merchant in Orlau; Protector: Josef Wechsberg, Doctor in Polish Ostrava	Prom. 6. March 1897 Me. Troppau, 26. June 1896	VIII., Neudeggergasse 4
Wechsberg, Joseph	Orlau, 10. March 1837		Prom. 1. February 1859 Me. Leutschau	–
Wechsberg, Julius	Polich Ostrava, 21. December 1879	Emanuel Wechsberg, Merchant in Polish Ostrava	Prom. 13. March 1903 Me. 24. June 1897	IX., Türkenstrasse 9
Wechsberg, Leo	Orlau, 25. May 1875	Ferdinand Wechsberg, Merchant in Orlau; Protector: Josef Wechsberg, Doctor in Polish Ostrava	Prom. 16. Juny 1899 Me. Vienna, 14. July 1893	VIII., Neudeggergasse 4
Wechsberg, Max	Orlau, 10. March 1870	Albert Wechsberg in Moravian Ostrava	Prom. 13. July 1895 Me. Bielitz, 26. Juny 1889	IV., Schleifmühlgasse 20/I/14

(continued)

Name, first name	Birthplace, date	Father's name, Father's profession	Study Rigorosum A-Level	Student's address in Vienna
Weinstein, Emil	Brosdorf, [Bronsdorf?] 26. April 1870	Jacob Weinstein, Merchant and Gastroenter in Brosdorf	Prom. 15. June 1895 Me. Teschen, 4. July 1889	VIII., Schlösselgasse 16/ II/14, IX., Berggasse 17/I/9, Alser Strasse 16/ II/1/17
Weissberger, Bernhard	Jablunkau, 11. April 1869	Marcus Weissberger, Gastroenter in Jablunkau	Prom. 15. July 1893 Me. Lieblitz, 8. July 1887	IX., Roßauer Lände 11/2/5
Werner, Richard	Freiwaldau, 22. July 1875	Sigmund Werner, Merchant in Freiwaldau	Prom. 7. July 1899 Me. Weidenau, 22. July 1893	VIII., Wickenburggasse 3/III/1
Wiener, Leopold	Troppau, 13. March 1876	Gustav Wiener, Merchant in Troppau	Prom. 20. July 1901 Me. Troppau, 16. July 1895	I., Rauhensteingasse 7/II
Wolfert, Adolf	Wigstadtl, 12. May 1861	–	Rigor. 24. July 1891 Me. Troppau, 5. July 1881	–
Seisler, Josef	Bielitz, 7. October 1858	Isak Zeisler, Town council in Bielitz	Prom. 3. July 1882 Me. Troppau, 26. March 1877	I., Salzgries 41
Ziffer, Adolf David	Roppitz, 15. October 1864	Emanuel Ziffer, Tradesman in Kotschau	Prom. 28. April 1888 Me. Prerau, 20. July 1882	II., Novaragasse 16/II/15
Ziffer, Ferdinand	Marklowitz, 1. December 1849	Moritz Ziffer, Merchant in Polish Ostrava	Prom. 23. March 1878 Me. Troppau, 12. July 1869	II., Theresiengasse 3/ III/18
Ziffer, Josef	Friedek, 29. July 1860	Protector: Moritz Spitzer, Doctor of Medical Sciences in Friedek	Prom. 19. November 1897 Me. Teschen, 21. September 1877	IX., Schwarzspanierstrasse 15, Waisenhausgasse 20
Zwillinger, Jakob	Troppau, 19. July 1876	Bernhard Zwillinger, Proprietor in Wal. Meseritsch	Prom. 2. April 1900 Me. Moravian Weisskirchen, 24. July 1894	III., Geusaugasse 21, VIII., Bennogasse 18/8

[a]Vienna city district

3. Jewish students from Silesia at the medical faculty of the University of Vienna between 1914 and 1938

Name, first name	Birthplace, date	Father's name, Father's profession	Study Rigorosum A-Level	Student's address in Vienna
Auerbach Berta, (1925: Münz, Berta)	Teschen,1. July 1898	Berthold Auerbach, Court official in Teschen	Prom. 6. Junie1924	VIII.,Josefstädterstr. 87/22
			Me. Teschen, 4. July 1918	VI., Webgasse 4/23
Bachrach, Artur	Troppau, 11. February 1892	Ignaz Bchrach, Upholsterer in Oderfurt and Moravian Ostrava	Prom. 31. January 1921	IX., Türkenstrasse 21/II/23
			Me. Moravian Ostrava, 7. July 1910	
Barber, Ferdinand	Deutschleuthen, 1892	Hermann Barber, Industrialist in Bielitz	1918 in the third semester	IX., Müllnergasse 6/14
Bauer, Alfred	Friedek, 29. August 1892	–	Prom. 11. March 1924	–
			Me. Vienna, 10. July 1912	
Beck, Arthur	Benisch, 1894	Hermann Beck, Gastroenter in Benisch	Prom. 20. February 1922	–
Beck, Heinrich	Benisch, 20. December 19	Vorm. Hermann Beck, Gastroenter in Benisch	Rigor. 3. June 1922	V., Kompertgasse 6/4
			Me. Freudenthal, 17. June 1919	
Bernfeld, Anna, geb. Salomon	Hirschberg, 1. October 1892	–	Nostrification doctoral degree (Berlin) – 21. December 1923	–
			Me. Breslau, 13. March 1913	
Bettelheim, Else Wilhelmine	Teschen, 1900	Eduard Bettelheim, Director of the Oldis Commission inVienna	Scholar, 1918	II., Taborstrasse 48
			Me. Vienna, II.	
Better, Fritz/ Friedrich	Bielitz, 1. September 1896	Karl Better, Merchant in Bielitz	Prom. 27. January 1922	IX., Müllnergasse 15/13
			Me. Bielitz, 8. July 1914	VIII., Breitenfeldergasse 17/17

(continued)

Name, first name	Birthplace, date	Father's name, Father's profession	Study Rigorosum A-Level	Student's address in Vienna
Bock, Gustav	Troppau, 1897	Karl Bock, Kaufmann in Moravian Ostrava	Scholar, 1918 Me. hung. Brod	IX., Bleichergasse 8
Borger, Richard	Bielitz, 5. February 1894	–	Prom. 27. January 1922 Me. Bielitz, 9. July 1913	–
Bornstein, Alfred	Bielitz, 14. March 1902	Josef Bornstein, Merchant	Prom. 5. April 1927 Me. Bielitz, 24. June 1921	VI., Ägidigasse 12/7 XIX., Grinzing, Baracke 33/7
Bornstein, Otto	Bielitz, 14. April 1903	Josef Bornstein, Merchant	Prom. 5. April 1927 Me. Bielitz, 24. June 1921	VI., Ägidigasse 12/7
Brand, Dorothea	Bielitz, 14. October 1889	–	Prom. 29. January 1916 Me. Vienna 8. July 1910	–
Brenner, Eugen	Reichwaldau, 20. May 1899	Josef Brenner, Merchant in Reichwaldau	Prom.7. April 1924 Me. Teschen, 3. March 1917	IX., Schlagergasse 11/15
Brenner, Max	Dombrau, 29. July 1900	Samuel Brenner, Merchant in Dombrau	Rigor. 15. February 1922 Me. Teschen 4. July 1918	VIII., Breitenfeldergasse 17/44, IX., Hebragasse 1/7
Dattner, Bernhard	Ustron, 7. July 1887	Adolf Dattner, Merchant in Ustron	Prom.10. November 1911 (Doctor of law) Prom. 16. June 1919 (med.) Me. 2. July 1906	IX., Lackierergasse 7/6
David, Israel	Deutsch Wartenberg?, 10. July 1902	Protector: Mother Chaje David	1930 in the ninth semester	II., Ybbsstrasse 22/22
Dörfler, Hans	Bielitz, 6. September 1881	Dr. Stefan Dörfler, Gymnasium professor in Brünn	Prom. 30. July 1918 Me. Vienna, 4. July 1910	IX., Eisengasse 28/19
Eisner, Emanuel	Niedek, 7. January 1896	Josef Eisner, Guest house owner in Niedek	Prom. 11. May 1922 Me. Teschen, 28. September 1914	IX., Porzellangasse 54/II/25

(continued)

Name, first name	Birthplace, date	Father's name, Father's profession	Study Rigorosum A-Level	Student's address in Vienna
Epstein, Kurt	Breslau, 1903	Max Epstein, Magistrate's employee in Breslau	1926 in the eighth semester, before: University of Breslau	XVII., Jörgerstrasse 41/I/6
Feuerstein, Ludwig	Bielitz, 11. August 1903	–	Abs. 9. December 1938	IX., Türkenschanzstrasse 23, VIII., Wickenburggasse 3, IX., Währinger Strasse 46
Fischbein, Moritz	Kopitau/ Oderberg, 12. October 1891	Markus Fischbein, Synagoga servant in Oderberg	Prom. 26. July 1921 Me. Teschen, 13. July 1914	IX., Hahngasse 17/5
Fränkel, Elly	Alexanderfeld/ Bielitz	Bernhard Fränkel, Director of the company in Saybusch	Scholar, 1918	III., Erdberger Lände 20/20
Gessler, Ernst	Freudenthal, 22. July 1898	Heinrich Gessler, Private owner in Freudenthal	Prom. 27. June 1924 Me. Freudenthal, 1. December 1916	XIX., Chimanistrasse 22
Glesinger, Bernhard	Woikowitz b. Friedek, 9. August 1888	Anton Glesinger, Economist in Woikowitz	Prom. April 1919 Me. Friedek, 13. September 1905	IX., Liechtensteinstrasse 41/II/12
Gold, Emmerich	Teschen, 20. January 1888	Leopold Gold, Merchant in Poznan	Prom. 23. January 1914 Me. Poznan, 18. February 1908	IX., Lustkandlgasse 29/13
Goldberg, Ernst	Bielitz, 16. September 1894	Moritz Goldberg, Bank officer in Bielitz	Prom. 20. February 1923 Me. Bielitz, 8. July 1914	VIII., Josefstädterstrasse 23, Langegasse 50
Gross, Erich	Dzieditz, 1903	Heinrich Gross, Doctor in Dzieditz	Scholar, 1922	IV., Weyringergasse 29/16
Grossmann, Arpad	Polich Ostrava, 1. October 1898	Alexander Grossmann, Official in Moravian Ostrava	Rigor. 15. June 1920 Me. Moravian Ostrava, 4. July 1917	IX., Bleichergasse 8

(continued)

Name, first name	Birthplace, date	Father's name, Father's profession	Study Rigorosum A-Level	Student's address in Vienna
Grünberger, Ernst	Dzieditz, 1895	Jeremias Grünberger, State Railway Council in Dzieditz	Scholar, 1914	II., Pazmanitengasse 28
Grünspan, Samuel	Leipnik/Bielitz, 11. December 1901	Hermann Grünspan,Tax collector in Biala/Bistritz	Prom. 24. July1925 Me. Bielitz, 2. July 1919	IX., Wasagasse 23/28
Haas, Rudolf	Königsberg/ Troppau, 12. October 1898	Adolf Haas, Baker in Königsberg	Prom. 23. July 1923 Me. Troppau, 11. November 1916	–
Hauser, Ernst	Breslau, 26. December 1918	Fritz Hauser, Industrialist in Meran	1937 in the first semester,before: Theresian Academy, Abs. 26. August 1938	I., Operngasse 4/16
Hirsch, Paul	Jägerndorf, 2. Januar y 1906	–	Prom. 10. July 1931 Me. Vienna, 25. June 1924	–
Hornung, Leo	Ostrava, 2. August 1902	Isidor Hornung, Merchant in Moravian Ostrava	Scholar, 1922	XX., Staudingerstrasse 10/19
Jungmann, Martin	Teschen, 24. September 1895	Samuel Jungmann, Merchant in Bielitz	Prom. 22. March 1921 Me. Bielitz, 11. July 1913	IX., Glasergasse 10
Kaluki, Julius	Liegnitz, 1906	Mose Kuluki, Merchant	1926 in the fifth semester, before: University of Breslau	IX., Pichlergasse 1/6
Kempler, Paul	Martinau, 1898	Julius Kempler, Merchant in Vienna II.	1918 in the first semester, before: Sophien Gymnasium	II., Ausstellungsstrasse 45/9
Knöppfelmacher, Anna	Troppau, 21. Januar 1900	Dr. Julius Knöpfelmacher, Chief Financial Officer in Moravian Ostrava	Rigor. 22. March 1923 Me. Bielitz, 3. July 1919	IX., Garnisongasse 7

(continued)

Name, first name	Birthplace, date	Father's name, Father's profession	Study Rigorosum A-Level	Student's address in Vienna
Kurzmann, Rudolf	Teschen, 24. November 1894	Isidor Kurzmann, Merchant in Vienna	Prom. 24. June 1921, Vienna Me. Vienna, 8. July 1913	VII., Lindengasse 37
Landau, Arthur	Bielitz	Israel Landau, Gastroenter in Bielitz	Scholar, 1922	VII., Mariahilfer Strasse, VIII., Laudongasse 26/15
Laufer, Leopold	Polish Ostrava, 19. February 1897	Jakob Laufer, Merchant in Polish Ostrava	Rigor. 18. May 1920 Me. Moravian Ostrava, 11. June 1915	Ayrenhoffgasse 1
Lenzner, Heinrich	Groß Knezendorf, 1895	Karl Lenzner	1918 in the second semester	II., Kaiser-Josef-Strasse 36/17
Lichtwitz, Otto	Troppau, 1. March 1893	–	Prom. 3. November 1922 Me. Vienna, 12. July 1911	–
Loebel, Robert	Bielitz, 12. October 1893	Fabrikant Protector: Dr. Alfred Reisz in Vienna	Prom. 30. July 1919 Me. Bielitz, 12. May 1912	IV., Technikergasse 5
Löwy, Alfred	Hruschau, 10. April 1898	Simon Löwy, Merchant in Hruschau	Prom. 11. March 1924 Me. Moravian Ostrava, 16. November 1926	VIII., Wickenburggasse 2/5
Moller, Alberta	Troppau, 31. January 1903	Adolph Moller, Merchant in Vienna	1925 in the seventh semester, before: University of Prague Me. Troppau	I., Gonzagagasse 1/4, IX., Alserstrasse 10/14
Morgenstern, Wilhelm	Polish Ostrava, 1900	Emil Morgenstern, Merchant in Vienna II.	1918 in the first semester Me. Moravian Ostrava	II., Ybbsstrasse 28/25
Müller, Isidor	Mosty/Teschen, 15. January 1884	Heinrich Müller, MErchant in Teschen	Prom. 20. December 1917 Me. Friedek, 11. July 1906	I., Wollzcilc 21, III., Bechardgasse 13
Münz, Stella	Bielitz, 28. February 1895	Ludwig Münz, Military doctor	Prom. 28. February 1921 Me. Lviv, 25. May 1914	VIII., Pfeilgasse 30/17/III

(continued)

Name, first name	Birthplace, date	Father's name, Father's profession	Study Rigorosum A-Level	Student's address in Vienna
Oständer, Erich	Bielitz, 22. October 1903	Isidor Oständer, Official in Bielitz	Prom. 5. April 1927 Me. Bielitz, 25. April 1921	II., Ferdinandstrasse 27/4/III
Perls, Else	Waldenburg, 1904	Wilhelm Perls, Doctor of Medical Sciences in Breslau	1926 in the eighth semester before: University of Freiburg/ Breisgau	VIII., Breitenfeldergasse 25/I/15
Pollak, Eugen	Bielitz, 12. June 1890	Bernhard Pollak, Professor in Vienna	Prom. 9. December 1914 Me. Bielitz, 14. July 1909	I., Grillparzerstrasse 7
Popper, Hans	Orlau, 1897	Adolf Popper, Doctor of Medical Sciences in Bautsch in Moravien	1918 in the first semester, Me. Moravian Ostrava	–
Quittner, Helene	Troppau, 11 or 22. April 1891	Theodor Quittner, Manufacurer in Vienna	Prom. 3. April 1925 Me. Vienna, 3. July 1918	I., Hohenstaufengasse 10
Reichert, Gerhard	Teschen, 22. November 1898	Dr. Samuel Reichert, Doctor in Teschen	Prom. 21. December 1923 Me. Teschen, 23. October 1916	IX., Porzellangasse 20/2
Reichmann, Moritz	Polish Ostrava, 3. December 1897	Samuel Reichmann, Merchant in Moravian Ostrava	Prom. 23. October 1924 Me. Moravian Ostrava, 15. October 1915	II., Ausstellungsstrasse 53, XVI., Geroldgasse 7, IX., Hahngasse 5
Rhoden, Margaret(h)e	Oderberg, 5. September 1892	–	Prom. 31. January 1919 Me. Teschen, 11. July 1913	–
Rittermann, Sami Josef	Bielitz, 12. May 1898	Protector: Erna Rittermann, Professor in Lviv	Prom. 23. December 1922 Me. Prague, 8. May 1916	IX., Rotenlöwengasse 7/23
Roger, Herbert	Stauding, 1900	Benno Roger, Broker in Troppau	1922 in the seventh semester, before: University of Prague	Weidlingau-Hadersdorf, Mauerbachstrasse 30
Rosenbaum, Erich	Bielitz, 1897	Adolf Rosenbaum, Cutter in Bielitz	Scholar, 1918 Me. Bielitz	VII., Burggasse 100

(continued)

Name, first name	Birthplace, date	Father's name, Father's profession	Study Rigorosum A-Level	Student's address in Vienna
Rosenberger, Günther	Emanuelsegen b. Pless, 25. November 1912	Adolf Rosenberger, Official in Cottbus	Scholar, 1938 Abs. 27. June 1938	IX., Alser Strasse 10/14
Rosenthal, Dr. Alfred	Lautsch, 15. October 1893	Josef Rosenthal, Merchant in Vienna	1934 in the second semester	VI., Mariahilfer Strasse 109
Rosenthal, Robert	Odrau, 28. May 1896	Josef Rosenthal, Merchant in Vienna	Prom. 25. July 1922 Me. Vienna, 8. April 1915	VI., Mariahilfer Strasse 109
Rosner, Erwin	Schumbarg, 25. November 1893	Sigmund Rosner, Merchant in Schumbarg	Prom. 6. June 1924 Me. Moravian Ostrava, 12. Juli 1912	Hotel zum Hirschen
Scheftel, Valerie	Troppau, 4. November 1911	Toni Scheftel, Store directress in Troppau	Prom. 4. March 1938	II., Franz Josefs Kai 29
Schleuderer, Isidor	Teschen, 2. August 1899	F. Schleuderer, Merchant's widow in Teschen	Prom. 21. November 1924 Me. Teschen, 28. December 1918	VIII., Blindengasse 36
Schmidt, Friedrich	Groß Hernsdorf, 26. September 1891	Julius Schmidt, Chief Forester in Odrau	Prom. 8. November 1917 Me. Friedek, 2. July 1910	VIII., Auerspergstrasse 17
Schneid, Otto	Jablunkau, 1900	Jakob Schneid, Official in Bielitz	Scholar, 1918 Me. Bielitz	–
Schneider, Hugo	Freistadt, 19. June 1897	Israel Schneider, Professor in Teschen	Prom. 17. March 1922 Me. Vienna, 30. June 1916	IX., Lackierergasse 6
Scholzel, Hans	Breslau, 1905	Otto Scholzel, Middle school teachers in Breslau	1926 in the sixth semester, before: University of Breslau	VIII., Piaristengasse 11/2
Schramek, Walter	Teschen, 21. June 1898	Samuel Schramek, Merchant in Teschen	Prom. 20. May 1924 Me. Teschen, 3. May 1916	II., Ob. Donaustrasse 33, XVII., Hernalser Hauptstrasse 37
Schrötter, Bruno	Ellgoth/Bielitz, 31. May 1899	Jakob Schrötter, Grain exporter and merchant in Bielitz	Prom. 10. July 1924 Me. Bielitz, 3. November 1918	VII., Kirchengasse 43/II/17, VIII., Feldgasse 12/II/22

(continued)

Name, first name	Birthplace, date	Father's name, Father's profession	Study Rigorosum A-Level	Student's address in Vienna
Schrötter, Siegfried	Ellgoth b. Bielitz, 18. June 1894	Jakob Schrötter, Merchant in Bielitz	Prom. 6. April 1922 Me. Bielitz, 12. July 1913	IX., Nußdorfer Strasse 61/II
Sigmund, Robert	Dziedzitz, 5. April 1897	Jakob Sigmund, Doctor in Dziedzitz	Prom. 6. April 1922 Me. Bielitz, 7. July 1915	IX., Pichlergasse 6/5
Silbermann, Isidor	Bielitz, 1. March 1899	Adolf Silbermann, Metal founder in Bielitz	Prom.18. December 1924 (and Nostrification) Me. Bielitz, 21. December 1917	X., Hebbelgasse 2 (k.u.k. Reservespital)
Silberstein, Maria Margarete	Vienna, 22. May 1918	Friedrich Silberstein, Professor in Vienna	Abs. 24. June 1938	XVIII., Anastasius-Grün-Gasse 100
Silberstein, Moritz	Bielitz, 30. May 1899	Dora Mährer, Seamstress	Prom. 24. June1927 Me. Bielitz, 8. November 1917	IX., Bleichergasse 9/III, II., Ob. Donaustrasse 43/17, XIX., Pokornygasse 15
Silbiger, Arthur	Bielitz, 2. September 1899	Silbiger in Bielitz	Rigor. 4. May 1923 Me. Bielitz, 25. February 1918	I., Franz-Josefs-Kai 45
Sobel, Siegmund	Bielitz, 24. May 1900	Wilhelm Sobel, Merchant in Bielitz	Prom. 23. June 1925 Me. Bielitz, 4. July 1918	XX., Mathildenplatz 5/14, IX., Porzellangasse 45/17
Socha, Otto	Ustron, 11. August 1894	–	Prom. 6. April 1922 Me. Teschen, 14. July 1914	–
Soltau, Hans Jehuda	Hruschau, 17. May 1899	Jean Gerson Soltau, Official in Vienna	Prom. 11. March 1924 Me. Vienna, 7. March 1917	IX., Hahngasse 30/14
Spitzer, Ida geb. Tramer	Bludowitz, 17. April 1881	Geza Spitzer, since the beginning of the war in the field	Prom. 7. April 1924 Me. Oberhollabrunn, 9. July 1917	XX., Pappenheimgasse 6/I/21
Steiner, Frieda	Bielitz, 1. January 1899	Prof. Dr. Max Steiner, Rabbi in Bielitz	Prom. 7. April 1924 Me. Bielitz, 2. July 1917	VIII., Josefstädter Strasse 34, II., Körnergasse 7

(continued)

Name, first name	Birthplace, date	Father's name, Father's profession	Study Rigorosum A-Level	Student's address in Vienna
Stern, Robert	Skotschau, 3. August 1895	Berthold Stern, Veterinarian in Skotschau	Prom. 21. December 1923 Me. Bielitz, 10. July 1914	IX., Schulz-Straßnitzki-Gasse 6, Porzellang. 20/27, I., Elisabethstrasse 24/25
Strauss, Edwin	Teschen, 12. June 1885	Philipp Strauss, Merchant in Teschen	Prom. 24. July 1914 Me. Teschen, 28. June 1904	VIII., Wickenburggasse 3/32
Thieberger, Richard	Bielitz, 8. August 1896	Bernhard Thieberger, Cutter in Moravian Ostrava	Prom. 23. December 1921 Me. Moravian Ostrava, 15. Juli 1914	IX., Porzellangasse 9/II, XVIII., Karl-Ludwig -Strasse 1, II., Taborstrasse 22/II/36
Tramer, Erwin	Bielitz, 1900	Heinrich Tramer, Baker in Bielitz	Scholar, 1918 Me. Bielitz	II., Ob. Donaustrasse 43
Traubner, Jakob	Wirbitz/ Freistadt, 11. May 1898	Samuel Traubner, Merchant in Wirbitz/Freistadt	Prom. 20. July 1927, before: University of Bern Me. Moravian Ostrava, 11. June 1918	I., Sonnenfelsgasse 15/I
Vogl, Alfred	Bielitz, 23. May 1895	Th. Vogel, Music teacher in Vienna	Prom. 3. June 1921 Me. Vienna, 7. July 1914	VIII., Albertgasse 49
Weiss, David Adolf	Breslau, 18. September 1889	–	Prom. 30. October 1915 Me. Breslau, 5. March 1909	–
Werber- Müller, Olga	Czechowitz, 14. June 1897	Josef Werber, Miller in Czechowitz	Prom.. 23. July 1923 Me. Bielitz, 2. July 1917	VIII., Josefstädterstrasse 34
Zanker, Arthur	Oderberg, 22. July 1890	Josef Zanker, Merchant in Oderberg	Prom. 19. December 1914 Me. Olmouc, 14. July 1909	II., Valeriestrasse 4b

Abs. Absolutorium, *chir.* surgeon, *Me* matriculation examination, *med.* medical doctor, *Nat.* Nationale, *Prom.* doctorate, *Rigor.* Rigorosum, *Spon.* bürgschaft

Bibliography

Alexander, M. (2008). *Kleine Geschichte der böhmischen Länder*. Ditzingen: Reclam.

Bauer-Merinsky J. (1980). Die Auswirkungen der Annexion Österreichs durch das Deutsche Reich auf die medizinische Fakultät der Universität Wien im Jahre 1938: Biographien entlassener Professoren und Dozenten, University dissertation, Wien.

Beck von Mannagetta L., & von Kelle C. (1906). Die österreichischen Universitätsgesetze. Sammlung die für die österreichischen Universitäten gültigen Gesetze, Verordnungen, Erlässe, Studien- und Prüfungsordnungen usv. im Auftrage des k. k. Ministeriums für Kultus und Unterricht mit Benützung der amtlichen Akten. Wien: Manzsche k.u.k. Hof-Verlags und Universitäts-Buchhandlung.

Conrads, N. (1994). *Schlesien. Deutsche Geschichte im Osten Europas*. Berlin: Wolf Jobst Siedler Verlag.

Doležal, H. (2007). Oberschlesische Studenten an der medizinischen Fakultät der Universität Wien in den Jahren 1818 bis 1938. In J. M. Dyrda & J. Miozga (Eds.), *Górny Śląsk – dokumentowanie dziejów lecznictwa ze szczególnym uwzględnieniem ziem dawnych księstw cieszyńskiego i opawskiego* (pp. 154–195). Katowice: Śląski Uniwersytet Medyczny.

Doležal, H. (2010). Oberschlesische Pharmazeuten und Doktoren der *Chemie* in den Dekanats-Rigorosen-Protokollen der Medizinischen Fakultät der Universität Wien in den Jahren 1751 bis 1872. In W. Kaczorowski & J. M. Dyrda (Eds.), *Wkład Górnego Śląska w dzieje europejskiego lecznictwa* (pp. 127–144). Opole: Wydawnictwo Uniwersytetu Opolskiego.

Doležal, H. (2012). Schlesier an der Medizinischen Fakultät der Universität Wien 1752–1938. *Archiv ostdeutscher Familienforscher, 20*, 121–150.

Drimmel, H. (1959). Die Hochschulreform von Thun-Hohenstein. *Österreich in Geschichte und Literatur, 3*, 1–7.

Gall, F. (1965). *Alma Mater Rudolphina 1365–1965. Die Wiener Universität und ihre Studenten*. Wien: Austria Press.

Gedenkbuch für die Opfer des Nationalsozialismus an der Universität Wien 1938. http://gedenkbuch.univie.ac.at. Accessed 24 May 2015

Haslinger, P., & Kreft, W. (Eds.). (2010). *Historisch-topographischer Atlas schlesischer Städte/Historyczno-topograficzny atlas miast śląskich/Historicko-topografický atlas slezských měst*. Marburg/Wrocław: Wissenschaftsforum/Verlag Herder-Institut.

Höflechner, W. (1988). *Die Baumeister des künftigen Glücks. Fragment einer Geschichte des Hochschulwesens in Österreich vom Ausgang des 19. Jahrhunderts bis in das Jahr* (p. 1938). Graz: Akademische Druck- u. Verlagsanstalt.

Jordak, K. (1965). *Die Universität Wien 1365–1965*. Wien: Bergland.

Mühlberger, K. (1993). *Dokumentation "Vertriebene Intelligenz 1938". Der Verlust geistiger und menschlicher Intelligenz an der Universität Wien von 1938 bis 1945*. Wien: Archiv der Universität Wien.

Rexroth, F. (2015). Planskizze für ein Luftschloss. Die rudolfinische Stiftungsurkunde von 1365 und die Entstehung der Wiener Universität. In H. Rosenberg & M. V. Schwarz (Eds.), *Wien 1365 eine Universität entsteht* (pp. 14–27). Wien: Christian Brandstätter Verlag.

Tuisl, E. (2014). *Die Medizinische Fakultät der Universität Wien im Mittelalter. Von der Gründung der Universität 1365 bis zum Tod Kaiser Maximilians I. 1519*. Göttingen: V&R unipress.

Uiblein, P. (1965). *Überblick über die Geschichte der Universität Wien: 600 Jahre Universität Wien*. München/Wien: Universität Wien.

von Bruch, R. (1998). Die Universitäten in der Revolution 1848/1849. Revolution ohne Universität – Universität ohne Revolution. In W. Hardtwig (Ed.), *Revolution in Deutschland und Europa 1848* (pp. 133–160). Göttingen: Vandenhoeck und Ruprecht.

Weczerka, H. (Ed.). (1977). *Schlesien. Handbuch der historischen Stätten*. Stuttgart: Kröner.

Weinrich, B. (1990). *Niederösterreichische Ärztechronik*. Wien: O. Möbius.

Žáček, R. (2004). *Dějiny Slezska v datech*. Praha: Libri.

Chapter 11
Between 'Here' and 'There': The Dual Identity of Dr. Izrael Milejkowski

Naomi Menuhin

Abstract Dr. Izrael Milejkowski (1887–1943), a member of the Jewish-Polish *inteligencja,* visited the Land of Israel in April 1936 for the First World Congress of Jewish Physicians. At that point, he closely examined his position on Zionism and, at the same time, being a Polish patriot. He struggled to be a citizen of equal rights in Poland while maintaining his Jewish-national pride and working for his own people in several areas. The article discusses mainly Milejkowski's worldview and ideas. He maintained a split identity—Polish and Jewish—a duality that represented a new direction for his people. The article focuses on the relations between the Diaspora and the Land of Israel in Dr. Milejkowski's worldview.

Keywords Milejkowski · Physician · Intelligencia · Identity · Warsaw · Z.L.R.P

11.1 Introduction

According to the *Centralne Archiwum Wojskowe*, (Central Military Archives) Milejkowski was born on July 17th 1887, in the town of Krewo in Oszmiana County—then a part of Vilnius County in Lithuania and currently part of Belarus. At some point, his family moved to Warsaw, where Izrael, his brother, and sister acquired higher education. He married Bella Sovalsky and had one child, Yanina, born in 1913 (*Główna Biblioteka*).

After graduating from the University of Warsaw's School of Medicine in 1914, Milejkowski joined the Russian Army to serve as a military physician. He joined the Polish Army at the end of World War I for a service he completed in 1923. He then started working as a physician in the Dermatology and Venereology Department of Warsaw Jewish Hospital and started a private practice at home. Z.L.R.P, the Jewish physicians' organization in Poland, was formed in the 1920s (Ohry–Kossoy and Ohry 2012, 455–462), and Milejkowski soon became an active member of the orga-

N. Menuhin (✉)
Jewish History Department, Ben-Gurion University of the Negev, Beersheba, Israel
e-mail: menuhinn@post.bgu.ac.il

M. Moskalewicz et al. (eds.), *Jewish Medicine and Healthcare in Central Eastern Europe*, Religion, Spirituality and Health: A Social Scientific Approach 3,
https://doi.org/10.1007/978-3-319-92480-9_11

nization, and was even one of its leaders in the 1930s. Through the years, Milejkowski published articles in both the Zionist and professional press, all the while serving as a physician and a public activist. In his Z.L.R.P capacity, Dr. Milejkowski visited the Land of Israel in April 1936 for the First World Congress of Jewish Physicians. At that point, he closely examined his position on Zionism and, at the same time, being a Polish patriot. This stage in his life raises numerous questions, the answers to which may shed a light on the world of Milejkowski's character. Milejkowski returned to Poland and continued his public work. When World War II started, all of Warsaw's Jewish residents were forced into a ghetto where Milejkowski was a prominent member of the *Judenrat*: head of the ghetto's Medicine and Sanitation Department. Later, he initiated the Hunger Study (Apfelbaum 1946) and took part in the activity of the Faculty of Medicine in the ghetto (Ofer 2015; Balinska 2012). Like many other Warsaw Jews, Milejkowski died in the January 1943 Action. The little research existing on Dr. Izrael Milejkowski focuses on the work he performed at his last place of residence: the Warsaw Ghetto, (Ofer 2015; Roland 1992). These works discuss mainly his Hunger Study on the pathology of hunger and the human body's starvation mechanism, conducted by a group of physicians led by Dr. Milejkowski.

11.2 The Visit to the Land of Israel and Subsequent Writings

The First World Congress of Jewish Physicians (Friedman 1936) opened on April 23rd, 1936 in Tel Aviv. Held in order to discuss the status of Jewish physicians in the land of Israel and worldwide, mainly in Europe, the five-day congress was attended by 1126 people, including the Polish delegation of 68 members. The congress was a key point in the ongoing dialogue between European Jewish physicians and the Jewish community in the Land of Israel in the interwar period, a dialogue that significantly expanded in the 1920s and the 1930s. Individual Jews' viewpoints on Zionism were reflected in their social circles, the newspapers they read or published in, and the schools their children attended.

The visit to the pre-state Israel was momentous in Milejkowski's life. It allowed him to examine and reflect on both his Polish and his Zionist-national identity. It later became apparent that it was a turning point in his life, for if he chose to stay in Israel and not return to Poland, his fate would probably have been different. Milejkowski took part in the congress as a representative of the Organization of Jewish Physicians in Poland. He arrived in the Land of Israel in early April 1936, traveled the land, met its residents, took notes, attended meetings, and took an active part in the congress.

After the congress, he returned to Warsaw, bearing the notes he had taken during his visit, which included his thoughts and feelings from before the visit. The documents tell the story of a writer with a strong desire for personal expression and a highly developed historical consciousness. His notes were published under the title "With the World Congress of Jewish Physicians in the Land of Israel—Characteristics and Reactions." Most of them were published in *Der Mament,*[1] a Polish newspaper in Yiddish. The first article was published on May 15th, 1936, and the last one in September.

Milejkowski's articles depict his personal impressions and feelings on one or two specific subjects (List of Articles 1936). He wrote mainly about the Land of Israel and his feelings related to its vistas; the first encounter with the Land of Israel and its colors, the sunlight, the beach of Tel Aviv, the streets of the first Hebrew city and its people. He wrote about his encounter with Jerusalem, trips across the land, the revival of Jewish settlements, and Jewish and Arab residents. It should be noted that Milejkowski visited the Land of Israel during the 1936 Arab uprising against the British rule, and witnessed the decline in the Jewish-Arab relations. He also wrote about his impressions from the congress.

For Milejkowski, the Land of Israel was an unfamiliar world to learn about. For him, it comprised personal, national, historical, and philosophical layers. It was a place rich in details, colors, smells, and tastes, vivid and real, but also filled with glory and holiness. He found a place that had a message for the world. Milejkowski conducted a dialogue with the land and its sights, with his readers, and mainly with himself. At the end of each dialogue, he chose his personal stand, which remained outside the Zionist action.

When Milejkowski arrived in the land of Israel, the Polish ground was already burning under the Jews' feet. The situation of Jewish physicians was deteriorating. As a public figure, a publicist, and a physician, Milejkowski shared his personal experiences from the Land of Israel with Poland's Jews in general and with his peers in particular. His personal experiences were very emotional, packed with descriptions of the local nature and filled with images and rich language. His experiences were published in a series of articles that constituted a "continuous diary." It may be assumed that the notes were written in real time, though they were published in *Der Mament* only after he returned to Poland under the title "With the World Congress of Jewish Physicians in the Land of Israel—Notes." Still, only a few articles were dedicated to the congress itself. Most of them revolved, as mentioned above, around his personal experiences.

The first article published was "A Soul Without a Body." It began with the following explanation:

> I feel that it is just impossible to deal with the picture without its Israeli frame… The picture loses its deepest meaning and content without the Israeli frame. It would be like a soul without a body, it would be powerless, fictitious. The picture would lose all of its deepest meaning and content." (Milejkowski, *Der Mament* 15.5.1936)

---

[1]The daily newspaper *Der Mamant* (The Moment) was established in 1910 by Tzvi Prilutzky and his son Noach. Eliezer Zilberberg, and Chaim Prozansky worked for the paper as well. The privately owned paper was a-political and represented the part of the Jewish society that strove to acculturate Jews into the rest of the Polish society. With time, the paper became pro-Revisionist and carried articles by the Revisionist Movement's leading ideologists, Ze'ev Jabotinsky and Uri Tzvi Greenberg. The paper was proactive in maintaining Jews' national rights and interests in Poland. It newspaper responded strongly to all major political events and addressed issues from the point of view of the typical Polish Jew. (Levinson 1953; Cohen 2001).

In other words, though the congress was of utmost importance to him, it was still only a part of the larger whole he called Israel.

Milejkowski reached the Land of Israel by ship (Polonia). Those who travelled with him, took a train from Warsaw to the Constances Port in Romania, to the Black Sea, and from this port, most of the members took a ship for Jaffa. For Milejkowski, the trip was filled with excitements and never-ending wonder. He was preoccupied with many feelings and thoughts as the ship made its way to the Land of Israel. "As we came closer and closer to Israel, our feelings strengthened. What can I compare it to? I guess this is what a young groom feels as he travels for the first time to meet his future wife." (Milejkowski, *Der Mament* 22.5.36). This is excitement filled with wonder and anticipation: "I heard so much about her from dad and mom, from the matchmaker, and from all those surrounding me! They said that she's as pretty as gold, as fine as a diamond, and as quiet and gentle as a dove." Yet, there is fear as well: "And this feeling that the Land of Israel has awakened in all of us on the ship was curiosity mixed with fear, closeness and deep, tight-knit relationships, still unfamiliar…" (*ibid*).

The sight of the Land of Israel from the ship left him speechless: "When finally, in the distance, like in a fog, we could see the Land of Israel, we all stood on the deck, frozen, in silence, speechless, as our gazes turned to the vague outline of the distant view that was still unclear. Some of us even shed a tear or two, filling the sea with a few more drops" (*ibid*).

His writings are filled with family images, starting with the description of travelling by ship to the Land of Israel as a groom before his wedding, travelling to meet his soul mate whom he had never met before, but had heard so much about. The image of the Land of Israel as a bride is interesting as itself, for in those years Milejkowski was a widower. Another example is the description of the physicians' attitude towards the congress as parents looking at their newborn baby.

The use of family motifs by Milejkowski was so prominent insofar as he used them to draw conclusions and generalizations—from the individual in a family to the national "togetherness," which is a big, expanded family. Thus, for example, referring to the Congress, Milejkowski devoted a large amount of words to the "togetherness" motif, hoping for the creation of a worldwide alliance that would take care of the interests of Jewish physicians everywhere. His soul was sensitive to spiritual experiences, and he tended to find holiness in the sights, people, and nature of the Land of Israel. The Jewish residents are described in superlatives of bravery and self-sacrifice. He was astonished by the strength of their belief in and grip of the Land of Israel, their work and contributions to the development of the land.

The decisive demand that Jews speak the Hebrew language in the Land of Israel, in everyday life as well as in scientific gatherings, had Milejkowski in a disadvantage. In Poland, Milejkowski was a public figure, a leader, a publicist and a physician—positions in which language is a fundamental tool. In Israel, he lacked that tool. This might have been an obstacle for his stay here, though as someone who was proficient in many languages, he could make an effort and learn it, if he truly wanted to. He was not frustrated by the use of Hebrew, but by the "fanatics" who promoted it. He expected them to treat Yiddish speakers with more respect.

The encounter with the Land of Israel and its sights confronted Milejkowski with the reality of the Diaspora. He, who was a member of the Zionist Party and its representative on the Council of the Warsaw Jewish community, was about to test his Zionist believes. This examination was reflected in the constant comparisons he drew between "here" and "there," "east" and "west," "modernity" and"tradition," and "us" and "them." The encounter with the Land of Israel provided him with a sort of mystical experience, as if nature itself rose to receive him: the beaches, the mountains, the fields, the valleys—all welcomed him together with the streets of Tel Aviv, Jerusalem with its holiness, the Izrael Valley, and even the northern settlement of Yesod Hama'ala.

Before to the Congress started, early meetings were held at one of the fancy cafés along the beach (Milejkowski, *Der Mament* 28–29.5.1936). "We fell in love with them, with our beachside meetings… I was a regular guest in the magnificently built part of our Jewish city."[2] The use of the words "our city" shows a strong sense of belonging: this city belongs to us too. Milejkowski may have lived in Warsaw, but he owned Tel Aviv as well. And yet, along with a sense of belonging, Milejkowski started slowly feeling estranged and alienated, emphasizing the fact that, despite all, he did not truly feel he belonged in the Land of Israel: "As one who has just been cursed, I would walk among singing and joyful faces, and like the wind I would stand on the beach on the far end of the sea. Here is the 'Land of Israel' and there are 'the people of Israel' and, unfortunately, the bridge connecting the two is still too narrow, tough, and long… I would look at the teens around me and it appeared to me that, together with me, millions of eyes looked at them, eyes filled with tears and longing… as well as feeling of joy, combined with sadness that engulfed me. As I strolled the beaches of Tel Aviv, I would see and feel them so strongly…." (Milejkowski, *Der Mament* 28–29.5.1936).

Here, Milejkowski changes his pattern of thought: Tel Aviv becomes the center in his mind and Warsaw, the periphery. With these words, he expresses the double entendre that characterized his experiences while making his way to the Land of Israel and his encounter with it. He loved the land and yet was not truly a part of it; he was excited about all that was done in the land, but remained an outside observer. His body was on the beaches of Tel Aviv, but his eyes were the eyes of all the Jews of the Diaspora. The small moments he experienced on the streets of Tel Aviv appeared in his eyes as enchanting "dimples," the title he gave his first article, dated 5.6.1936 (Milejkowski, *Der Mament* 5.6.1936).

He was amazed with everything he saw, even with a young woman that sold bus tickets because in Tel Aviv, even riding the bus was a special experience: "In order not to lose time, a young, uniformed woman sold us bus tickets. She personally approached every passenger. The bus soon arrived at the station and the people waiting for it—men, women, and children—boarded it. But How? One by one, quietly

[2] Interestingly, many who visited or immigrated to the Land of Israel in those days describe it in a fashion that closely resembles Milejkowski's writings. They speak of encountering the land's vistas and strong sunshine as compared with the lands they had left, fear of the unknown, and so on. See Miron 2004.

and peacefully, without pushing and shoving, without noise. No one was pushing with their elbows; no one stepped on anyone's toes. When the bus was full, those who remained outside kept standing without moving. Nobody begged for mercy. No one tried to sneak in…" (*ibid*).

He must have compared this experience with the one from Warsaw, the one he was familiar with. After all, he had a point to make: "And I, who have travelled often, for years now, on the 'Jewish' bus, here in Poland, have received an unusual impression of the Israeli bus system." (*ibid*). He explained the order with a "magic spell" that the Land of Israel casts upon its people:

> How very orderly it was—the result of self-control and deep emotion, a calm understanding of the world around them…. What a wonderful change this spell of the Land brings upon people! It looks quite simple—Jews riding a bus here and there—and yet I still see the magic in it. The deep meaning behind this phenomenon: here, on our Jewish streets, it stands before me like a bitter wrinkle reflecting our lives. There again, in the streets of Tel-Aviv, it lights up like dimples on the sweet face of a young woman, shining on the Jewish city… (*ibid*).

This is a Zionist text. Interestingly enough, the Zionist vocabulary replaced the vocabulary used by the Diaspora Jews, with phrases such as "the magic spells of the Land of Israel" or "the sunbathed Jewish city" that take on a magical sort of meaning and connotation. Milejkowski had to draw the comparisons. Having encountered the young Hebrew city, he felt the pain of the powerless and living hard lives Jews of the Diaspora. One of the things he noticed, and that particularly struck him, was the absence of police officers in Tel Aviv, as compared with Warsaw: "Here is another minor thing. I walk along Allenby Street in the twilight hours. It is large and packed with people. Bicycles and cars move back and forth. It is already past working hours and people spill out onto the street. There is much movement, and yet no rush. People stride at a walking pace along the sidewalks of the city, and those who are in a hurry, take a ride. I make my way through the densely populated area. Suddenly a thought pops into my head. What is this? I do not see a single police officer! A police-less city, I think to myself, and yet all seems to be functioning normally and safely, without fear or worry." (*ibid*).

It would seem he was tempted to leave everything behind and stay in Israel, but Milejkowski was conflicted: he was proud of the Zionist enterprise and wanted to be a part of it, but all along the way there were things that reminded him that this was not his place.

11.3 Here and There

As mentioned, Milejkowski drew constant comparisons between "here" and "there," between Poland (Warsaw) and the Land of Israel. For example: here, in Tel Aviv, the sunlit streets of the white city—there, the greyness of Warsaw; here you wait in line for the bus and the driver is welcoming, whereas there, people are rushing and shoving. It was as if Tel Aviv was the center and Warsaw, the periphery. Here, Jewish people speak Hebrew, whereas there, they speak Yiddish. Here, there are no policemen on the streets, whereas there, the police are visibly present on the streets of Warsaw. Milejkowski loved the Land of Israel, but was not a part of it. He was excited by the work done here, but remained an outsider. His body was on the shores of Tel Aviv, but his eyes were the eyes of a Diaspora Jew.

> Here is a small anecdote. A group of visiting physicians traveled to a northern settlement where they met a local farmer. At some point, they asked her: What do the residents of the settlement do with gifted children? "Thank God," she replied, "so far, the Lord has kept us from having gifted children." (Milejkowski, *Der Mament*, 19.6.1936).

This reply made the visitors laugh, but Milejkowski felt that it was the essence of the difference between the viewpoint of the Jews of the Diaspora and that of the Jews in the Land of Israel. He wrote:

> We, the Jews of the Diaspora, only want to view our children as gifted. It follows from the internal struggle for the right to live and for a place to live in. But in Israel, it is different. Life, particularly rural life, makes completely different demands. Here, children must have healthy arms and legs and be fit for hard physical work in a limited space, close to the land—the land on which they must live their lives and go through in a rural-Jewish fashion. (Milejkowski, *Der Mament*, 19.6.1936).

The words of the farmer strengthened his feeling that the Land of Israel was not his place. The Land of Israel needed farmers to work the land and fighters to defend its strong men and women. The farmer's response aroused laughter among the listeners, but in fact, they had to face a new point of view, quite disparaging the idea of the Jewish society in the diaspora. The "new man" of which the Zionist dream wished to create, appeared to be not a learner, an educated, a physician, but one connected to the land.

11.4 Language Barriers

As mentioned before, the Hebrew language issue was indeed a difficult one for Milejkowski. The fact that Hebrew had to be used on a daily basis including the medical meetings in Israel put Milejkowski at a clear disadvantage. He spoke several languages fluently. He had stated that he was versed in Polish, Russian, French, and German (though oddly enough, he did not mention Yiddish). Apparently, he had no Hebrew, which frustrated and troubled him. A few days after the Jewish physicians arrived in the Land of Israel for the conference, a general get-acquaintance meeting of physicians was held in Tel Aviv. It took place at the Health Center, a community institution that sought to promote the local health and hygiene (Milejkowski, *Der Mament*, 5.6.1936). "The leaders of the Congress Committee of Palestine talked, and their language was fluent Hebrew," he remarked, adding that he felt frustrated:

> Finally, it is my time to speak, and I have been given permission to talk, I walk onto the stage and suddenly feel tense. This is the first time I am publicly standing in the Land of

> Israel and, in addition, I am faced with language questions and issues. It is hard for me to speak their language, our language, Hebrew; and at the same time use our mother tongue—Yiddish. (*ibid*).

With time, Milejkowski learned to distinguish between thc language of the people and the language of the land. The language of the people was Yiddish, a Diaspora language that was the common tongue of the Jews of Europe; the language of the land was Hebrew—the language associated with Israel's history. This is his main stance in regards to the language issue:

> That damned language question! Putting aside philosophical reasoning and questions, with the spontaneous strength of a natural legal phenomenon, the Israeli reality answers that question. Hebrew *is* the language of the Land of Israel. It grew organically with traditions that are thousands of years old. The language is filled with clean and clear air, and it grows and blossoms on every grass blade and flower of it. This is the truth of the Land of Israel, but unfortunately not the truth of the People of Israel… (*ibid*).

The fact that some of their colleagues spoke Hebrew made other guests feel begrudged. Milejkowski regretted that most of the Jewish people were not fluent in Hebrew, but rejected the grudge: "Those who resented the fact that their friends in the Land of Israel spoke mostly Hebrew are most definitely wrong" (*ibid*).

Discussions on the use of Hebrew were prominent and intense in many circles at the time. At the conference, there was even someone who told the visiting physicians that they must learn Hebrew before deciding to visit the Land of Israel.[3]

According to Milejkowski, Hebrew belongs to the Land of Israel and there "could be no other way," but as a guest, he made other demands: "We wish to understand the Israeli reality, but at the same time, we have the right to want the people in the Land of Israel to understand us." By saying "us" Milejkowski referred to the Diaspora Jews, again making a noticeable distinction between "us" and "them," "here" and "there."

11.5 One Nation?

What are the relations between the Diaspora Jews and the Jews living in the Land of Israel? Are they truly one nation? Are they brothers or rather distant cousins? Again, Milejkowski used family metaphors:

[3] The use of Hebrew occupied physicians in Israel at the time. In 1932, the Tel Aviv Branch of the Medical Association instructed its members must deliver lectures in Hebrew only, but the daily *Davar* reported (1935, 1 January, p. 4) that physicians who worked for the local Sick Fund still delivered addresses in German, and that many were pleased with the public use of non-Hebrew languages. Some argued that if certain physicians find the instruction to speak in Hebrew publically too difficult, they should have their lectures translated or have others speak for them in Hebrew. Others claimed that physicians are making an effort to follow that instruction, but there is no need to go to extremes and that they should be allowed to speak in other languages as well. The Hebrew University of Jerusalem (in a board meeting dated 30 November, 1936) decided that it would hire only individuals who speak Hebrew "to a certain extent."

> …I especially felt it when I watched the beautiful, blonde-haired Israeli children; those joyful and purposeful children. What would become of them in the future? Are they siblings from one household, brothers, blood related? Or are they only related by means of living in the same universe, somewhat strangers and distant from one another? This problem is one with which Israel must cope: this is by all means an educational issue. All children must be shown the bridge that connects us to them already in schools. (*ibid*).

Thus, Hebrew, and the way it was treated, were indicative of the relations between the Jews living in the Land of Israel and the Diaspora Jews. Milejkowski eventually believed he found a solution that helped him tie all loose ends together. It was respectful of the Hebrew-speaking physicians, but set them straight and kindly reprimanded them. Milejkowski talked about the Fifth Commandment: "Honor thy mother and thy father" (Leviticus, 19: 3). In other words, children must not ignore, humiliate, or cause their parents pain. They should care for their parents as they age before caring for themselves. Milejkowski referred to Yiddish-speakers Diaspora Jews in Europe as the parents of Israeli Jews. He took the podium and settled the debate, in Yiddish:

> I feel, my dear colleagues, that here, upon this land where I am right now, we must speak Hebrew, but I also believe that my Yiddish should not sound strange to your ears. It is, after all, the language that your mothers and fathers spoke—and 'honor thy mother and thy father' is so sacred to us (Milejkowski, *Der Mament*, 5.6.1936).
>
> Again Dr.Milejkowski used familial terminology, depicting the relationship between the diaspora Jews and the Jews in the land of Israel. He demanded to fulfill the deed "respect your mother and father" and asked: what is the relationship between the diaspora Jews and the Jews in the land of Israel? Are they both one? Are they close siblings, or distant relatives.

He did not elaborate on his final choice of language, leaving that to his listeners and readers.

Touring Israel, the physicians reached the northern settlement of Yesod Hama'ala (Milejkowski, *Der Mament*, 3.7.1936), where they were hosted by the settlement's physician. There, on the shores of the Hula swamp, the local physician related that he and his children contracted malaria, but he would not leave the settlement despite the high personal toll that exacted of him. The local physician's moral-medical set of values suited Milejkowski's own values. There he was—a man who lived in the metaphorical swamp of anti-Semitism, watching the strengthening of anti-Semitic movements in Poland in those years—standing on the shores of the Hula swamp, trapped between two swamps, and yet he felt his duty was not to abandon his community and, therefore, he must return to Poland.

This may have been an indication of his future position as chief medicine and sanitation officer of the Warsaw Ghetto. If Milejkowski wanted to, he could have used his personal and professional ties to save his life, but he chose to stay—just as the physician from Yesod Hama'ala did in the face of adversity—to the bitter end, in his case. Milejkowski always felt he had to return to his community, his familiar surroundings, and the people he represented. As a public figure, Milejkowski felt committed to return to his people. Did rational considerations prompt his return to Poland, or perhaps emotional or other considerations served as the decisive factors? Perhaps it was all of the above. We will never know.

11.6 Conclusion

The Congress protocols serve as a testimony for Milejkowski's perception of himself as a Jewish scholar who wished to fit in and live as a citizen with equal rights in his country (Poland), not as a Jewish physician who serves and treats only Jewish patients. The protocols shed a light on the insights regarding the tension he felt between his Jewish and his Polish identity, and on the difficulty to settle both. The fact that he was a Jewish physician in Poland created a complicated picture: facing anti-Semitism, racist laws, and the difficulty of making a living, he nevertheless desired to remain a Polish citizen, preferably with equal rights. In fact, he firmly stated, "I am Polish." His notes from the Land of Israel testify that even though he was deeply impressed with the land and its vistas, with the Zionist project and the use of Hebrew, his heart remained in Poland, and that it never made him feel inferior. He appreciated the reality of the Land of Israel, but he knew where his place was: he was a leader of his people, and his people were in Poland; this is where he had to go.

Milejkowski never rejected his Judaism, and certainly not Zionism. He lived in both worlds—the world of a Jewish nationalist and the world of a Polish scholar—simultaneously. His logical statements at the Physicians' Congress, as well as in his journalistic articles, reflected the strong empathy he felt—like many other Diaspora Jews—towards his country of residence—Poland, in his case. This feeling did not come at the expense of his Jewish identity, as evidenced by his well-known activities for the benefit of the Jewish community in general and Jewish physicians in particular.

When the Congress ended, Milejkowski returned to Poland. In 1936, political reality in Poland was deteriorating. The lives of the Jewish people in Poland became more and more difficult, especially for the Jewish physicians. A person who could adequately understand reality would take action and go to the Land of Israel, especially if he were an enthusiastic of the Zionist project, or maybe emigrate to a different country.

Milejkowski was killed in the Nazi *Aktion* of January, as a Zionist, although his notes from his visit in the Land of Israel give the impression of a person who is torn between identities. On the one hand, he was excited with the beauty of Israel, its sights, its people and its language, but on the other hand his place was not there, but in Poland, as a Jewish-Polish scholar. His notes from his visit to the Land of Israel indicate that even though he expressed great empathy, excitement, and astonishment with the Zionist project, he could not personally undertake the Zionist mission. He was standing all along in a position of distance. His notes reflect a vibrant romantic person with a sensitive soul and a great attention to details, a man who humanized nature and the sights, who felt empathy towards the people of the land he visited, and yet his feeling that his place was not here can be felt throughout. He remained an outsider, a Polish scholar. His place was not "here." He knew that he was rooted in the sights of his country, Poland. His belonged in Poland, where he knew he had to serve the public he represented. It seems that these two opposite directions characterized Milejkowski's adult life. He was a member of the Zionist Movement, a publicist, a public figure. In his final days, he stated the Land of Israel was the only solution for the Jewish population, but in his own personal life, he aspired to fit into the Polish society and be a part of it.

It seems that Milejkowski's image is inconsistent with the distinctions that split the Jewish society in Poland in the interwar period into Zionists, Orthodox Jews, and members of the Bund Movement. His image may create an intermediate space, an image that represents a split identity, a dual phenomenon that probably symbolized a new direction. He was fascinated by the Zionist enterprise, but only watched it from the outside, without taking an active part in it, as he aspired to attain professional and public fulfillment in his country, Poland.

The encounter with the land of Israel helped Dr. Milejkowski see himself as a Polish and a national Jew. He came back to Warsaw, where he felt home. There was his culture, language, community, readers, patients, the people of which he served. As many others of the Jewish intelligentsia, Dr. Milejkowski likewise found a way to integrate in the Polish society without losing his Jewish identity or his national identity, and did not find any contradiction between his Jewish identity, the long history of the Polish Jews and his Polish identity. Albeit, his identity changed during the commencement of World War II—he become convinced that the land of Israel is the only solution for the Jewish question.

Bibliography

Apfelbaum, E. (Ed.). (1946). *Choroba Głodowa; Badania Kliniczne Nad Głodem Wykonane W Getcie Warszawskim Z Roku 1942*. Warszawa: American Joint Distribution Committee.

Balinska, M. A. (2012). The medical school in the Warsaw Ghetto, 1941–1942. *Polin, 25*, 463–466.

Centralna Archiwum Wojskowe (CAW): Główna Karta Ewidencyjna W.B.H poz 919.

Cohen, N. (2001). Jewish daily journalism in Poland. In I. Bartal & I. Gutman (Eds.), *Existence and breaking—Jews of Poland throughout the generations* (Vol. 2, pp. 301–326). Jerusalem: Zalman Shazar Center.

Friedman, D. A. (1936). *The first international congress of Jewish physicians' book*. Tel Aviv: Harfuah [Hebrew].

Główna Biblioteka Lekarska Dział Starej Książki: 307–15 -"Fragen zur erstmaligen Meldung der Heilberufe".

Levinson, A. (1953). *History of the Jews of Warsaw*. Tel Aviv: Am Oved [Hebrew].

Marcus, J. (1983). *Social and political history of the Jews in Poland 1919–1939* (pp. 263–291). New York: Mouton Publishers.

Milejkowski, I. (1936a, May 15). With the world congress of Jewish physicians in the land of Israel—Characteristics and reactions. *Der Mament* [Yiddish].

Milejkowski, I. (1936b, May 22). With the world congress of Jewish physicians in the land of Israel—Characteristics and reactions. *Der Mament* [Yiddish].

Milejkowski, I. (1936c, May 28–29). With the world congress of Jewish physicians in the land of Israel—Characteristics and reactions. *Der Mament* [Yiddish].

Milejkowski, I. (1936d, June 5). With the world congress of Jewish physicians in the land of Israel—Characteristics and reactions. *Der Mament* [Yiddish].

Milejkowski, I. (1936e, June 12). With the world congress of Jewish physicians in the land of Israel—Characteristics and reactions. *Der Mament* [Yiddish].

Milejkowski, I. (1936f, June 19). With the world congress of Jewish physicians in the land of Israel—Characteristics and reactions. *Der Mament* [Yiddish].

Milejkowski, I. (1936g, July 3). With the world congress of Jewish physicians in the land of Israel—Characteristics and reactions. *Der Mament* [Yiddish].

Milejkowski, I. (1936h, July 10). With the world congress of Jewish physicians in the land of Israel—Characteristics and reactions. *Der Mament* [Yiddish].

Milejkowski, I. (1936i, July 24). With the world congress of Jewish physicians in the land of Israel—Characteristics and reactions. *Der Mament* [Yiddish].

Miron, G. (2004). *German Jews in Israel* (pp. 119–124). Jerusalem: Magness Press.

Ofer, M. (2015). *White coats in the Ghetto*. Jerusalem: Yad Vashem [Hebrew].

Ohry-Kossoy, K., & Ohry, A. (2012). Dedicated physicians in the face of adversity. *Polin, 25*, 455–462.

Roland, C. G. (1992). *Courage under siege: Starvation, disease and death in the Warsaw Ghetto*. New York/Oxford: Oxford University Press.

Chapter 12
A Doctor's War Testimony: The Four Incarnations of "Dr. Twardy"

Monika Rice

Abstract This chapter critically compares the various embodiments of four versions of wartime diaries of Dr. Julian Aleksandrowicz, famous Polish-Jewish hematologist, humanist, and ecologist, throughout the four decades of them being written and republished. It analyzes them from the perspective of the shifting Polish-Jewish identities through which the author worked in order to achieve a consensus of his own final identity. Various aspects of Dr. Aleksandrowicz's war activities are emphasized in different ways in the consecutive editions of his memoirs, and his choices of emphases signal a dynamic work of subjective memory in his creation of an identity for himself. By referring this development of identity to contemporary stages of the socio-political and historical reality of postwar Poland, this chapter traces the complex path that followed in the wake of the process of assimilation and acculturation that might have typified a medical section of Polish-Jewish intelligentsia remaining in Poland after the war.

Keywords Postwar Jews · Polish-Jewish identities · Polish-Jewish doctors · Jewish intelligentsia in Poland · Jewish memoirs · Doctors' memoirs

This chapter traces continuities and modifications found in four versions of war memoirs written by one of the best known Polish-Jewish physicians of the last century, Dr. Julian Aleksandrowicz. By comparing specific themes and tropes that take on prominence, or, conversely, seem to retreat into the background in consecutive versions of the memoirs, this analysis strives to understand the process of creating a national-cultural identity by a famous representative of Polish intelligentsia.

Dr. Julian Aleksandrowicz (1908–1988) was a well-known Polish hematologist, and a propagator of ecological, preventive medicine. Among other achievements, he investigated the characteristics of the anticoagulant heparin, he invented a substance

M. Rice (✉)
Holocaust and Genocide Studies Programs, Gratz College, Melrose Park, PA, USA
e-mail: mrice@gratz.edu

M. Moskalewicz et al. (eds.), *Jewish Medicine and Healthcare in Central Eastern Europe*, Religion, Spirituality and Health: A Social Scientific Approach 3,
https://doi.org/10.1007/978-3-319-92480-9_12

that reinforces immunity in patients with blood disorders, and he patented a mechanism for the extraction and conservation of blood, which popularized blood donation. He published several dozen important medical textbooks, several hundred scientific articles, and many popular books on ecology, ethics, and preventive medicine. Dr. Aleksandrowicz was also famous for the inclusion of art in the process of a patient's healing, and for pioneering research on the role of magnesium in the human diet.

Born into an assimilated Jewish family in Kraków, he earned his medical degree in 1933 from the Jagiellonian University. During the Second World War, he fought in the September Campaign, before serving the Underground as a doctor in the forests of Radom-Kielce area in southeast Poland. Between the periods of his being a soldier, and a partisan, Dr. Aleksandrowicz and his family were forced into the Kraków ghetto, where he headed one of four Jewish hospitals in the Nazi-restricted area. He also hid in several hiding places organized by a series of Polish rescuers (whom he called "true people;" some of whom were awarded the title Righteous Among the Nations by Yad Vashem; for example, Józef Adamski; YVA, M.31.2/2104.2).

Ten years after the war, Dr. Aleksandrowicz sent a Polish-language memoir entitled *Kartki z dziennika doktora Twardego* (Pages from the Diary of Doctor Twardy [Dr. Twardy was his nom-de-guerre]) to Yad Vashem, suggesting that it should be translated into Hebrew and published as either a book, or a series of journal articles. Nothing, apparently, came of this offer. Seven years later, in 1962, the memoir was first published in Poland under the same title, and then republished four times (in 1967, 1983, 2001, and 2014), each time, by Kraków's Wydawnictwo Literackie. The book has enjoyed enormous popularity in Poland, commensurate with Dr. Aleksandrowicz's fame as a humanist-physician whose Clinic of Hematology in Kraków was a place of not only physical recovery, but also psychological and artistic healing for many a patient. Until his death, in 1988, Dr. Aleksandrowicz, therefore, prepared for publication three slightly different Polish versions of his memoirs; these, in turn, also differed from the manuscript deposited in Yad Vashem in 1955.

My goal in tracing continuities and alterations in four versions of the memoirs is, firstly, to establish the extent to which they might differ, and, secondly, to analyze how recreating his war memoir enabled Dr. Aleksandrowicz to reconfigure each time his own identity as a Polish-Jewish doctor. In my analysis, I am following Paul Ricoeur's concept of narrative writing as an identity creating process. Drawing on Aristotle's idea of emplotment, Ricoeur identifies an integrating process that "gives a dynamic identity to the story recounted: what is recounted is a particular story, one and complete in itself" (Ricoeur 1991, p. 21). By creating an intentional synthesis of heterogeneous events or incidents, the narrator actuates a plot that can be "at once concordant and discordant," but where concordance has ultimate primacy over discordance (Ricoeur 1991, p. 22). Applying this understanding to autobiographical writing, orienting life events in such a narrative creates a meaningful story of self-understanding and, thereby, creates an interpretation of one's life.

There are two dynamic aspects to "autobiographical emplotment": the first consists in the fact that constructing a narrative identity (the "self" of the text) involves the multiple social roles the narrator plays. Therefore, according to Monika Fludernik,

> Identity should … be used in the plural—identities—to acknowledge the multiplicity of roles and their contextual relevance… . identities are constituted in the interplay of individuals with other people in social contexts of family, work, study, leisure activities, etc. Although narrators generally believe they have a clear identity, that identity is an accumulation of performative stances and memories of past experiences which creates a continuity of self-understanding between roles and between contexts (Fludernik 2007, p.261).

The second dynamism refers to the relationship between the text and the reader. According to Ricoeur,

> the process of composition, of configuration, is not completed in the text but in the reader, and, under this condition, makes possible the reconfiguration of life by narrative (Ricoeur 1991, p. 26).

This dialectical relation between "the world of the text" and "the world of the reader" is essential when considering the historical reception of famous autobiographies.

By emplotting his own narrative of war experiences, Dr. Aleksandrowicz performed his dual ethnic–cultural identity, emphasizing certain experiences while discarding others, or qualifying some by interpretative comments. It is my goal to unravel which life episodes—and in what way–might have been eliminated, added, or edited in the consecutive narratives, in order to determine the character of reconstructing of Dr. Aleksandrowicz's identity. Through this exercise, I intend to contribute to the broader subject of the postwar Polish-Jewish identities of the intelligentsia, or to an "identity group portrait," to the extent that it refers to narrators of this particular social group.

As this research concerns the recreation of a mixed ethnic identity, these passages referring to Polish-Jewish relations will be of primary interest. It is important to note, however, that in accordance with the multiple roles that individuals play in society, and, consequently, the multiple identities derivative from these roles, the war story created by Dr. Aleksandrowicz could have been analyzed through other lenses and themes.

12.1 The Yad Vashem Manuscript ("YVM"), 1955

The first known version of the memoir is a 74 page Polish manuscript, sent to Yad Vashem in 1955 (YVA O33/190), consisting of 15 chapters and author's introduction. In the introduction, the narrator assumes an "objective," third-person position, to introduce "stories" that are "authentic events, which filled the life of a doctor, existing in Poland in the middle of the twentieth century" (YVA O33/190, "From the Author," unnumbered).[1] After that declaration, the narrative shifts into the first-person to announce a positivistic perspective:

> I looked at the people of this period with an eye of a naturalist. I observed how the period was shaping its characters. I saw reactions of different individuals to similar impulses. I perceived a rich scope of their actions, from heroism to extreme degradation. Many reflections were born out of these memories.

The proper memoir opens with the account of military duty in the September Campaign. On a retreat mission to Kielce, a distraught lieutenant imagines that "Dr. Twardy" is a spy. Recognizing a psychotic episode in the soldier, who panicked about the safety of his family, "Dr. Twardy," who was almost ordered to be shot, managed to expose the lieutenant's condition. The "adventure" tellingly introduces close-calls with death, rampant during the war, and not only due to "ordinary" war activities.

Following a general order, "Dr.Twardy" moves eastward, together with other men of combat age, beyond the new Polish-Russian border. Finding shelter among his family in Lviv, the doctor, nevertheless, soon attempts to return home - a pattern followed by many Polish refugees who could not decide which occupation, German or Soviet, was likely to be more or less lethal. After a death-defying escape on foot, having stumbled on a human frozen corpse, "Dr. Twardy" and his wife (who had joined him) finally reach German-occupied Poland in the deep night and spot a light in a hut. Risking everything, they decide to ask for shelter. To their utter shock, a group of women welcomes them, singing Christmas carols while spinning; the doctor and his wife had been expected. The saving light that led them to the hut, became "the light of good people." In spite of this heart-warming story, and of the new hope that Polish rescuers provided to the Aleksandrowiczs, the chapter ends with a deadpan conclusion, accentuating the tendency of targeted victims to trust in their fear rather than in other people: "Sometimes, fear has a quick ear" (YVA O33/190, p. 10).

Proper diary entries are found in the chapter from which the document took its title (*Pages From the Diary*), and which covers the pre-ghetto and ghetto periods (January 1940 – June 1942). It seems that, overall, these were authentic diary entries, written during the war, and, possibly, edited only afterwards. Before the ghetto, the doctor worked in the hospital on Skawińska Street; with other Jews, he was also forced to a weekly humiliating physical labor. On February 22, 1940, he openly describes a hostile Polish crowd that he encountered while shoveling snow:

> We are studying psychology doing this. What a wonderful study: a group of street people jeers at us, throws snow and horse's manure at us. An eight-year-old boy joins them. He is just about to throw a snow ball at us. His mother runs to him. She whispers something to his ear. The kid hangs his head low and leaves in shame (YVA O33/190, p. 13).

The mob is contrasted with "Councillor Mroziński," an old patient, who joins him in clearing the streets while saying,

[1] All translations from the Polish are my own.

> Let me [do it], doctor, at least for a moment; it is my internal need for demonstration. I have to show you, to show you and the others what I feel (YVA O33/190, p. 13).

When reporting on a Nazi scalping beard off an old Jew, the narrator also discusses Jewish collaborators as well as "good Germans," striving for objectivity:

> What precious evidence to a theory of what decides about human being are these observations: it is neither a world view, nor religion, nor belonging to a state, nor race, but a collection of biological factors that form themselves into the notion of "a character" (YVA O33/190, p. 15a).

Describing a moment at the creation of the ghetto, in March 1941, "Dr. Twardy" chooses to focus on a tram that runs through it, allowing the Gentile Poles to drop food packages to starving Jews (YVA O33/190, p. 16). "Dr. Twardy" presents Polish-Jewish interactions of the time through images like this—of Poles riding through the isolated Jewish area out of concern for Jewish survival. Even so, his perception of the situation is complex; it does not allow to ignore shameful Polish reactions. The ghetto's wall being built—the same wall that, as the narrator observes, acquires a shape of a "grave monument"—is looked upon by the "cold, emotionless faces of the bricklayers" (YVA O33/190, p. 17). An old prostitute decides to aid the wall builders with a theatrical gesture: "Let me also contribute to the extermination of this tribe," she announces pathetically, and lifts one brick, while the crowd joins in her laughter. Helpful heroes (the "good people") coexist with those who are actually quite pleased with Hitler's project of Jewish ghettoization. In the end, while there are no crowds to applaud those who drop food packages while riding through the ghetto, there are always crowds to be entertained by those who hurt the Jews. "Dr. Twardy" explains this polarization of the bystanders' attitudes by the occupier employing a "biological method," through which the indifferent become eventually evil. Paradoxically, however, this method also allows for goodness to shine: the only "Aryan" in the ghetto, the pharmacist Tadeusz Pankiewicz, a great rescuer of the ghetto population, and the author of the famous *Cracow Ghetto Pharmacy*, appears to have "found the meaning of life." In the place of military struggle, he has "infected" people with Goodness and Faith: "With the glow of his hot heart, he inspired the spirit of the unfortunate—and he won. He brought back the faith in man to many, and that is why many people survived" (YVA O.33/190, p. 17). Goodness and evil, then, appear to be equally distributed, and can be found anywhere: among Germans, Jews, Poles. This is one point that "Dr. Twardy" seems to be making throughout his diary's narrative.

At the end of May 1942, the news of another deportation threatens the relative peace in the ghetto. "Dr. Twardy" hides several hundred people in a small hospital on Józefińska Street with an official capacity of 50 patients. An informant denounces "Dr. Twardy," who receives a directive from the Gestapo to "resettle (read: to murder)" 470 patients; the doctors are to make the selection themselves. Dantean scenes follow, as patients offer gold and jewelry to doctors in attempts to save themselves or their relatives. "Dr. Twardy" assures us that no bribes were accepted. In the end, the doctor's ingenuous wife suggests a regrouping of a column of sick people that allowed them to mislead the inspecting Germans. The doctor states, "I prolonged the lives of my patients until November 1942."

A moral dilemma is suggested in the chapter *He Killed His Sick*, in which "Dr. Twardy" shares the repulsion he felt at witnessing a colleague giving poison to patients to "save" them from certain, violent death during the final action of the liquidation of the Kraków ghetto. On this day (March 13, 1943), "Dr. Twardy" moves with his wife and son to the "Aryan side," leaving his parents forever. His mother wants to join the fugitives, but "Dr. Twardy" manages to convince her to stay with her husband.

The next chapters depict events on "the Aryan side": hiding at the Adamskis', a visit from extortionists that fatefully resulted in an unsuccessful suicide attempt by the doctor, who swallowed a poison, and in his subsequent placement in the hospital.

After that, the narrative follows "Dr. Twardy" to the Underground unit of "Huragan," a charismatic officer. While on a mission to recapture a Polish residence, now occupied and used as a German arms magazine, the doctor encounters a Polish lady who supports the Underground despite working for the Germans. Years later, while seeing her as a patient in his office, he hears that, like him, she was on "Aryan" papers, and so her deception was double: she was a secret member of the Underground as well as a secret member of the Jewish people.

At some point, the atmosphere in the partisan camp sours: "Huragan" appears to spy on "Dr. Twardy" who, in breach of all conspiratorial etiquette, reveals to him his real name. In a most bizarre coincidence, "Huragan's" name also turns out to be Aleksandrowicz. This realization, rather than putting him at ease, makes him suspect that the doctor is playing some sort of "psychological game." The memoir is full of meaningful stories like this one, illustrating the paradoxes of life lived *quasi in extremis.* Another episode describes a brutal killing of captured German prisoners by a Polish soldier. The doctor is disgusted; later, some men from the division meet to discuss the murder with a chaplain. They conclude that the reading of "adventure" books may have normalized brutality for young, impressionable minds. While reflecting on the morality of war, they are surprised by the news that the Germans, in revenge, have begun burning the neighboring village.

Similar to the experience of life in the ghetto, the experience of the partisans is full of contradictions, verging on the heroic and the ridiculous, with moments both comic and tragic. A concentration of Underground troops before the Polish Uprising of August 1944 is depicted, for example, with irony:

> Everyone with a command of conspiratorial location, everyone with a password. What a wonderful, psychological study, what types of people. In uniforms straight from a needle, armed often strangely. Some of them with uhlan's sabers. Many had spurs, and there was even one in a shako of the uhlan Belin. Oh, imagination, incorrigible imagination! Yet, so much moving charm radiated from these people that were marching out into an unknown tomorrow, or, rather, into a more certain death, with a whole weight of romanticism. Uhlan's shako, amaranth, uhlan's sabres, girl's sash, all accessories of last years' poetry were found in our forest (YVA O.33/190, p. 56).

These appurtenances of a Polish revolutionary tradition, however, are juxtaposed to the admiration for what this vainglory symbolizes: during a surprise attack, a young soldier Zulik becomes injured in one eye. "Dr. Twardy" operates immediately, without anesthetic, while the patient bravely jokes, and, as soon as the eye is patched, joins the battle.

> Thanks to [the] disdain [for] bodily injuries that I saw in Zulik, in the face of a task, the fulfillment of my medical duty gave me full satisfaction. I learned a lot from Zulik. Thank you for that lesson that you gave me, a wonderful boy, corporal Zulik! (YVA O.33/190, p. 60).

The penultimate chapter, *Scholar and Sage*, is a tribute to all of the Aleksandrowicz's rescuers, and, most importantly, to the Adamskis. The deep friendship that formed between the doctor's father and a simple, decent bricklayer Józef Adamski did not cease when the Aleksandrowicz family was enclosed in the ghetto. Adamski, then, smuggled food packages, and paid for it with his life. Caught and beaten by a German soldier, he died from his wounds.[2] When "Dr. Twardy" returned to Kraków right before the war was over, it was the Adamskis who welcomed him like their own son. The story stands in contrast to the bitter encounter with a philosopher, a former "friend," who renounced all association with him and refused him assistance. "Dr. Twardy's" father's words that "not every sage is a scholar, and not every scholar a sage," close the chapter.

In the last chapter, *Return*, "Dr. Twardy" describes his rejoinder to his family's hiding place in Wieliczka. He feels trapped, like he is suffocating, as he tries to mimic his family's hiding protocol: existence under the bed is juxtaposed, in his mind, to life on the fresh green grass of the forest, and to the freedom of fighting, soldierly ethos. He managed to remain in the hiding place for 10 days, after which the calling of the forest proved too strong to resist. The manuscript closes with his dramatic "escape," in October 1944, from the uncomfortable safety of the hiding place.

12.2 First Published Version ("B 1962")

The first published memoir is graced with a cover, in a style reflecting contemporary aesthetic of the 1960s, with a sketch of two trees: one with a leaf, the other with a Star of David. The edition was preceded with a dedication to the Associations of Psychological Health, and to particular rescuers – the "true people."

The work has been significantly extended: in place of fifteen, there are now 19 chapters. As some excerpts from the YVM have been combined or placed in another chapter, there are, actually, seven new chapters in B 1962. Six of them describe in a more fully way the actions of the Underground, depict some particularly brave personalities of the Polish resistance movement, and, generally, sketch a heroic image of Poland in its struggle against Hitler. The last chapter, also new, treats of

[2] Only at the end of his life Dr. Aleksandrowicz confessed that Adamski was beaten by a Polish policeman. He explained withholding this truth by "falsely conceived patriotism" (Aleksandrowicz 1992, p. 95).

"Dr. Twardy's" return to the professional medical life in Kraków. Some content has also been reduced: the first two chapters – *Spy* and *The Light of Good People* – do not appear in B 1962.

More interesting are certain changes in emphasis: the authorial introduction juxtaposes more starkly the contrast between being "'a slave' of Kraków's ghetto" and a "'free man' from the forest" (Aleksandrowicz 1962, p. 7). While such juxtaposition is gently implied in YVM, neither the term "slave" nor the term "free man" appears there. The distance of a "naturalist" has, likewise, become augmented by the use of the third person throughout the whole section, in which the author devotes much more space to moral questions surrounding the pervasiveness of suffering, still framed in an evolutionary perspective:

> ... evil is what contradicts the development of human consciousness... . Evil is what regresses us back to what is unconscious.... we cannot be the passive subjects of evolution, as we are wasting it. We must direct its progress and, through this, the plight of the world. It is most possible today, especially today, when, thanks to science, we have gained a skill that allows us to shape environment in order for people to get along, so that, through this, their sufferings could be less frequent. Conscious evolution—means acting according to a principle that [the] man of the future will be what the man of today will make him to be (Aleksandrowicz 1962, p. 9).

The extended introduction seems to place the memoir in a universalist plane of moral literature, in synchrony with the way the Holocaust was perceived and studied during the first three decades after the war, starting with Kogon's *The Theory and Practice of Hell*.

In the chapter *Pages from the Diary*, which opens B 1962, the order of entries has been rearranged, but what might seem surprising is what remains intact: the entry of February 22, 1940, for example, still openly describes a hostile Polish crowd jeering at Jews (the narrator among them) who are being forced to shovel snow. In fact, this time, to emphasize the tragic lot of his coreligionists, "Dr. Twardy" adds Mordecai Gebirtig's poem "It hurts" ("It hurts/It hurts so terribly ... /when it is not the stranger – the enemy/but the one who is captured/the same as us, /a young Polish boy and a Polish girl/from the agony of our blood ... when they laugh at it, when the old men are beaten"). While YVM rather laconically describes the happenings, B 1962 contains a more developed interpretative framework, which is even less congratulatory to its Polish audience. Is it, perhaps, because "Dr. Twardy" does not have any ethnic axe to grind that he offers a universalist critique of human behavior, and refuses to become disillusioned in human nature after witnessing hostile Polish reactions? As he explains: "But the bitterness of these words did not weaken my faith in man. I always thought there is something contemptible identifying any improper attitude with the view of an entire social group" (Aleksandrowicz 1962, p. 16).

When describing the deportation of May 1942, in which "Dr. Twardy" was forced to "reduce" the number of his hospital patients, the narrator represents himself as more defiant towards the *Judenrat* (more negatively depicted) as well as toward the German authorities. Although, in YVM, Gutter (the *Judenrat* chairman) simply asks him the question and politely thanks him for fulfilling the task, in B

1962 Gutter "asks with a sharp, unpleasant voice"; he is rude and suspicious. "Dr. Twardy" answers him "insolently," after which he is asked to repeat the report to the Germans. He does so ("I have nothing to lose, anyway… ," he concedes), reflecting on their brutal, guttural laughter (Aleksandrowicz 1962, p. 32).

Another poem, "It Burns," by Gebirtig is added when "Dr. Twardy" provides a kaleidoscopic overview of the getto's political attitudes while "looking into windows" on July 6, 1942 (Aleksandrowicz 1962, p. 36).

In an entry in January 1943, "Dr. Twardy" describes his old friend, Dr. Jan Lachs, the famous obstetrician and historian of medicine who "had welcomed him into the world," and who, in the ghetto, reintroduced an ancient method for the reversal of circumcisions. "Dr. Twardy" was somewhat appalled and incredulous upon witnessing the demonstration of this method (which derived from a method employed by Jewish athletes competing with the ancient Greeks, and included a systematic increase of weight placed on the foreskin until it was removed); Dr. Lachs, however, is adamant that gruesome wartime events and the reasons behind them form a part of the rich texture of the total experience of life:

> You think that the meaning of life consists only of joys and pleasures. No, the meaning of life consists in the richness of what one experiences, and in the way one experiences it. There is no sharp distinction between beauty and ugliness, happiness and unhappiness, between health and illness. You should measure the value of life with the intensity of sensations and experiences, and creative action, and you will draw conclusions like a naturalist, while you will know the value of action from acts that will be beneficial also to other people (Aleksandrowicz 1962, pp. 38-39).

This entry was not part of the YVM; most probably it was written after the manuscript had already been sent to Yad Vashem. Perhaps, the memory of Dr. Lachs may have served as a proper canvas to sketch "Dr. Twardy's" peculiar "naturalist's" perspective that afforded him the distance with which he struggled to observe his own life as well as the behaviors of others. It provided a moral context for a passage that followed and that was, likewise, absent in the YVM, a passage that describes without excuses the behavior of those cynical extortionists who force unfortunate Jews on the "Aryan" side to demonstrate that their genitals are also "Aryan."

In a disturbing passage describing a doctor who provided his patients with cyanide tablets prior to a brutal deportation, the narrator is very explicit and direct that the martyrdom these patients were going through was a Jewish martyrdom, one which signified the absolute end of a Jewish presence in Kraków, and that had lasted through seven more or less glorious centuries. There is no universalization or "Polonization" of the victims, who would be dragged down the stairs or shot on the spot. These are the "Kraków Jews" (Aleksandrowicz 1962, p. 47).

When the narrator describes the escape from the ghetto, the impression of the mother that the reader receives, changes. Although, in YVM, she opted to leave her husband, in order to escape, in B 1962, she appears to be more heroic. There is no mention of the fact that she expressed the wish to leave; here, she just "stayed with the father" (Aleksandrowicz 1962, p. 49).

The new chapters concerning the Underground, and the family's Polish rescuers, while seem to be written specifically for Polish audience, contain what their style suggests, the greatest, most noble memories that "Dr. Twardy" salvaged from the war.

In the chapter *From Melina [hideout] to the Green Forest*, "Dr. Twardy" writes about the experiences of enjoying freedom, fighting back against Germans, and meeting his rescuers. He describes the soldiers' camaraderie as well as the trust the new doctor had to earn; testing him, the young soldiers who are to deliver him to "Huragan" leave him alone in the forest. Later, as a revenge, "Dr. Twardy" innocently informs the leader of the pranking fighters that his laxatives are aphrodisiacs.... The revenge on the soldier, who is now stuck for hours in the latrine, tastes sweet.

After his suicide attempt during the visit from extortionists once the family transits to the "Aryan" side, "Dr. Twardy" ends up in a hospital. Upon leaving the hospital, he is in need of finding another shelter, so he goes around to former acquaintances and friends in search of help. First, he visits "Dr. H.'s family," where he had left some goods before; he hopes to retrieve some of his things, or, possibly, to borrow money. He receives a rather surprise greeting from Dr. H.'s son, who assures him that they might try to send him some money, but that no cash would be forthcoming at the moment. "Dr. Twardy," then, decides to go to the artist Wodzinowski, who, however, turns out to be dead; his daughter Wicula, though, welcomes him very warmly, and introduces him to her friends (mostly members of the Polish Workers' Party), who happen to be having a meeting there. They all swiftly come to his aid. Hiding places are arranged, *Kennkarten* (false identity papers), registration cards, and *Arbeitskarten* (work papers) are printed. "Dr. Twardy" finds himself living close to the apartment of the famous poet Zuzanna Ginczanka. He is out of ideas how to help his parents, and whether he should even try to help them, considering that his mother is able to provide a minimum of provisions for his father in the ghetto. Observing and benefiting from helpful members of the resistance makes "Dr. Twardy" expressly long for his forest and his partisans; he becomes obsessed with the idea of rejoining the Underground forces. From "Biuletyn," the official organ of the Polish Underground State, he learns about General Sikorski's call to all citizens to join the clandestine army. However, "Dr. Twardy's" multiple attempts to be recruited again all fail due to the antisemitic attitudes of Underground superiors. The doctor does not mince his words of anger and disgust at the prejudices that prevent his precious medical expertise, extremely needed in the forest conditions, from contributing to the war against Hitler. Only the intervention of a highly influential friend solves the situation. Finally in April 1944, the AK (Armia Krajowa – *the Home Army*) sends him to Radom. Right before leaving for his underground mission, he manages to have his manuscript on the pathology of blood vessels hidden with a friend, who buries it in his garden. In Radom, the doctor is delirious with excitement at being part of the active struggle again: "But it is a colored film, vivid, in contrast to the gloomy scenes I experienced so far" (Aleksandrowicz 1962, p. 73). "Dr. Twardy" has found his freedom and his goal; he is in his element, as the last sentences of the chapter show:

> From now on, the days were passing in various ways: life was a mixture of idyll and terror, careless laziness, and drudgery, fight, and rest. I covered the wounds of people I had argued with the day before. I was getting to know new, valuable, interesting comrades, to whom I would say goodbye forever the next day. But one great feeling would not leave me in those days, the most beautiful and most human feeling: the joy of freedom. I was aware that my life was now in my own hands, that it was me—not the enemy—who could dispose of it. That I could die only in struggle, and not as a defenseless, hunted animal (Aleksandrowicz 1962, p. 78).

In *Pole and Hungarian Cousins Be*, "Dr. Twardy" describes an episode of cooperation with Hungarian soldiers, who, accidentally, came into possession of a cache of AK documents. "Szczesny," one of the companions, whose German is superior to everybody else's, volunteers to recapture the archives, and succeeds, with great bravery. At the end of the chapter, the narrator reveals that "Szczesny's" name is Artur Spindler, and that, since 1943, he has been hiding from "racist discrimination." The doctor mentions some other heroic actions of Spindler, for which he received a battlefield commission as an officer, together with the Polish Cross of Valor. The fact that Spindler was Jewish is not explicitly mentioned, which parallels "Dr. Twardy's" own hiding of his Jewishness in the Underground. Enumerating Spindler's successes aligns itself with the overall idea that partisan struggle redeems the ignominies of the situation of Polish Jews in hiding, that it is a noble goal to spend the war in an "active fight against the Hitlerian occupier" (Aleksandrowicz 1962, p. 155). As "Dr. Twardy" adds, Spindler's fate would take him, after the war, to Australia. In a nostalgic burst of rhetoric, the narrator asks whether the hum of Australian palms would remind him of the "speech of oaks and beeches in the Konecki forests?"

Commandant Maryska, the last of the new chapters on the Underground, is an account of a controversial, Underground commandant with an inflated ego, who is very brave, and effective at killing the Germans, but also vengeful and disrespectful toward his superiors. This commandant might have even employed antisemitic slurs, as "Dr. Twardy" seems to suggest when mentioning that "Maryska" liked to sprinkle his speeches to his soldiers with expressions from "fascist propaganda" (Aleksandrowicz 1962, p. 168). "Maryska," apparently, was eventually killed by his fellow Polish soldiers. This chapter adds a gray hue to the heroic image of the Underground; together with the story of a moral dilemma provoked by the brutal killing of German prisoners by a young Polish soldier, this story provides a context for more diverse, even problematic, reflections on the morality of war.

Another potentially significant change occurred in the chapter that describes capturing of a residence taken over by Germans and used as weapon magazine. An elegant Polish woman who was "captured" when "Dr. Twardy's" people attacked the residence is not revealed as a Jewess, as the information that she was also on "Aryan papers" is absent. Is it because Dr. Aleksandrowicz did not want to reveal that Prof. Maria Ichnowska (as the footnote informs) was actually of Jewish origin? Another small difference in the same chapter concerns money; in the YVM Polish woman's money is mentioned as being taken by "Dr. Twardy's" soldiers along with a host of other useful things. Money, though, is, not mentioned at all in B 1962.

Perhaps money was something, less than noble, that should not be associated with heroic partisans?

Other changes consist, mostly, of added historical information. In the chapter *Strange Meeting in the Forest*, for example, there is a longer passage concerning "Huragan's" Underground achievements. There are also some poems written by a young war companion, conveying a nostalgic, heroic image of the forest partisan life.

The chapter *Scholar and Sage*, dedicated to the rescuers of "Dr. Twardy's" family, and to all those who risked their lives to help those in danger, contains a few minor changes, among which one of the more noticeable is that the narrator's calculation that there are "fifteen percent people like Adamskis" in the society, becomes less measurable—"even a small percentage of people"—as if a statistical projection could no longer be operative. The passage as a whole, however, proves that one could speak quite openly about the ghetto, its risks, the help that was given, and the unwillingness of some to help.

In *Return,* the narrator adds that Soviet troops "liberated" Kraków a few days later, a circumstance that is not mentioned in YVM. The word "liberation," of course, was the only permitted interpretation of the communist take-over of Poland after the war.

In the last chapter *Surprise*, which is likewise new, the narrator depicts his first postwar visit to the general meeting of Kraków Medical Association. He feels awkward and out of place; although some of his colleagues are friendly, others appear hostile or indifferent. He declares his personal philosophy, which is always to focus on the good people, while erasing from memory the hostile ones. Dr. Ferenc approaches him with a wonderful surprise. "Dr. Twardy" confesses that, following the murder of his parents by the Nazis, the only thing that would make an impression on him was human kindness. His joy was, nevertheless, enormous when Dr. Ferenc produced the scholarly manuscript that "Dr. Twardy" had given to him for safekeeping. The memoir ends with a kind of summary of his war losses and gains:

> Emotionally and intellectually, I experienced several years of Hitler's occupation, which fate allowed me to survive. My experiences were sealed by an episode from a memorable meeting of the Medical Association. Enriched with this knowledge about man, I found a path for further life (Aleksandrowicz 1962, p. 191).

In summary, B 1962 is significantly extended with additional information on "Dr. Twardy's" experiences in the Underground. In fact, the events related to his clandestine military activity, together with various portraits of significant partisan figures, take up 13 of the 19 total chapters, so that they appear to dominate the whole narrative. Although "Dr. Twardy" fought actively "only" in the September Campaign of 1939, and then rejoined the struggle (after several unsuccessful attempts impeded by antisemitic prejudice) for about 7 months between April and October of 1944, this period lends its tone to the entire document. The two years spent in the Kraków ghetto are covered by the longest chapter of the book, the chapter entitled *Pages from the Diary of Dr. Twardy*.

12.3 Second Published Edition ("B 1967")

The second edition brings a new "Introduction" by Professor Władysław Bartoszewski, a famous Catholic Underground fighter and member of the Home Army who was active (from September 1942) in the Provisional Committee for Aid to Jews, as well as in its successor, Żegota—the only governmental organization in occupied Europe devoted exclusively to the rescue of Jews. Bartoszewski was also a prisoner in Auschwitz. Then, like the thousands of other resistance fighters who had fought for an independent Poland on behalf of the Polish-Government-in-Exile (London), Bartoszewski suddenly found himself "on the wrong side of history," once the Soviet Army "liberated" Poland from the remnants of German military might, and took political control of the country. After the war, repressed by the security service, he was twice imprisoned on suspicion of being a spy, serving a total of 7 years, which prevented him from continuing his studies. In 1954, on the cusp of the wave of de-Stalinization, his sentences were deemed "wrong." After regaining his freedom, Bartoszewski was able to finish his studies, after which he worked as a writer, a journalist, and a historian.

During the 1960s, Bartoszewski established himself as an authority on Polish-Jewish relations. Awarded the Polonia Restituta medal for his wartime assistance to Jews (in 1963), he was also invited by the Yad Vashem Institute to reside in Israel, where he was bestowed the title Righteous Among the Nations. In 1967, together with Zofia Lewinówna, he published a collection of Jewish testimonies of Polish rescuers, *Ten jest z Ojczyzny mojej* (published as *Righteous Among Nations* in 1969 in London).

Bartoszewski places *Pages* within the Polish memoir-writing tradition, in particular, the most recent war memoir genre. Mentioning Polish and Jewish authors including Dr. Ludwik Hirszfeld and Dr. Zygmunt Klukowski, Dr. Aleksandrowicz seems to fit "naturally" into this "Polish" intelligentsia, especially given that his writings reveal "varied attitudes and a diversity of received experiences—from heroism, to indifference, to small-mindedness, until maliciousness" (Aleksandrowicz 1967, p. 7). Then, there is this helpful list of "Polish" achievements of Dr. Aleksandrowicz:

> [a] reservist officer, a member of the September Campaign in the 74th Infantry Regiment… ; a doctor in the Kraków Ghetto; a soldier of the AK partisans in Kozienice and Konecki forests; presently, a world-famous scholar… . The order of events as in many Polish biographies, but, at the same time, an extraordinary example of an active and optimistic attitude to life, people, the tasks one puts in front of oneself, one's occupation (Aleksandrowicz 1967, pp. 7–8).

Nevertheless, the larger part of the introduction is devoted to these aspects of life that are not typical of "many Polish biographies": life in ghetto, hiding on the "Aryan" side, and witnessing the deportations. All of these lead Bartoszewski to expand on the widespread phenomenon of assistance rendered by Poles to their Jewish neighbors: first, individually, then through various organizations. Bartoszewski recapitulates his concept of Poles coming *en masse* to the rescue of

Jews under threat, and places Dr. Aleksandrowicz among those who benefited from Polish help; he also admires the doctor's ability to "draw lasting values of moral conclusions from the events that he describes, and the author's faith in "the ability of nations and people to coexist, unscathed by his tragic experiences" (Aleksandrowicz 1967, p. 15). Those Jews who found themselves in the tragic situation of Nazi exclusion from the world could thus always count on their helpful Polish neighbors. Polish Jews—especially prominent scholars who had rather typical "Polish biographies"—could keep their "faith... unscathed by tragic experiences." The glory of an honorable Polish ethos was preserved, while the noble, "Polish" Jews emerged from the Holocaust "unscathed." As we shall see, in this version of his memoir, "Dr. Twardy's" self-portrait fits closely that optimistic and consoling, Polish interpretation.

As far as the text of the memoir itself is concerned, the editing changes are subtle and few; most of the chapters remain unaltered, and they appear in the same order as in B 1962. Only in the chapter *Pages From the Diary...*, is there a significant extension of material, consisting of entries concerning 1939. The first two chapters of the YVM that were missing in the 1962 version (*Spy* and *The Light of Good People*) are now incorporated in *Pages* under the 1939 entries. One telling change in *The Light of Good People* is found in the description of the doctor's agonizing return home across the German-Soviet border together with his wife; the YVM depicts a moment in which the wife stumbles across a corpse frozen in the snow. That gruesome detail is absent from B 1967, where it is replaced, significantly, by a moment in which the doctor notices the footprints of a rabbit upon the snow. This find inspires him to search for hope:

> Even such a faint trace of life becomes something heartily intimate in such a waste. The imprinted traces of rabbit's paws gave us new strength. Again we jump into arduous wading in the snow (Aleksandrowicz 1967, p. 36).

The entries from 1939, absent from the 1962 version, fill the reader in primarily on the events of the August 18th mobilization and the September Campaign. The description of the first bombing on the night of August 31st to September 1st is depicted in short, rushed sentences that powerfully affect imagination, allowing one to empathize with the shock of realization that the war being discussed is something real. Following with a detailed description of the first successful military actions, and of the retreat toward Kielce, the author brings in the episode of being suspected of being a spy by a group of soldiers with whom he was about to travel (an episode missing from the 1962 edition). There are cosmetic changes to the YVM: in the manuscript, "Dr. Twardy" appears to be speechless, and he confesses, out of fear of being shot, to experiencing a sudden need to urinate. In the 1967 version, this physiological reaction is not described; instead, "Dr. Twardy" appears to be more aggressive and faster to assert his self-defense against the absurd suspicion. This shift could be explained simply by the need to get to the point quicker, as the entire episode is abbreviated in the later version.

On September 17th, when the Soviet troops invaded and occupied Eastern Poland, "Dr. Twardy" reports, in accordance with pro-Soviet communist interpretation of history that was obligatory at the time in Poland, that "the Soviets are coming to help us. A new spirit entered us" (Aleksandrowicz 1967, p. 31).

Other changes to B 1962 are minor: the author omits a hilarious, if ribald, story of revenge on one young soldier, for whom "Dr. Twardy" recommended laxatives as aphrodisiacs. Perhaps, it would not have suited the decorum of a serious subject matter. A further reflection on "Dr. Twardy's" regained dignity, following his joining the partisans, is developed. Additional sentences also emphasize the doctor's felt need to contribute to Poland's struggle:

> The feeling of a regained human dignity, for which I fought with the help of true people, was a recompense for yesterday's humiliations. Again, I am a soldier, yet with a multiplied awareness of my service, my social role—as compared to September 1939 (Aleksandrowicz 1967, p. 111).

The new chapter on the Underground might have subversive undertones: it describes a group of Italian anti-fascist patriots deported to occupied Poland. While the story itself is poignant, the fact that it is told by a legendary Underground fighter (Commandant "Grab"), who did not let himself be cowed or silenced once the communists took over, is most significant. In fact, Commandant "Grab" kept the memory of the anti-communist struggle alive in his commemorative speeches. Selecting this particular story teller might suggest a hint of Dr. Aleksandrowicz's true political sympathies, and might thus counterbalance his obligatory interpretation of the Soviet takeover of Poland in September 1939 as "military help." The new final chapter (*The Healthcare Personnel of Occupied and Underground Kraków*) depicts Polish physicians and nurses, like Dr. Ludwik Żurowski (a recipient of the Righteous Among the Nations Award), who were protective toward Jewish colleagues and patients. It describes the German organization of the local healthcare system, and even some examples of "good" Germans. It also mentions a racist, Polish newspaper. Finally, it describes the underground medical school and healthcare system, with the separate hospitals for Polish and Jewish patients, and it closes with the moral judgment that "Kraków's medical world, beyond a few exceptions, came out of the hard trial of characters—victorious" (Aleksandrowicz 1967, p. 226).

In the newly published *Epilog*, the doctor reminisces about meeting a certain Dr. Heilmeyer, who distinguished himself during the occupation as one of the "good Germans," one of those who showed "a true human attitude against the background of the fascist-perverted nation" (Aleksandrowicz 1967, p. 226). Dr. Aleksandrowicz encountered him, more than 10 years after the war, at an international congress on hematology. In spite of the acknowledgment, earlier in the text, that Dr. Heilmayer "showed a great deal of heart [for] the suffering society, and left behind heartfelt memories of himself," he feels a certain awkwardness when meeting him again face to face:

> For a moment, the figure of an officer in a *Luftwaffe* uniform appeared in my thought; the same face looks up at me from the hat with swastika—the face of an enemy. Now we shook our hands (Aleksandrowicz 1967, p. 226).

Answering Dr. Heilmeyer's question as to where "Dr. Twardy" was during the war, he responds: "I was where you would, probably, not come out alive…." He comments that a "forced smile" accompanied their conversation. Following the description of this strange meeting, "Dr. Twardy" closes his book with a reflection on whether his and Dr. Heilmayer's generation managed to pass on the truth about their times to a new generation, and on whether they were able to pass on their knowledge and experience, so that the youth would draw conclusions of their own which would "organize the world better." He ends with a positivistic conclusion:

> the advancement of the science in all fields and the constant, multifaceted raising of human consciousness to higher and higher levels, will eventually lead to the place of violent aggression and genocide being replaced by such actions that contribute to the positive development of man and humanity (Aleksandrowicz 1967, p. 227).

Altogether, B 1967 contains fewer changes and additions in comparison to B 1962, than B 1962 contains in comparison to YVM. The most significant changes in B 1967 testify to the stronger embracing of Dr. Aleksandrowicz's identity as an Underground soldier. He achieves this end via the addition of the forward by Bartoszewski; the entries referring to the September Campaign, which emphasize the longer status of Dr. Aleksandrowicz as a soldier; and some minor, stylistic changes that underlie certain political trends. The political realities implied include a need to remain, at least on the surface, faithful to the new communist version of Polish history. At the same time one might work to subvert it, and to emphasize a stronger connection with the ethos of the Polish Underground, of which Bartoszewski serves as a prominent example.

Another change to this edition consists in 16 photographs that depict scenes including "Dr. Twardy" and his co-partisans in the forest, Nazi offices and Nazi officers in the Kraków ghetto, an SS defilade, Dr. Aleksandrowicz's false documents, and Dr. Aleksandrowicz's rescuers. One photograph depicts the ghetto's wall and the ghastly scene of an empty street with bags and random objects scattered dramatically on it after a deportation.

12.4 Third Published Edition ("B 1983")

The third Polish edition, from 1983, came after the period of martial law in Poland. Bartoszewski, who had signed the August 1980 intellectuals' letter in support of the protesting workers from the Polish coast, was detained under martial law, and released in April 1982. Nevertheless, his introduction to the memoir is present, although now placed at the end of the book. It is also significantly expanded, with an exposition of certain differences between Poles and Jews under German occupation, and of slow realization of the final goals of the occupier. Bartoszewski introduces certain personalities responsible for the rescue of Jews, and emphasizes a policy of eliminating those traitors who had harmed Jews in hiding (Aleksandrowicz 1983, pp. 178–186).

The physical appearance of the book testifies to a Polish nostalgia for the Underground: the cover presents a black and white photo of a porous surface (forest? stone?) with a Home Army Medal encircled by the words: Radomsk-Kielce Region 1939–1945.

In addition to his previous introduction, Dr. Aleksandrowicz wrote a new introduction to the third edition explaining what was added. Primarily, the new material included not only portraits and descriptions of people connected with the Underground, but also reflections on life and death, on suffering, and on the meaning of history. In it, he expresses the conviction that life continues after death—in culture, ideas, literature, history, and memory of other people. Suffering is an existential necessity, for it liberates intellectual strength in the subconscious, and it inspires what the doctor calls "illumination." Dr. Aleksandrowicz confesses that some of his own inspiration for medical work came during the darkest war times, when he was under threat. The ultimate meaning of existence of the human race is the continuation of its existence. The doctor believes that what he has experienced might become a message that will contribute to the building of a New Order.

B 1983 follows B 1967 fairly closely. There are some stylistic changes, as well as six new chapters, five of which deal with the Underground. Among the stylistic changes, in the section *Pages from the Diary of Doctor Twardy*, there is a significant transformation of the narrator's person describing the tragic events of the day of the liquidation of the Kraków ghetto: in all previous editions, the events in the hospital are described from the perspective of the narrator's "friend," whose thoughts, feelings, and actions are recalled in the third person. This is a "friend" who is distressed by another "action," and who decides that it is better to offer his incurably ill patients a poison, to the disbelief of the narrator as well as that of the nurses administering the poison. The earliest (YVM) version ends most dramatically:

> we ran out of the hospital. From that moment on, I lost him from my sight. He—the doctor, led by mercy, who killed his sick (YVA O33/190, p. 30).

B 1962 and B 1967 do not contain these condemning words. In them, there is rather an emphasis on the doctor's own distress, and on his attempts to get some answers (approval, perhaps?) from the narrator, who, however, remains speechless. B 1983 reveals that the mental processes previously projected on a "friend" are, in fact, the narrator's own reflection. In B 1983, it is now "Dr. Twardy" who orders the nurses to give the poison to his patients. The tragic evolution of his moral judgment (from clear moral condemnation, to dispassionate reporting, to acknowledgment of responsibility) appears over the course of the four versions of the memoir.

Another transformed text is that of the "Mother's Narrative" of the escape from the ghetto. In YVM, the doctor's mother is ready to leave her husband and follow her son in search of safety. In B 1962 and B 1967, the intention of the mother, which could be interpreted egoistically, is not revealed; there is only the laconic statement, "Mother stayed with the father." By B 1983, the narrator reveals that he is the one who convinced his mother to stay, thus returning to the original version that put the mother in a light that might not be seen as the most sacrificial.

Another potentially meaningful change is found in the placing of additional emphasis on the deeds of a notorious Nazi collaborator, Ignacy Taubman, a former schoolmate and neighbor of Dr. Aleksandrowicz, who displayed a particular determination to harm him (Aleksandrowicz 1983, p. 58).

One salient feature of each published version has been a significant expansion of the Underground narrative. This time, the description of events and portraits of prominent partisans are intertwined with reflections on the ecological balance that governs the world of man and his environment, as well as the doctor's own medical achievements in addressing diseases of civilization.

When discussing the history of medicine under the occupation, the doctor adds a plea to commemorate the rescuers; together with his colleagues from the Polish Society of Mental Health, he asked Yad Vashem to honor the rescuers of Jews, as a positive background for the condemnation of the genocide during the Eichmann Trial. The letter took too long to be useful during the trial, but Yad Vashem did initiate honoring the rescuers by planting trees in their name, which would become a seed leading to the creation of the Righteous Among the Nations Award.

12.5 Conclusions

Over the course of the four incarnations of his memoir, Dr. Aleksandrowicz created a strong identity as a heroic Polish freedom-fighter, one of the "boys from the forest" who wanted to belong to the tradition of Polish insurrectionary struggle. This theme takes progressively more space in the successive editions, in which the doctor's desire to join the fighters is expressed more strongly, while descriptions of partisan life, and the treatment of achievements of important personalities of the Polish Underground expand. Although the story of a "Polish soldier" is a prominent category of identification, however, there are at least two other secondary identities that Dr. Aleksandrowicz constructs in his memoirs.

The first "secondary identity" is that of a Jewish doctor in the Kraków ghetto, who observes and shares in increasingly discriminatory policies against his fellow Jews, and who tries to prevent some of the most brutal deportation actions. This aspect of identity becomes, then, a vehicle for moral reflection, which brings forth an unexpected revelation in the last version of the memoir: only in B 1983 does Dr. Aleksandrowicz admit that he was, in fact, the doctor who, in the words of YVM, "killed his patients." This disturbing confession is accompanied by thoughts that attempt to find "balance" in his life's message, and to formulate a sort of "end-of-life sagacity" that could be passed on to the people of the future.

Another strand or "tertiary identity" that struggles and undermines the primary identity of the "Polish soldier" appears with regard to the experience of antisemitism, especially among his fellow soldiers. The frustrations accompanying his initially failed attempts to join the Underground, as well as his colleagues' bigotry and opportunism in the face of the Nazi treatment of the Jews, are explicitly expressed in the texts. The impossibility of combining the image of a noble Polish soldier with the dissonant notion of antisemitism results in the creation of a universalist "sage" who understands that there are good and bad people in every nation and that it is one's "character" that is responsible for one's behavior, and not merely one's ethnic or cultural conditioning. As these universalist thoughts become more prominent in each version, the moral assessment of the doctor's own choices undergoes reevaluation: from an earlier embrace of a traditional moral perspective that would condemn any harm done to patients (even in the name of saving them from a more brutal imminent death) to a later confession of uncertainty, but a less doubtful one—as to whether his action may have been, rather, an act of mercy.

One interesting, less expected feature of Dr. Aleksandrowicz's process of building his self-portrait through the memoirs, is the openness with which he discusses the more shameful examples of Polish antisemitic behavior, which creates an uncomfortable mirror for Polish readers. Considering the enormous popularity of the *Pages*, one may wonder how these difficult passages must have been received at the time when a heroic image of Polish conduct during the war, and, in particular, a false paradigm that considered Poles only as helpers and rescuers of Jews (a paradigm to which Bartoszewski's book only contributed) was an unquestioned assumption. The memoirs appeared both before Jan Błoński's ground-breaking article *The Poor Poles Look at the Ghetto*, which is considered as a starting point for Polish public "soul-searching," and before the historiographical revolution introduced by Jan Gross' *Neighbors*. Perhaps, the critical fragments in the *Pages* were, simply, overlooked, and ignored.

References

Aleksandrowicz, J. (1962). *Kartki z dziennika doktora Twardego*. Kraków: Wydawnictwo Literackie.
Aleksandrowicz, J. (1967). *Kartki z dziennika doktora Twardego*. Kraków: Wydawnictwo Literackie.
Aleksandrowicz, J. (1983). *Kartki z dziennika doktora Twardego*. Kraków: Wydawnictwo Literackie.
Aleksandrowicz, J. (1992). *Tyle wart czlowiek*. Lublin: Wydawn. Lubelskie.
Fludernik, M. (2007). Idenity/alterity. In D. Herman (Ed.), *The Cambridge companion to narrative* (pp. 260–273). Cambridge: Cambridge University Press.
Ricoeur, P. (1991). Life in quest of narrative. In D. Wood (Ed.), *On Paul Ricoeur* (pp. 20–33). London/New York: Routledge.

Chapter 13
"Ich bin ein Koszaliner"? Struggles with Belongings in Borderlands. Leslie Baruch Brent's Autobiography *Sunday's Child? A Memoire*

Miłosława Borzyszkowska-Szewczyk

Abstract The stipulated topic of reflection is an attempt at following the identity project contained on the pages of Leslie Brent's autobiographical texts, especially in his autobiography *Sunday's Child? A Memoire* (London 2009). Leslie Brent is a British professor emeritus, a renowned scientist in the field of immunology. He is the co-discoverer - with Peter Medawar and Rupert Billingham - of acquired immunological tolerance (Billingham RE, Brent L, Medawar PB. Nature 172:603–606, 1953). He was born in 1925 as Lothar Baruch in the city of Köslin and escaped the Holocaust with one of the "Kindertransports" travelling from Berlin to Great Britain – the operation that saved 10,000 Jewish children from extermination. Köslin – which with the shifting borders after 1945 became the Polish city of Koszalin – is located in Pomerania, a region spanning the historical and cultural Slavic-Germanic, and now also the administrational, Polish-German borderland. Since 1989 Brent has regularly visited the town of his birth, actively taking on the role of "emblematic Jew" in the local project of reviving Jewish memory.

The intention of this article is to answering the following question: How does the author position himself as an individual in charting the borders of homeliness and strangeness laying between the everydayness of the pre-WWII German Köslin – marked by gradual exclusion of its Jewish community and the reality of Great Britain, in the context of his assimilation/acculturation process, as well as the realities of the post-1989 Polish Koszalin – the time of his first return journey to the hometown and subsequent visits?

Keywords Jewish identity after Holocaust · Pomerania · Koszalin · Köslin · Borderland · Regaining memory in Central-Eastern Europe · Leslie Baruch Brent

---

M. Borzyszkowska-Szewczyk (✉)
University of Gdańsk, Gdańsk, Poland
e-mail: mborzyszkowska@ug.edu.pl

M. Moskalewicz et al. (eds.), *Jewish Medicine and Healthcare in Central Eastern Europe*, Religion, Spirituality and Health: A Social Scientific Approach 3,
https://doi.org/10.1007/978-3-319-92480-9_13

My reflection touches on the borderlands in a multitude of ways, starting with the fact that Leslie Baruch Brent as an immunologist represents a field that encompasses medicine. He comes from Pomerania – one of the regions spanning the historical and cultural Slavic-Germanic, and also the administrational Polish-German borderland. He was born as Lothar Baruch in 1925 in pre-war Köslin. The Jews could be perceived as a "virtual" space in-between, a 'Gedächtnisgemeinschaft', situated between national options and identifications (Loew et al. 2006, p. 10). Analyzing Brent's autobiographical work (borderland of documentary and autofiction) with an identity project in mind, one may approach this task, as among others, a case study viewed in the context of Polish-German borderland's cultural entanglement, war fates and post-war German-Jewish Pomeranians, as well as in the light of post-war Polish-Jewish-German dialogues. The text, its author, as well as their analysis also fit into the process of "regaining memory" in Central-Eastern Europe (Diner 2003, p. 11), especially in Pomerania.

The stipulated topic of reflection of this article is an attempt at following the identity project expressed in Leslie Baruch Brent's autobiography with my deliberations generally concentrating on the following issues: What model of shaping Jewish identity is being constructed by the author in the tangle and tension of experiences connected with, among others, anti-Semitism – including persecution in the Third Reich, the experience of the Holocaust, internal Jewish divisions, forced emigration and integration, including his career path of a scientist-immunologist in Great Britain? Whether and just how is the identity project constructed? Are the experiences falling "in-between" the culturally defined identities also recorded and/or reflected on? How does the author position himself as an individual in defining the borders of homeliness and strangeness, between the everydayness of the pre-WWII German Köslin – characterized by gradual exclusion of its Jewish community, and the realities of Great Britain in the context of his assimilation/acculturation process, as well as the realities of post-1989 Polish Koszalin – the time of his first return journey to the hometown?

Analyzing Leslie Brent's autobiographical texts as an narrative identity project with the reception of his texts and person in mind, this tasks may be approached in a number of ways: It is a case study of Jewish identity following the Holocaust, viewed in the context of Polish-German borderland's cultural entanglement and processes, pre-war migration of German Jews from Pomerania, as well as from the perspective of post-WWII Polish-Jewish-German dialogues on the past, viewed as a process of shaping cultural memory of the future. Both the texts and their analysis also fit within the process of "regaining memory" in Central-Eastern Europe (Dan Diner 2003[1]), with one of its Polish catalysts being the so called Jedwabne-debate after 2001 (Michlic and Polonsky 2004) connected with the country's Jewish past. This article also constitutes an example/a case study on how Jews and the local societies of present-day Polish cities experience spaces of Jewish heritage and work

[1]Especially see chapter "Schuldiskurse und andere Narrative. Epistemisches zum Holocaust", pp. 180–200.

on their preservation and restoration, contributing to the process of reconstructing the Jewish Past in Eastern Europe.[2]

The expression *Ich bin ein Koszaliner* contained in the title of my article, to which I have added a question mark, was articulated in 2005, during Brent's third visit to Koszalin, taken on the invitation of city authorities. It was then, among others, that Brent was presented with the title of an Honorary Member of the Koszalin District Chamber of Physicians.

13.1 The Author and His Text

Leslie Baruch Brent's education, professional career and private life had all unfolded in Great Britain where he arrived with the first of Kindertransports on 1st December 1938, joined the army in 1943 as a 'friendly enemy alien' and adopted the name of Leslie Brent. Mr. Brent holds the British title of Professor Emeritus and is referred to in book reviews in professional magazines written as "one of Britain's foremost immunologists" (Batchelor 2009).[3] He is the author of an autobiographical work *Sunday's Child? A memoire* (London 2009). Mr. Brent was a member of the team whose leader, Peter Medawar, was awarded the 1960 Nobel Prize for discovering the phenomenon of immune tolerance – the discovery which laid the foundation for the development of transplantation surgery.

Until his retirement in 1989, Leslie Brent had for 21 years held a professoriate at the Saint Mary's Hospital Medical School in London, where he was the head of Department of Immunology, created of the Clinical Immunology Laboratory and co-founded the local Allergy Clinic. Between 1976 and 1978, Mr. Brent was the President of the (International) Transplantation Society. One year earlier, 1975 he was invited to conduct lectures in Poland (Warsaw, Cracow, and Wroclaw) by the founder of Polish nephrology Tadeusz Orłowski (1917–2008) (Kokot 2009). It was then that he visited the Auschwitz-Birkenau Memorial and Museum.

In his autobiography[4] Leslie Baruch Brent deals with the subject of *Where do I come from?* which is to a lesser degree expanded by the question of *What world did the Holocaust take away from – not just – me?* – primarily in the context of the lost Jewish world of Köslin. Therefore it cannot be taken exclusively as a memorial book on the remembrance of collective trauma (Adamczyk-Grabowska and

[2] The article is connected with a wider research project, which main aim is to reconstruct Jewish topographies of Pomerania based on autobiographies of authors of Jewish origins from Pomerania (after 1945), on which the author is working currently.

[3] See also Monaco, A. P. (2009): Sunday's Child? A Memoir By Leslie Baruch Brent, *The New England Journal of Medicine*, 16.7.2009, 317. Monaco presents Brent as "a world-renowned immunologist and transplantation biologist".

[4] The German edition of the book: Leslie Baruch Brent (2010): *Ein Sonntagskind? Vom jüdischen Waisenhaus zum weltbekannten Immunologen,* Berlin: Berliner Wissenschaftsverlag. Work in progress is Polish translation of Brents memories, what should be published in the series of the Kashubian Institute.

Kopciowski 2011). Brent's story revolves first and foremost around the questions of *Who am I today? Where have I come to? How have I got here?* His autobiographical text does not quite escape classification. Therefore, an interpret was spared the pleasure (and toil) of deciphering the inter-genre games, characterizing autobiographical texts. *Sunday's Child? A Memoir* is at its heart an autobiography of a scientist in the classic sense of the word, where the question *Who am I?* is located at the very center. The narrator's focus – even when directed at other characters and institutions of significance to the biographical pathway portrayed in the form of essay recollections in the structure of the text – do not stray far from their relation with the *Self* (or *Me*) which is the dominant subject and object of auto-reflection. This question will be expanded with Brent's comments and views on global political and social situation, such as racial intolerance, the Israeli-Palestinian conflict or global warning.

Starting from the position of a refugee in Great Britain, Brent reconstructs in detail the stages of his education and the pathway of professional career as an immunologist. This perspective is shifted in the German edition, as clearly evident in the subline added to the title, in which the departure point is positioned in his childhood experience of being under the care of Jewish Orphanage in Berlin – Pankow. The book's review published in the bulletin "Deutsche Gesellschaft für Chirurgie" aptly interprets this subtitle as both "a program and a table of contents of this moving book" (Trede 2009). The author also shares his reflections on the causes of Holocaust, founding of the State of Israel and Israeli policy on the Palestinians, Jewish identity, Zionism and anti-Semitism. Complement to so profiled life-story, are his accounts of post-war "returns" to his native town – once German Köslin, the present-day Koszalin in Poland.

13.2 Köslin of the 1930s Through the Eyes of a Jewish-Home Child

During the interwar period, Köslin was a city in the German province of Pommern (Jaroszewicz and Stępiński 2007), populated by about 28 thousand people, out of which around 170 were of Jewish faith (Salinger 2006, 2/III, p. 482). The region which was referred from the perspective of the day as the "backwaters of the Reich, a town off the main track, well maintained and clean but with a provincial smell to it" (Grzebałkowska 2015, p. 87). The author of these accounts was born into a family of "German Jews or Jewish Germans", whose head was a commercial trader with a non-orthodox approach to the question of religion: "My parents [...] practiced their religion with conviction though without orthodoxy" (Brent 2009, p. 7).

The chapter dealing with the period of the Nazis coming to power in Germany in 1933, is entitled *Childhood by the Baltic*. It contains Arcadian images-recollections "of happy, secure, uncomplicated time in a loving family" (Morris and Morris 2010, p. 269) on the eve of the coming Holocaust, during the incubation of Nazism. The memoirs of Leslie Brent paint a picture of a (petty)bourgeois home with no academic traditions, liberal in its cultivation of Jewish rituals. The religiousness of the family is summed up by the autobiographer in the following sentence: "Judaism, although not of the orthodox kind, played an important part in our life" (p. 14). It points to the forms of participation in the city's Jewish community – regular visits at the synagogue, grandfather's practice of orthodox Judaism, social contacts, memories of celebrating holidays of the Jewish calendar. As the showpiece of this "small but vigorous community" he presents "the absurdly grandiose synagogue" (p. 14).[5]

13.3 Berlin-Pankow (1936–38). Stay in Jewish Orphanage

The turning point of 1933 is indicated by a telling statement "Paradise was irretrievably lost." (p. 16). In the provinces, the anti-Semitism of neighbors and institutions took on clear form earlier than it had in cosmopolitan Berlin. One of the commentators of the book points out that in such translucent communities the spirit of German nationalism had found prompt expression in its aversion towards Jewish co-citizens (Fleischkacker 2010). In 1936, after their only son was expelled from school (*Mittelschule*) as a consequence of a provocation, the parents decided to send the boy to the Jewish Orphanage of Berlin-Pankow. This is where, along with 7 or 8 other children, Lothar was chosen to leave on the "kindertransport" – a rescue effort organized following the "Kristallnacht" prior the outbreak of WWII, aimed at relocating Jewish children (10, 000 persons) from Germany, Austria, the Czech Republic, Poland and the Free City of Gdańsk to Great Britain (Benz et al. 2003).

Brent's parents and older sister as many other Jews from the German provinces moved to Berlin in 1938 hoping for the anonymity of a metropolis. In 1942 they were deported from Berlin to Riga, where they were shot immediately after their arrival. The last letter which Lothar received in 1942 contained a coded message "Wir verreisen" [We are going on the journey]. On one hand, during the narration there is a re-occurring awareness of the randomness of human fate, especially in connection with surviving the Shoah, on the other, his story continues to repeat one of the central questions of Holocaust-survivors i.e. *Why me?* (Lammel 2009).

[5]A new synagogue at Am Kleinen Wall near to Lazarettstrasse had been opened 1885. Jewish society counted to this point of time the highest level – 303 persons. The temple was built on square, flanked with dome, on it the inscription appeared: *From the rising of the sun to the place where it sets, the name of the* Lord *is to be praised* (Psalm 113,3). In the interior has been placed an organ. The synagogue was destroyed during the November Pogrom 1938 (Salinger 2006, 481–482).

13.4 Autoportrait – A Documentalist of His Own Successful Assimilation

Based on the autobiography of Leslie Brent, there forms an image of a man of action – a scientist who at times takes on the role of a writer, led – as he admits – by the obligations he feels as a Shoah survivor towards the fate of his family and the community he comes from. The author adopts the attitude of a witness of history, striving during adult life to shift his position from being an object of history to being that of its subject, completing the sequential challenges of a scientist-immunologist. The flow of the story does not quite carry the reader away, mostly due to its form rather than its content, which is a consequence of the adopted style described by a British reviewer as "powerful, but dispassionately told, story" (Morris and Morris 2010, p. 269). The bulk of the narrative is conducted in a testimonial manner, listing the results of painstakingly collected and ordered dates, notes, recollections, facts – summing up or rather giving a sense to the twists and turns of life and the choices made. The narrator does not break in any place the "autobiographical pact" (referring to Philippe Lejeune concept: 1975, Polish Edition: 2001), appearing to be a documentalist of his own life, interweaving it into the historical-social context: "extraordinary life story is powerful, engrossing and timely" (The New England Journal of Medicine, 2009). At this point, one is reminded of the words of Georges Gusdorf, the French philosopher and social historian, that an autobiographer "displays his image on the background of this which surrounds him" (Gusdorf 2009, p. 20) – society, generation, discourse, time and space, while at the same time underlining the uniqueness of own fate. The autobiographer Leslie Baruch Brent deliberately constructs the order of his own biography, closing its central axis in the play of words: *From Intolerance (Racial) to Tolerance (Immunological).*

Constructing the cause and effect chain of his life-history, the figure of the mentioned *Sunday's Child* has been introduced as an element of the narrative framework with its assigned symbolism of a child of fortune – the lucky one in many levels, not only as a Holocaust-survivor. Brent recalls it both in the introduction and conclusion of the story and title of his memoir. It comes directly from a fragment of the last letter he received from his parents which they had managed to send through the Red Cross a few days before their deportation from Berlin to Riga: "Du bist eben ein Sontagskind!" [After all, you are a Sunday's child!]. The words of his parent's final message are interwoven by Brent – the author, as intended to equip the boy with a good measure of self-confidence. The autobiographer sums up the story of his life, as follows:

"[…] But my father was right in pointing out that I was 'ein Sonntagskind'. I have been extraordinarily fortunate at various critical points of my life and I have succeeded in building my life on the foundations laid so lovingly by my parents in my first 11 years. "(p. 293)

Sunday's Child? with a question mark in the autobiography's title could be read in the same context, as a telling self-reference to a Holocaust survivor who had lost his entire family.

13.5 The Returns? "Ich bin ein Koszaliner"?

The recollection of receiving a fax from Koszalin in July 1999 inquiring whether David Brent, whose matzeva was in the collection of a local museum is a relative, opens the tale of Brent's own life (David turned out to be the author's uncle). The report from the 2005 visit to Koszalin closes that story. In this way, the Koszalinian motifs were staged as the second narrative framework of autobiographer's life story.

Do the visits to the place of birth and childhood require an interpretive-explanatory comment *per se*? Probably not. As their interpreter, I am rather intrigued by what meaning or juxtaposition is assigned to them by the author from the perspective of time and space. Since his first trip to the post-WWII Koszalin in 1989 just prior to the fall of the communist system in Poland, Leslie Baruch Brent has visited his hometown a number of times and through private contacts has actively participated in the life of the city. He published the impressions on his subsequent stays in Koszalin in the British "Journal of the Association of Jewish Refugees". He gives them titles such as: *How I came to believe in miracles* (2005) and *More miracles of Poland* (2006). "Miracle" is the keyword chosen by Brent to present his impressions, being placed in titles of both articles. His positive experiences in Poland are interpreted as miracles, i.e. phenomena beyond rational perception and expectations. This interwoven meaning may be interpreted as a response to the stereotype of post-war reinvigorated, omnipresent Polish anti-Semitism (Chodakiewicz: 2003, 11 et seq.).

13.6 Hybrid Identity Project?

Identity issues are coped with in a separate chapter of the autobiography, aptly entitled [*Chapter Twenty-Two: A Question of Identity?* (pp. 267–274). The identity project of Leslie Baruch Brent is of a periodic character with a multi-level structure, shifting along with changing social-political milieus: from the assimilation model to that of a shared identity combining elements of various contexts he is a part of. The author himself points the process as follows: "My persona and my voice have developed subconsciously, no doubt strongly influenced by the different environments to which I have been exposed [...]" (p. 271).

The main factor is his Britishness (not Englishness – as the autobiographer underlines) with experience of migration: "I certainly was, once, a German-Jewish refugee" (p. 271). British culture nearly completely dominated Brent's experiential horizon. He explains his situation quite metaphorically, borrowing terms from transplantology when he clarifies: "I was transplanted at the age of thirteen from a German-Jewish environment into England and, although mindful of my roots, I adapted to life in Britain with, ultimately, a British name" (p. 267).

Employing Stephen Bochner's typology of coping with an unfamiliar culture, the autobiographer seems to be dominated by the assimilation attitude (Bochner 1982) as the immigrant who has successfully (and completely) mastered British cultural codes, pushing aside/rejecting for number of decades his German and de facto Jewish socialization. Therefore, in the tale of his own life, Leslie Baruch Brent does not make any considerable pause at the experience of alienation, quite the opposite – he underlines his (inborn) ability to adapt, as well as the help he had received from third parties (other people), interpreting its character through his own positive approach to life. He points to no longer considering himself a refugee and after a chain of arguments concludes: “all this makes me feel that I am truly embedded in the fabric of British society” (p. 271). The attribute “nearly” placed next to the declared British identity refers to a period of time following his retirement – which is connected with renewing contacts with various circles from his youth, among others, taking part in reunions of Kindertransport children, the pupils of the Jewish Orphanage in Pankow in 2001 and the Bunce Court School in 2002, as well as forming links with present-day Koszalin.

German language and 11 or 13 years of German socializing are definitely assigned to the childhood period, with emphasis put on feelings of loneliness and isolation experienced in German school vulnerable to constant abuse as the only Jewish child in his class, deepened further by the trauma of abrupt withdrawal of friendship by his Christian neighbours’ children (see p. 16).

When it comes to language, Leslie Brent stresses that at home it was *Hochdeutsch* and not Yiddish that was spoken. By naming *Hochdeutsch* – standard German – he provides further explanation: “the German spoken by educated people” (p. 17), in this way pointing to the adopted model of Western-European Jewish emancipation, as well as the cultural standards of a *Bildungsbürgertum*. The message is made even stronger by listing interior decoration elements of the first flat (for example grand piano) or memories of a bourgeois summer idyll on the Baltic coast in Grossmöllen – the postwar Polish Mielno (p. 13). Those objects memorisied/recalled in the autobiographies, could be interpreted as expression of an aspirated cultural pattern of *Bildungsbürgertum* and awareness of familiar continuity and cross-generational amalgamation (see Shalgros 2007, p. 135). This originally homogenous message about the language used by the family, which unequivocally sets the narrator on the axis of Western-Eastern cultural models of Jewishness, supplements and differentiates the complement message that individual words in Yiddish were used at home. He also states that along with a group of other Jewish children he attended Hebrew lessons as part of religious education, although he underlines that his parents were not Zionist-oriented. From the perspective of time, he expresses astonishment at this fact and in the next sentence laconically justifies the situation: “They were rooted too deeply in German culture to consider this to be an option” (p. 20).

Germaneness is accordingly part of the past, marked with bitter-sweet memories, it does not play any role either as *Heimat* or as *Vaterland*. In contrast to the memoirs of authors of Jewish descent from Gdańsk,[6] one may not speak here of any developed regional or local identification (Borzyszkowska-Szewczyk 2015).

---

[6]G. Dworetzki (1985): *Heimatort. Freie Stadt Danzig. Von Danzig nach Gdansk*, Düsseldorf: Droste Verlag; F. Meisler (1996): *On the Vistula Facing East*, London: André Deutsch; E. Pintus

Presenting himself as an agnostic and a secular person, the author gives three main reasons for abandoning Jewishness: "the omnipotent Jewish Got had not protected his chosen people" (p. 270), transplantation into England and secular education at the Bunce Court School. Therefore, he does not extend his sense of belonging out of religion. It seems to be functioning as an element of the "collective of memory" ('Gedächtniskollektiv' due to Diner 2003, p. 198): "I believe that my Jewishness is strongly linked to my childhood experiences and the fate of my family in the Holocaust, which I continue to feel so acutely and painfully" (p. 274). Therefore, he affirms his Jewish identity be reconnecting with the home town and not by adopting Israel as the new home. His critical views of the State of Israel and its policies towards the Palestinians, manifested by his joining the group "Jews for Justice for Palestinians" and consciously refraining from visiting Israel, constitute a decisive emphasis of choosing to live in the diaspora.

The "Ich bin ein Koszaliner" of the title, should be understood as a rhetorical figure. The aim of this expression, a paraphrase from the 1963 West Berlin speech of J.F.Kennedy, is to establish contact with its addressees – the modern-day Koszalinians. It is a consciously interwoven and at the same time clearly manifested expression of empathy, as well as (just how consciously?) of slinging a bridge between own experience and that of present-day inhabitants of Koszalin. In the autobiographical narration we will not find any references or reflections confirming a firmly set attitude of this type.

Why should the modern Koszalinian theme constitute an important complement to his own life story? According to Stephen Bochner, consequences of the mentioned attitude of assimilation may take the form of intercultural conflicts (see Thomas: 2003, p. 282), including those of internal character. The role of an emblematic Jew in the reception of modern day Koszalinians, seems to allow Brent to fill the breach caused by leaving behind his Jewish culture which returns with force after his professional career ends. In this way it reconciles – through words – a life mutilated by the Holocaust. The confirmation of this thesis may be found in the act of writing his memoirs with the aim of establishing continuity and the adopted narrative construction itself which is opened and closed by an image-recollection of Koszalin. As Fani Oz-Salzberger and Amoz Oz pointed out in their essay *Jews and Words*, Jewish continuity has always depended on words, those being spoken out and those written down, being dependent on a growing labyrinth of interpretations, discussions and disputes (Oz and Oz-Salzberger 2014, p. 9). Yet another argument is his returning to the family name of Baruch in the early 1990s, which the author laconically describes as a reconnecting with the past (he inserted the name Baruch as his middle name).

(2005),: *Moje prawdziwe przeżycia / Meine wahren Erlebnisse*, ed. by J. Borzyszkowski, Gdańsk: Instytut Kaszubski; M. Ryczke Kimmelman (2005): *Life Beyond the Holocaust. Memories and Realities*, ed. and with introduction by G. G. Schmidt, Knoxville: The University of Tennessee Press; M. Ryczke Kimmelman (1996): *Echos from the Holocaust*. A Memoir. Knoxville: The University of Tennessee Press; E. Lichtenstein (1985): *Bericht an meine Familie. Ein Leben zwischen Danzig und Israel. Mit einem Nachwort von Günter Grass*, Darmstadt / Neuwied;: W. Kaelter & G. Cohn (1997): *From Danzig. An American Rabbi's Journey,* Malibu, Calif.: Pangloss Press.

There is not a single place where Brent – the narrator seems to be lost in time that is not his own or adrift in-between cultures.[7] The narrator does not emphasize the effort of reaching into hidden layers of memory containing buried traumas and records of amnesia/suppression, does not share any profound dilemmas of life's choices. It is quite intriguing that the art of recording experiences does not reflect the "reminiscence effect" – the returning of old memories, which can be commonly observed in old age (Draaisma 2010, p. 9). The autobiography also does not have the structure of a reminiscence, which forms when more attention is given to the early, initial periods of life (Draaisma, p. 86 et seq.). Such structure usually characterizes authors of ripe age. The reader has the impression that Brent's strategy is to describe episodes or opinions devoid of any emotional sharing with their reader.

Relations with today's Koszalin could be interpreted here rather as a result of strategy selected for synthesizing life-experience. Therefore, the project does not have any (postmodern) hybrid structure with third spaces (Bhabba 2004) but rather a carefully time-divided and segregated multi-contextual construction. Brent-the narrator presents his acculturations process as divided into contexts and periods, with each acculturation being represented by a different process model. His active professional life is marked rather by a single-dimensional conceptualisation of acculturation implied assimilation i.e. an immigrant who gradually gives up his identification with the culture of origin and moves towards identification with the culture of contact (Olmedo 1979). In this image home and host cultures are positioned as opposing rather than counterbalancing. For Brent, the breaking point seems to be the process of retiring from professional life combined with his first visit in the post-war Koszalin of 1989. This experience coincides with a start of his internal 'cultural mediation' (Bochner 1982), which includes Brent's Jewishness and evolves into an integrationist model of acculturation. This process allows for some sojourns aimed at synthesising more cultures and acquiring multicultural personalities.

13.7 Leslie Baruch Brent as the "Emblematic Jew" of Koszalin

The collective identity needs its guiding lights. Not only in Koszalin's local publications, Leslie Brent is consistently presented as the Koszalinian Noble Prize Winner, e.g. in the invitation to a meeting with the renowned immunologist held on 12th July 2005 held at the local theatre Teatr Propozycji Dialog or a Co-Winner of the Noble

[7]The text is not a representation of an attitude of deepened confession, one of the attitudes mentioned along withones of testimony and challenge in the concept of "autobiographical triangle" authored by the Polish literature theoretician Małgorzata Czermińska (Czermińska, 2000).

Prize (Pacholski 2006). This fact concerns also articles published in the collective volume on the history of Jews in Pomerania (Romanik 2007, p. 214).[8]

Nevertheless Brent's presence and Koszalin's present-day contacts with him constitute a catalyst for taking steps aimed at commemorating the pre-war Jewish community of Köslin – the Koszalinian Jewish memory revival for which the fall of communism was an important turning point. This process includes such local activities as tidying up the grounds of the Old Cementary in 2005, Marches of the Living, publications on the history of Jewish community by the local priest Henryk Romanik, engaged in the Christian-Jewish dialog in Poland (Romanik 2006), as well as commemoration and cleaning up of the grounds of the New Jewish Cementary (2006). This range of actions in which participated a variety of Koszalinian social circles drew wide attention by the regional and local media, as well as by British ones thanks to Brent's own articles. The majority of these activities have been initiated and supported also by Zdzisław (Zibi) Pacholski, an artist-photographer who has filled the role of a liaison between the local community and Leslie Brent. In his cultural engagement and in his art Pacholski regularly refers to the figure of Leslie Brent, for example in the exhibition *Tutaj – Here – Hier* presented in many places in Poland and abroad.

It seems that Leslie Baruch Brent constitutes a pretext for taking such steps, while also being a living *memento* – functioning in Koszalin as the emblematic Jew figure. His person bridges the past with the present consolidation of city's memory broken by the replacement of its population. His presence "certifies" the ongoing construction of the local landscape of memory referring to Jewish past without 'real' Jewish life. In an interview for the London Daily Mail after his first visit in Koszalin he said: "there is nothing there, not a trace of our presence remains, there are no Jewish cemeteries, I did not find any graves of my loved ones... only the woods around the city and at the seashore rustle in the wind as they did back then" [Pacholski, Posłowie].

Leslie Baruch Brent himself seems to accept this role. As he writes in his autobiography "I was happy to play an emblematic role in all this, even if the attention I received sometimes seemed well over the top" (p. 292). Brent reflects also the causes of assigning him to the role of the "emblematic Jew" by Koszalin's citizen as their confrontation with Polish anti-Semitic past:

> "It is legitimate to ask, why the Poles of Koszalin, who were in any way responsible for what happened in Köslin before the Second World War, have felt it appropriate to honour me in this way [referring to an official invitation by the City Mayor in 2005 – M.B.-S.] and to draw attention to the fact that the city had a small but thriving German-Jewish community before the war. I suspect that it has to do with a need on their part to confront their own anti-Semitic past and to reach out in a spirit of reconciliation." (p. 292)

[8] About Leslie Baruch Brents participation on world's first examples of acquired transplantation tolerance see for example Starzl T. E. (2006): Leslie Brent and the Mysterious German Surgeon, *Annals of Surgeory*, no. 244 (1), 154–157. See also chapter of Brents autobiography *From Intolerance (Racial) to Tolerance (Immunological)*, p. 108–136.

Present-day Koszalin is a medium-sized city located in the Pomeranian Province (population app. 109,000) and seems to be a town which is in the process of searching for a form of its collective identity story. Time will tell just how broad/integrating/pluralistic will this project turn out to be in combining diverse narratives about the city's past coming from different ethnic and cultural-social groups and whether it will also account for the non-simultaneousness of their respective experiences. The concept of the "times of memory" (Gedächtniszeiten) of Dan Diner contains an imagination of simultaneity of multiple narratives about the past (Diner 2003, p. 8.) It is not just about providing for the fact of Jewish presence in the memory budget of the local community "as part of nostalgic vision of colourful 'multiculturalism'" (Kapralski 2015, p. 149) or ethical response to the Jedwabne-debate following the publishing in the year 2000 of the Polish version of Jan Gross' controversially received book "Neighbours" (Michlic and Polonsky 2004). The goal seems rather to work out a heterogeneous memory perspective of a City populated by representatives of culturally diverse regions of pre-war Poland which have formed the post-war population of Koszalin – including ethnic Ukrainians. Similarly to other regions of Poland, the 'Jewish memory project' in Koszalin is a part of the local struggle for democracy. "Jewish heritage is both a vehicle to reclaim aspects of a prewar history of multiethnic coexistence across Polish terrain, as well as to envision a future Poland that may stand among peers in an increasingly multiethnic European Union" (Lehrer 2015, p. 172). No doubt, the dynamics of this process and its results depend on the vision, balance of power and participation of local authorities, activists and opinion-makers. The socio-political context – the *Zeitgeist* plays a role too.

One has to underline that the original addressee – the target audience of Leslie Baruch Brent – consisted of British, English speaking readers, hence the emphasis, for instance in the introduction, on the merit of immigrants and entwining/connecting his own fate with a series of immigrant biographies, e.g. for example the African writer and human rights fighter Olaudah Equiano, who arrived in England in the seventeenth century as a slave. Brent refers to his autobiography in the *Preface* to his story. Hence, the life history takes on the form of a *success story*, painting an image of a fully successful assimilation of a German-Jewish child from Kindertransport. We will not find in this narration any references to a portrait gallery of Jewish-German symbiosis. Through his autobiography Leslie Brent puts his personal life-story into the historical and social context of – first and foremost – his "adopted country" (p. 274) – the United Kingdom.

Jewishness is presented as a cultural tradition in which he had grown up in, until the moment of leaving Germany, it is clad in memories of an idyllic childhood, yet also burdened by the experience of stigmatization and exclusion from the German community. At the same time, as recorded in the memoir writings from the initial post-war visit to the city on the verge of democratic changes, his dominating impression was the total removal/erasing of Jewish presence from Koszalin's memory. The act of recording his memories was also fittingly motivated by establishing a continuity of memory about the Jews of Koszalin. His time-fluctuating identity project seems to be a kind of response to the dominant perception of our times – an increasingly liquid and heterogeneous world (Bauman 2000).

Neither Brent's autobiography nor his travel journal have a postcolonial attitude in the recognition of his place of childhood which usually dominates travel writings of the so-called German "homesick-tourists" revisiting post-WWII Polish Pomerania (Borzyszkowska-Szewczyk 2004, 2009). The narrator does not betray objections by using Polish and German topographical names. His Polish dialogue partners are presented by first and family names, and not as nameless representatives of a nation with an assigned set of stereotypes. The acquaintances and friendships made in Koszalin are maintained in subsequent years. The post-WWII status of the town as part of the Polish State is clearly interpreted as a an act of historical justice in his first visit to Koszalin in May 1989. As he points out in his travel journal: "I would not have come here if Koszalin would still be German – this is absolutely out of the question and impossible in every and each way" (Brent 1989, p. 56).

References

Adamczyk-Garbowska, M., & Kopciowski, A. (2011). Memorial books as a remembrance of collective trauma. In F. Tych & Adamczyk-Garbowska (Eds.), *Jewish presence in absence. The aftermath of the holocaust in Poland, 1944–2010* (pp. 503–540). Jerusalem: Yad Vashem.

Batchelor, R. (2009, May). Sunday's child? A memoir? *Special book review: Immunology News.*

Bauman, Z. (2000). *Liquid modernity*. Cambridge: Polity.

Benz, W., Curio, C., & Hammel, A. (2003). Die Kindertransporte 1938/39. In *Rettung und Integration*. Frankfurt a.M.: Fischer Taschenbuch Verlag.

Bhabha, H. K. (2004). *The location of culture*. Abingdon: Routledge.

Billingham, R. E., Brent, L., & Medawar, P. B. (1953). Actively acquired tolerance' of foreign cells. *Nature, 172*, 603–606.

Bochner, S. (1982). The social psychology of cross-cultural relations. In S. Bochner (Ed.), *Cultures in contact* (pp. 5–42). Oxford: Pergamon.

Borzyszkowska-Szewczyk, M. (2004). Erinnerung und Gegenwart. Reiseberichte aus Pommern in der Pommerschen Zeitung (1989–2001). In D. Albrecht, B. Neumann, & A. Talarczyk (Eds.), *Literatur, Grenze, Erinnerungsraum* (pp. 163–180). Würzburg: Königshausen u. Neumann.

Borzyszkowska-Szewczyk, M. (2009). Wanderungen zwischen Gestern und Heute (gen Morgen): Wiederbegegnungen mit der "alten" Heimat in den Reiseberichten des deutschen Adels aus Hinterpommern und Ostpreußen nach 1945. In N. Donig, S. Flegel, & S. Scholl-Schneider (Eds.), *Heimat als Erfahrung und Entwurf*. Münster: LIT (= Gesellschaft und Kultur: Neue Bochumer Beiträge und Studien, ed. by P. G. Klussmann at. al., vol. 7), pp. 147–161.

Borzyszkowska-Szewczyk, M. (2015). Die Abwesenheit des Anwesenden. Das Bild Danzigs in Alfred Döblins "Reise in Polen" *versus* in der Autobiographik von Danziger Autoren jüdischer Herkunft. In M. Brandt & G. Kwiecińska (Eds.), *Jahrbuch für Internationale Germanistik, Reihe A – Band 121: Internationales Alfred-Döblin-Kolloquium. Interkulturelle Aspekte im Schaffen Alfred Döblins* (pp. 67–90). Bern [et al.]: Peter Lang.

Brent, L. B. (1989). *Köslin/Koszalin – eine sentimentale Reise.* Manuscript of German translation of travel diary hold by Leslie Baruch Brent during the first visit in Koszalin, kindly provided by the Author, archive of Mirosława Borzyszkowska-Szewczyk.

Brent, L. B. (2005, December). How I came to believe in miracles. *Journal of the Association of Jewish Refugees*.

Brent, L. B. (2006, July). More miracles of Poland. *Journal of the Association of Jewish Refugees*.

Brent, L. B. (2009). *Sunday's child? A memoir*. London: Bank House Book.

Brent, L. B. (2010). *Ein Sonntagskind? Vom jüdischen Waisenhaus zum weltbekannten Immunologen*. Berlin: Berliner Wissenschaftsverlag.

Chodakiewicz, M. J. (2003). *After the Holocaust: Polish-Jewish relations in the wake of World War II*, East European Monographs./Polish edition: (2008): *Po Zagładzie. Stosunki polsko-żydowskie 1944–1947*, transl. by Anna Madej. Warszawa.

Czermińska, M. (2000). *Autobiograficzny trójkąt. Świadectwo, wyznanie i wyzwanie*. Kraków: Universitas.

Diner, D. (2003). *Gedächtniszeiten. Über jüdische und andere Geschichten*. München: C K. Beck.

Draaisma, D. (2010). *Fabryka nostalgii. O fenomenie pamięci wieku dojrzałego* (E. Jusewicz-Kalter, Trans.). Wołowiec: Czarne.

Fleischkacker, A. (2010, January/Feburary). Ein Sonntagskind? Vom jüdischen Flüchtlingsjungen zum weltbekannten Immunologen. *Antifa*.

Grzebałkowska, M. (2015). *1945. Wojna i pokój*. Warszawa: Agora S.A.

Gusdorf, G. (2009). Warunki i ograniczenia autobiografii. In M. Czermińska (Ed.), *Autobiografia* (pp. 9–46). Gdańsk: Słowo/obraz/terytoria.

Jaroszewicz, M., & Stępiński, W. (Eds.). (2007). *Żydzi oraz ich sąsiedzi na Pomorzu Zachodnim w XIX i XX wieku* [Jews and their neighbours in Pomorania in 19 and 20. Century]. Warszawa: Wydawnictwo DiG.

Kapralski, S. (2015). Amnesia, nostalgia and reconstruction. Shifting modes of memory in Poland Jewish' spaces. In E. Lehrer & M. Meng (Eds.), *Jewish spaces in contemporary Poland* (pp. 149–169). Bloomington/Indianapolis: Indiana University Press.

Kokot, F. (2009). Tadeusz Orłowski – The founder of polish nephrology. Memorial article. *Polish Archive of Internal Medicine, 119*, 285–286.

Lammel, I. (2009, November 10). Mit viel Glück überlebt. Warum Leslie Baruch Brent Schuldgefühle plagen. *Die Seite Drei*.

Lehrer, E. (2015). Jewish heritage. Pluralism, and Milieux de Mémoire. The case of Krakóws Kazimierz. In E. Lehrer & M. Meng (Eds.), *Jewish spaces in contemporary Poland* (pp. 170–192). Bloomington/Indianapolis: Indiana University Press.

Lejeune, P. (2001). In R. Lubaś-Bartoszyńska (Ed.), *O autobiografii. Wariacje na temat pewnego paktu*. Kraków: Universitas.

Loew, P. O., Pletzing, C., & Serrer, T. (2006). Zwischen Enteignung und Aneignung: Geschichte und Geschichte in den "Zwischenräumen Mitteleuropas". In P. O. Loew, C. Pletzing, & T. Serrer (Eds.), *Wiedergewonnene Geschichte*. Wiesbaden: Harrasowitz.

Michlic, J. B., & Polonsky, A. (2004). *The neighbours respond: The controversy over the Jedwabne massacre in Poland*. Princeton: Princeton University Press.

Monaco, A. P. (2009). Sunday's child? A memoir by Leslie Baruch Brent. *The New England Journal of Medicine, 361*, 317–318.

Morris, J., & Morris, P. J. (2010). Sunday's child? A memoir by Leslie Baruch Brent. *Transplantation, 3*(89), 267–269.

Olmedo, E. (1979). Acculturation: A psychometric perspective. *American Pychologist, 34*(11), 1061–1070.

Oz, A., & Oz-Salzberger, F. (2014). *Żydzi i słowa*. Warszawa: Czytelnik/English edition: (2012): *Jews and Words*. New Haven: Yale University Press.

Pacholski, Z. (2006). Leslie Brenta powrót do domu. *Rocznik Koszaliński, 33*, 135–138.

Pacholski, Z. (2012). *Posłowie* (manuscript of the epiloque prepared for the planed Polish edition of Leslie Baruch Brents autobiography), in personal archive of M.B.-S.

Romanik, H. (2006). Szkice do portretu Żydów koszalińskich: Jakuba Salomona Borchardta i Leslie Barucha Brenta. In M. Jaroszewicz & W. Stępiński (Eds.), *Żydzi oraz ich sąsiedzi na Pomorzu Zachodnim w XIX i XX wieku* (pp. 193–215). Wydawnictwo DiG: Warszawa.

Romanik, H. (2007). *O Żydach w Koszalinie. Szkice do portretu zbiorowego*. Koszalin: PerMedia.

Salinger, G. (2006). *Zur Erinnerung und zum Gedenken. Die einstigen jüdischen Gemeinden Pommerns. Teilband 2: T. III A-M*. New York.

Shalgross, B. (2007). *Rzeczy i Zagłada*. Kraków: Universitas.

Thomas, A. (2003). *Lernen und interkulturelles Lernen*. In A. Wierlacher & A. Bogner (Eds.), *Handbuch Interkulturelle Germanistik* (pp. 276–287). Stuttgart: J.B. Metzler.

Trede, M. (2009). *Ein Sonntagskind? Vom jüdischen Waisenhaus zum weltbekannten Immunologen*. "Deutsche Gesellschaft für Chirurgie. Mitteilungen" 4/09.

Part IV
Jewish Doctors in the Face of Terror and Extermination

Chapter 14
Jewish Doctors: A Place in Holocaust History

Ross W. Halpin

> *Perhaps it was not even normal the way this woman clung to life. As a physician I often admired her indomitable life-instinct – at a place where the sky was always aglow from burning human bodies; where life had so little value and security; where everyone yearned for death, this woman kept on being so frantically attached to life.*
>
> (Perl 1948, p. 101)

Abstract Since the end of the Holocaust the inexplicable actions of SS doctors and the interest in Nazi medicine in general have become a recognized part of Holocaust history. Although research has been conducted on the work of Jewish doctors in ghettoes, a facet of the Holocaust that until recently has been ignored is the history of the life and work of European Jewish doctors leading up to the Holocaust from the beginning of the twentieth century and the role they played and sacrifices made during the Holocaust in the concentration, labor and sub-labor camps. Jewish doctors were the antithesis of Nazi doctors and deserve to be recognized for their service to humanity in a time and under conditions of extreme adversity. The SS doctors turned from healers to killers based on an ideology of scientific racism. The reasons for the decisions and actions of the Jewish doctors during the Holocaust were far more complicated and were multi-faceted. They were influenced by their history, socialization, culture, religion, education and years of rejection and discrimination. This chapter is a brief history of the time leading up to and including the life and work of Jewish doctors during the Holocaust.

Keywords Jewish doctors · Antisemitism · German doctors · Medicine · Holocaust · Survival · Choiceless choices · Human condition · National Socialism · Law for the reconstruction of the civil services · *Halakha*

R. W. Halpin (✉)
University of Sydney, Sydney, NSW, Australia
e-mail: rhal3993@uni.sydney.edu.au

M. Moskalewicz et al. (eds.), *Jewish Medicine and Healthcare in Central Eastern Europe*, Religion, Spirituality and Health: A Social Scientific Approach 3,
https://doi.org/10.1007/978-3-319-92480-9_14

14.1 Introduction

Dr. Giselle Perl was born on the December 10, 1892 in Sighet, Transylvania. Her family was observant orthodox Jews. Her father was a successful businessman providing Giselle and her siblings with the comforts of life. In 1920 Giselle, by then a paediatrician, and Ephraim Kraus, a doctor, were married and by 1928 were parents to Ella and Imre. According to Giselle life for the Kraus family was happy and peaceful despite the occasional inference of antisemitism. They enjoyed their success living in a beautiful house once owned by Cornelia Priel, the great Hungarian artist.

By April 1944, Perl and her family, including her mother and father and extended family had lost all possessions, taken by the State and ironically neighbours and from a local transit camp began the infamous train journey of 8 days to Auschwitz:

> … the small children cried with hunger and cold, the old people moaned for help, some went insane, others gave life to their babies there on the dirty floor, some died and their bodies travelled with us… (Perl 1984, p. 32)

What happened to Perl is representative of the dramatic transformation in life from one of relative peace, freedom, success and hope to that of degradation, suffering, despair, hopelessness and unbearable loss. To the Nazis they became sub-humans or nonpersons. Paradoxically in the camps the doctors lived a dual life as a doctor attempting to uphold a commitment to do no harm and protect the patient and at the same time they had an innate drive to survive. Nevertheless, memoirs, diaries and oral testimonies are witness to the impact Jewish culture, history, religion and personal traits had in their attempts to honour the Hippocratic oath. But at times this was impossible.

The history of Jewish doctors during the Holocaust has not been researched or considered of importance compared to other groups, places and events of that period. This has been borne out by the imbalance of research and literature between that of Nazi medicine and SS doctors and Jewish doctors during the Holocaust. Admittedly this imbalance is being corrected however much more work needs to be carried out.

As historians we are obliged to ask questions and provide answers that will add to the body of knowledge of the life and work of the Jewish doctors in the ghettoes and camps. Did their early life experiences and culture prepare them for the hell they were to enter? Did the doctors face ethical dilemmas or choiceless choices? Did faith influence decisions and choices? Under the inhumane conditions and relentless pain, suffering and the constant threat of death what motivated them to keep going? Was it the inherent drive to survive or the commitment to the Hippocratic oath? Were they prepared to sacrifice their own life? Were they healers or killers? Or both? Did their status provide them with the means to survive longer than ordinary prisoners? Most of the doctors who survived were in the age range of 40–60 years and not young. Why and how? A large proportion of doctors who survived were women. Why? How and why did doctors survive an average of 18 months while at the same time the normal prisoner lasted approximately 12 weeks? Was it luck or a combination of factors? As humans trying to survive were the doctors prepared to act immorally to survive? Did Jewish doctors become important even essential to the continuing operation of the camps? If so? Why and how? Primo Levi wrote that those who survived were the worst of the prisoners. Did the doctor fit into this category? Where did the doctors lie in Levi's theory of the *Grey Zone*?[1] What lessons can be learned from the doctors' actions, behaviour and sacrifices? The answers to these questions go to the very heart of the role of the doctors and their place in Holocaust history. Their behaviour and actions will show they were ordinary people who faced extraordinary challenges, challenges that often presented a struggle between *crise de conscience* and survival.

At the beginning of the twentieth century Jewish doctors were part of the fabric of the medical profession in Europe. In the tradition of Jewish culture medicine was considered the most honourable of professions. According to John Efron, "few occupations are as immediately linked to a group as medicine is to the Jews ... and ... as a people, the Jews have enjoyed an intimate and deeply symbiotic relationship with medicine." (Efron 2001, p. 234) Irrespective of the Jewish doctors' success they were still outsiders, which was made clear from their student days when many were physically abused, bullied and humiliated. Despite discrimination and ostracism whilst students they gained respect and success as doctors that became evident by their dominance in their fields of medicine and demand for their services by both Jews and non-Jewish patients.

Although there was antisemitism in Germany, the genesis of the problem between German and Jewish doctors was basically financial. After the end of WWI when German and Jewish doctors returned home, Jewish doctors either opened new practices or found work more easily than German doctors. This caused financial stress for the German doctors and their families causing bitterness and resentment that in turn was to play into Hitler's hands in implementing his euthanasia, sterilization and eventual genocide program. Hitler promised the German doctors that if elected he would rid Germany of Jews, including of course Jewish doctors, thus guaranteeing German doctors' financial security. Once Hitler was elected he honoured his promise. Hence the German doctors were bound to enter into a Faustian pact with Hitler wherein the price of financial security and social standing was the commitment to implement Hitler's program of euthanasia and sterilization to cleanse and purify German blood. The SS doctor was also instrumental in implementing the gassing program that murdered millions of Jews, POWs, gypsies, homosexuals and disabled people.

[1] The concept of the grey zone is essentially a metaphor for moral ambiguity, a conceptual realm that Levi characterises as having ill-defined outlines which both separate and join the two camps of masters and servants. It possesses an incredibly complicated internal structure, and contains within itself enough to confuse our need to judge (Rosenberg 2005, p. 9).

14.2 Jewish Doctors: Culture and Tradition

Many Jewish doctors prior to National Socialism had assimilated into their respective social groups and considered themselves citizens of the countries in which they and their families were born and lived. However some, including doctors, converted to Catholicism or non-Jewish religions. For some, it was an attempt to escape from violent antisemitism; for others a calculated move to find a way into areas of employment not open to avowed Jews (Roland 1992, p. 30). On the other hand, as Roland contends, there were a middle ground held by the assimilationists who considered themselves good Poles or even Polish first and Jews second. This is rather insignificant as there were fewer than 1600 Christian converts in the Warsaw ghetto (Roland 1992, p. 32) in a population of approximately 450,000. On those numbers it could be assumed in the total population before the Holocaust conversion from Judaism to Christianity was inconsequential.

Memoirs, oral testimonies, diaries and biographies suggest that Jewish communities were close knit valuing relationships, family ties, education, religious observance and the adherence to and respect for the laws of the state. They were dependent on strong relationships and bonds to their families and their communities. There was emphasis on following specific traditions that encouraged the observance of Jewish life and law by following customs such as living within walking distance of a synagogue or *mikva* (ritual bath) and close to their extended family for the purpose of gatherings on the Sabbath, holy days and festivals. Most children from observant families were required to devote time each day, sometimes hours, to studying the Torah. There was a strong commitment to keeping the family together – for the young and aged to live together and for children to care for their parents and other members of the extended family – and for observant Jews to participate in Jewish organizations and associations, particularly ones that extended help to those in need (Vaughan 1994, p. 2).

In terms of the history of Jewish doctors two common threads pass through four distinct periods. The first thread is the cornerstone of the Jewish attitude towards medicine: "choose life – if you and your offspring would live" (Deuteronomy 30.19 Tanakh[2]) and the second thread the successes Jewish doctors achieved prior to the mid 1930s juxtaposed to the difficulties experienced and the final hell the doctors encountered after that period. There was four distinct periods between 1900 and 1945 with the first ranging from 1900 to 1920. This latter period was one in which the Jewish people's lives and the Jewish doctors' lives and their fate began to evolve from the hard won gains of emancipation back to the isolation and exclusion of the

[2] The primary Jewish sacred text is the Tanakh, whose name is an acronym of Torah, Nebi'im and Ketuvim (Law, Prophets and Writings). The Tanakh consists of the same books as the Christian Old Testament, although in a slightly different order and with other minor differences. The Tanakh should not be referred to as the "Old Testament" in the context of Judaism, however, as the term implies acceptance of the "New Testament." The Tanakh contains 39 books in all (if each of the Twelve Minor Prophets is counted as one and the subdivided books such as Kings and Chronicles are counted as two): 5 books of the Torah, 21 books of the Prophets, 13 books of Writings.

past; the second period from 1920 to 1933 when Jews including Jewish doctors were subject to vitriolic abuse and discrimination and it became obvious that National Socialism and antisemitism were sweeping the country; the third period from 1933 to 1939 was a period of loss, when Hitler's promise to rid Germany of Jewish doctors became a reality. The final phase was from 1939 to 1945 when Hitler and the German nation while waging total war to gain world dominance attempted to carry out genocide of the Jewish people.

14.3 1933–1939 – Hitler Fulfills His Promise to the German Doctors

Antisemitism existed in Germany before Hitler became Chancellor yet it was relatively mild compared to that suffered by Jews in neighboring countries such as Poland, Hungary, France, and Russia. Nevertheless, the pervading atmosphere of blatant discrimination created by the imminent coming to power of an antisemitic National Socialist regime gave local councils and municipalities, universities and other institutions within Germany, and even in other countries, license to begin discriminatory actions against Jews. The new government no longer adhered to legal norms, which had offered the Jews some measure of protection in the preceding years. Once the Law for the Reconstruction of the Civil Service[3] was introduced in 1933, even tenured civil servants of non-Aryan descent were legally dismissed. Local government officials in Berlin and Munich, Julius Lippert and Karl Fiehler respectively, of their own accord broke legitimate contracts when they dismissed all Jewish public health and welfare physicians (Kater 1989, p. 185.) In some German cities such as Hamburg the welfare system appears to have operated according to the tenure of National Socialism from the very beginning. The governments granting the license to discriminate against Jews and other minority groups was enthusiastically accepted and finally introduced by local authorities.

German universities, such as Göttingen and Berlin, in response to the new racial and discriminatory policies, not laws, sacked Jewish academics including those in the medical faculties. The actions of the universities and similar institutions and local authorities were not dictated by law nor did institutions seek approval from the National Socialist party. Kurt Klare, co-founder of the Nazi Physicians' League, wrote to a colleague Dr. Scheidegg, "that Jews and philosemites (sic) ought to take note of the fact that Germans are masters of their own house once more and will

[3] The Law for the Restoration of the Professional Civil Service (Gesetz zur Wiederherstellung des Berufsbeamtentums, shortened to Berufsbeamtengesetz), also known as Civil Service Law, Civil Service Restoration Act, and Law to Re-establish the Civil Service, was presented on 7 April 1933, immediately after Hitler became Chancellor. The law gave the government the power to dismiss tenured civil servants including opponents of the Nazi regime, undesirables and civil servants who were not of "Aryan descent". This meant Jews and political opponents would be forced to retire or would be dismissed. This included lawyers, teachers, academics, judges and all public servants.

control their own destiny." (Kater 1989, p. 183). Dr. Conti, also a co-founder of the NPL, proffered that it was reasonable and not surprising that the German doctors who had suffered for so long at the hands of Jewish doctors were now asserting themselves as the legitimate healers of the German people. Hitler effectively orchestrated the gradual disenfranchisement of the Jewish doctor in German medicine. He appointed Gerhard Wagner to oversee the reorganization and restructure of the German medical system and it was Wagner aided by Carl Haedenkamp and Alfons Stauder who initiated the process for dismissing Jewish functionaries from the national medical associations as well as regional and local groups (Kater 1989, p. 183).

Despite the early rush in 1933 to victimize and discriminate against Jewish doctors, their expulsion did not take place overnight. The process evolved over a five-year period through a combination of propaganda, intimidation and official laws. From May 17, 1934 Jewish doctors were no longer allowed to be members of and practice within Germany's state-supported health insurance program. This was, in fact, a sequel to the Law for the Reconstruction of the Civil Services of April 7, 1933 that effectively removed all Jewish physicians (*Amtsärzte*) from the civil service. During 1934 and 1935 pharmacists and dentists were targeted and had to show proof of Aryan descent. Legislation began to have a far-reaching and permanent impact on the Jewish physician when the Reich Physicians' Ordinance of December 13, 1935 was introduced. Gentiles were forbidden to be treated by Jewish doctors or to attend a Jewish hospital. Jewish doctors were forbidden to call themselves "physicians" and only allowed to use the more degrading term, "sick treaters" (*Krankenbehändler*). This policy applied in all Nazi occupied territories. In Germany, some Jewish doctors, such as *Mischling*[4] who were WWI war veterans were initially protected from the full force of Nazi persecution and discrimination. Inevitably this lenient policy came to a swift end. This progressive erosion of the recognition of Jewish doctors in the Reich reached its conclusion with the legislation of the Fourth Ordinance of the Citizen Law, enacted on July 25, 1938, which revoked the medical licenses of all Jewish physicians thus preventing them from practicing medicine in Germany.

Antisemitism and the subsequent damning effect on Jewish medicine in Poland was at a much higher level than in Germany or Austria. Joseph Tenenbaum argues that in the post-Great-War period a new Poland was borne on a wave of pogroms where Jew-baiting, Jew-beating, beard-cutting and economic discrimination was endemic. (Tenenbaum 1963, p. 144) According to Dr. Edmund Goldenberg, a survivor of Auschwitz, born in Terezin, Czechoslovakia, but educated in Poland, antisemitism was unrestrained and widespread throughout Poland. During the 1930s when Hitler came to power many European countries such as Poland took a keen interest in the developments of the Jewish situation in Germany under National Socialism and were attempting to implement similar policies with similar outcomes

[4]Mischling – A person with two Jewish grandparents was considered to be either a Jew or a Mischling of the first degree. A person with only one Jewish grandparent is considered to be a Mischling of the second degree.

to temper the dominance of Jews particularly the Jewish doctors similar to the situation in Germany. The pre-Holocaust toxic relationship between the Jewish and Polish doctors was a forerunner to what was to occur in camps. Jewish doctors not only feared the Nazi SS doctors, they also were subject to bullying, humiliation and sometimes beatings by non-Jewish prisoners and doctors, particularly Polish doctors.

14.4 The Ghettoes

By the early 1940s the camps were well established however ghettoes and transit camps were still in use as holding stations. With overcrowding and shocking conditions hundreds of thousands died as a result of epidemics, starvation and the elements. The ghettoes introduced the Jewish doctors to conditions that were beyond their comprehension. Epidemics of illnesses foreign to many doctors, particularly young graduates were rampant. Medicine and equipment were almost non-existent, and many doctors were not qualified in the fields of medicine most needed. Starvation was at the core of almost every illness. The quantity and quality of food and water was so poor as to cause massive weight loss and emaciation, which led to the breakdown of the immune system, which in turn led to epidemics and chronic diarrhea. The population of the Warsaw ghetto was approximately five hundred thousand Jews and the Lodz ghetto had a population of approximately three hundred and twenty thousand. According to Dr. Adina Blady Szwajger, who commenced work as a nurse in the Jewish Children's Hospital in the Warsaw ghetto, the children and elderly were the first affected by the conditions presenting impossible challenges to the doctors "... babies, emaciated with hunger, died of a "gluteal infection" which struck them down one by one. No treatment or medication helped; nor did special nourishment or dressings of the trophic skin changes. They all died quickly, practically without crying" (Szwajger 1990, p. 29).

Both the Jewish doctor and the hospital were revered in the ghetto, the former as a healer and protector and the latter as the nerve center of the ghetto. Unlike the hospitals in the concentration camps which were considered transit houses to the gas chambers, the ghetto hospital was the *hekdesh*, a refuge for the sick, the destitute and the old.

Nonetheless as time passed the conditions became worse, leading to increased starvation and epidemics and obviously higher death rates. According to Szwajger,

> On the streets, there were more and more ragged paupers begging for a crust of bread. In the hospitals more and more flea-ridden, lice-infested, fungus-diseased children. More children emaciated from hunger with the eyes of adults; more and more tuberculosis. (Szwajger 1990, p. 30)

Dr Ludwik Hirszfeld, a serologist, of Jewish origin, (he had converted to Catholicism) who served on the Health Council of the Warsaw ghetto, was convinced that the sanitary conditions in the ghetto were an intentional prelude to the murderous intentions of the Nazis "...when one concentrates 400,000 wretches in one district, takes everything away from them, and gives them nothing, then one creates typhus." (Weindling 2000, p. 280.) It is within this context that Jewish doctors with an overwhelming number of sick and undernourished patients and virtually no resources were required to work.

In addition to their role as healers and carers, the Jewish medical community provided services that were indicative of their selflessness and dedication to humanity. In the Warsaw ghetto research was conducted and medical education was provided in the form of the continuation of university studies. Apart from the obvious value, these clandestine activities were a form of spiritual resistance. The governor of the *Generalgouvernement*[5] Hans Frank, predicted unequivocally that the Nazis had condemned 1,200,000 Jews to death by starvation (Roland 1992, p. 101). In the face of malnutrition and starvation as the primary infirmity, the medical community of the Warsaw ghetto saw the opportunity to carry out research on the disease. As Dr. Izrael Milejkowski, one of the originators of the hunger disease project explains:

> Hunger was the most important factor of everyday life within the wall of Warsaw ghetto. Its symptoms consisted of crowds of beggars and corpses often lying in the streets covered with newspapers. Mortality data on hunger and its two companions, tuberculosis and typhus, were collected from orphanages and refugee centres and from specific hospital material. (Winick (ed.) 1997, p. 40)

The results of this study, smuggled out of the ghetto, are still used as a reference source today. Further experiments of great value were carried out on other diseases, such as typhus.

14.5 1939–1945 – A Life in Hell

Dr Sima Vaisman, a survivor of Auschwitz, described her trip and arrival at Auschwitz.

> …the doors to the car are sealed; we settle ourselves in the dark as best we can and the journey to the unknown begins… Auschwitz, land of death… (Vaisman 2005, p. 28)

She expressed her feelings later in her memoirs when she felt so helpless watching women and their children entering the gas chambers, she said she "…fell into the heart of hell…" (Vaisman 2005 p. 52).

[5] Political administrative entity comprising the central part of Poland occupied by Nazi Germany in 1939–1945 but not incorporated directly into the Third Reich. The capital of the Generalgouvernement was Cracow. It was created by Adolf Hitler's declaration of 12 October 1939 and was essentially a German colony with a totalitarian regime and only minimal rights for the local population. The governor was Hans Frank. The Nazis persecuted the Poles and sent hundreds of thousands to Germany to work in the domestic and war industries as forced laborers. They repossessed the homes and properties of the Polish people that were given to German families who migrated to the newly occupied land. The area of 95,000 square kilometres composed the cities of Kielce, Krakow and Lublin.

The final phase of the history of the Jewish doctors took place from 1939 until the end of the Holocaust in 1945. The story of their life and work in the camps is absorbing and confronting for the reader because of the raw evil and inhumanity of the SS. It was a tragedy for the Jewish doctors because the price of their survival was either directly or indirectly the abandonment of the Hippocratic oath no matter how one looked at the matter. Due to circumstances beyond their control whatever decision they made was eventually at the cost of a life. As privileged prisoners they became members of what Primo Levi called the *gray zone:*

> with undefined contours, which both separates and connects the two opposing camps of masters and servants. It has an incredibly complicated internal structure, and harbors just enough to confound our need to judge. (Levi 1988, p. 27)

The majority of the doctors lived with prisoners in Birkenau or were assigned to sub-labor camps. They were regarded as "privileged" prisoners, a status that along with "personal traits" and "defense mechanisms" allowed the doctors the opportunity to live longer or survive the camp itself (Halpin 2014). Slightly larger portions of food, better quality clothing for protection against the elements and better sleeping arrangements, combined with personal traits of resilience, hardiness, strong self-esteem and defense mechanisms that provided respite and protection from the inhumanity that if worked in unionism could with luck save the life of the prisoner.

However, there was a dark side to these privileges. This included participation in selections mainly in the hospital and infirmaries. The Jewish doctors were regularly ordered to identify prisoners who could not return to work or under the instructions of an SS doctor were ordered to select a specific number of sick Jewish patients all of whom would be executed either by phenol injection or sent to the gas chamber. For the most innocuous reasons such as a patient making a noise, a doctor may be ordered to silence the prisoner. Dutch Jewish doctor, Elie Cohen, faced a dilemma when he was told to subdue a disruptive patient or he, Cohen, and all the patients in the ward would be executed. To save himself and the patients he chose to kill the disruptive patient. According to Lawrence Langer he made a choiceless choice rather than faced an ethical dilemma. To stay alive Cohen was forced to repeat killing in this manner, confessing that he suffered no ethical dilemma in repeatedly killing noisy patients, "Same old thing. We'll have to do it again. And we did too, and that man died as well." (Cohen 1973, p. 88). Due to the critical shortage of medication, doctors were constantly forced to make the decision of who would receive medication and who would not, thus, in effect the doctor became the executioner.

In each case the doctor was abandoning the Hippocratic oath of "do no harm", their personal principles of right and wrong and, most sacred of all, by not observing Jewish religious practices they were abandoning their Jewish religion and *Halakha,*[6] Jewish religious law. In the labor camps Saturday was designated a work

[6]The entire body of Jewish lawand tradition comprising the laws of the Bible, the oral law as transcribed in the legal portion of theTalmud, and subsequent legal codes amending or modifying traditional precepts to conform to contemporary conditions.

day despite the fact that according to the Jewish religion Jews are not allowed to work on the Sabbath. In hospitals and infirmaries in the extermination camps selections were often carried out on the Sabbath and major *"actions"*[7] were carried out on Holy days. The SS purposely chose Holy days to carry out atrocities against the Jews to humiliate them and inflict pain and shame because they as Jews were breaking sacred tenets of Jewish law.

Jewish tradition is fundamentally time-aligned such that time dominates both *Halakha* and life. This was a significant problem in the camps because of the difficulty in knowing the day of the week and the time of day. On a specific day, for example the Sabbath, certain rituals must be performed at given times. Many commandments could only be fulfilled at specific times such as reciting the *shema*.[8] Despite the threat of death some prisoners managed to preserve some Jewish traditions. Gad Goldman (Eliach 1983, p. 197) recalls that in his block in Auschwitz, someone "organized" (stole) a pair of *tefillin*.[9] At four o'clock in the morning prisoners would line up to put them on. Herman Weiss was in Buna (Auschwitz III) from May 1944 until January 18, 1945. During his 9-month period, he was privileged to use the *tefillin* once (Eliach 1983, p. 197). In most memoirs the Jewish doctors didn't comment on religious beliefs or matters regarding traditions and rituals.

14.6 Conclusion

At the beginning of the twentieth century in Western Europe, Jewish doctors along with many Jews, had assimilated into the general society. In Eastern Europe, such as Poland and Russia, assimilation was not as common. Emancipation initially gave the doctors opportunities to enrol in medical schools and they quickly played a dominant role in many fields of medicine and leaders in research receiving appropriate international recognition. Paradoxically, as the doctors became more successful, professionally and financially, open criticism and discrimination by non-Jewish doctors and the medical associations increased. The rise of Hitler and National Socialism assisted by the encouragement of the non-Jewish medical community began the gradual erosion of the rights and reputation of Jewish doctors.

The role of the Jewish doctor in the camps and ghettoes has a distinct and important place in Holocaust history. It begins after a long hard struggle for some recogni-

[7] "Actions" were a euphemism for a mass killing on any one occasion, for example, the killing of a number of pregnant women at any one time, the murder of gypsies or homosexuals or at the time of selections.

[8] *Shema* ("hear") is the Hebrew word that begins the most important prayer in Judaism. Hear, Israel, the Lord is our God, the Lord is One. *Sh'ma Yisra'eil Adonai Eloheinu Adonai echad.*

[9] *Tefillin* (sometimes called phylacteries) are cubic black leather boxes with leather straps that Orthodox Jewish men wear on their head and their arm during weekday morning prayer. Observant Jews consider wearing *Tefillin* to be a very great *mitzvah* (command).

tion and equality albeit grudgingly given by the rulers and non-Jews of Europe, but still with caveats. However the ingrain emotions of discrimination, antisemitism and hatred were unleashed under the license of National Socialism and the ascendancy of Hitler into absolute power. As time progressed life for the Jews became one of struggle, persecution, loss and despair. It then became a classic and unique story of how to survive extreme adversity by ordinary people. It is a source of information about the human condition when people are faced with ethical dilemmas in a world in which ethics and morality do not exist and what occurs when dedicated doctors are faced with choiceless choices that force otherwise good decent people to commit unmentionable acts to stay alive. It provides valuable data of the importance of specific personal traits such as resilience, self-esteem, empathy and it throws further light on the debate of nature versus nurture. In addition, it reveals that Jewish doctors particularly in industrial camps as Auschwitz played an unwilling role in human experiments and in treating injured and sick patients specifically to enable them to return to work that would help strengthen the German war economy.

The history of the Jewish doctors during this tragic period is the never-ending repeat of Jewish history. The European Jewish doctor, by the beginning of the twentieth century was lulled into a false sense of security and within a relative brief period of 45 years they and everything they represented, morally and ethically, had been decimated. Nevertheless, since the Holocaust, history has once again witnessed the re-emergence of the Jewish people none more so than the contribution made by Jewish medical scientists and researchers. Have we learnt the most important lesson from the tragic short history of the Jewish doctor in the space of 45 years? This century the alarming proliferation and mass appeal of extreme right wing parties, neo-Nazi groups and subtle endorsement of antisemitism and racial discrimination by elected governments and national leaders provides warning that history could repeat itself within a very short time frame.

Bibliography

Blady Szwajger, A. (1990). *I remember nothing more: The Warsaw Children's Hospital and the Jewish resistance*. London: Collins Harvill.

Cohen, E. A. (1973). *The abyss: A confession*. New York: W.W. Norton & Company.

Efron, J. M. (2001). *Medicine and the German Jews: A history*. New Haven: Yale University Press.

Eliach, Y. (1983). *Hasidic tales of the holocaust*. London: Oxford University Press.

Halpin, R. W. (2014). *The essence of survival: How Jewish doctors survived Auschwitz*. Sydney Jewish Museum: Sydney.

Kater, M. H. (1989). *Doctors under Hitler*. Chapel Hill: The University of North Carolina Press.

Levi, P. (1988). *The drowned and the saved*. London: An Abacus Book.

Perl, G. (1984). *I was a doctor in Auschwitz*. New Hampshire: Ayer Company Publishers.

Roland, C. G. (1992). *Courage under siege: Starvation, disease, and death in the Warsaw Ghetto*. New York: Oxford University Press.

Rosenberg, J. G. (2005). *East of time*. Blackheath: Brandl & Schlesinger.

Tenenbaum, J. (1963). The Jews in Poland. In L. Falstein (Ed.), *The Martyrdom of Jewish Physicians in Poland*. New York: Exposition Press.
Vaisman, S. (2005). *A Jewish doctor in Auschwitz: The testimony of Sima Vaisman*. Hokboken: Melville House Publishing.
Weindling, P. (2000). *Epidemics and genocide in Eastern Europe 1890–1945*. Oxford: Oxford University Press.
Winick, M. (Ed.). (1997). *Hunger disease: Studies by the Jewish doctors in the Warsaw Ghetto* (Volume 7 in the Wiley Series). New York: Wiley.

Thesis

Vaughan, L. (1994). *The Jews in London 1695 & 1895*. London: The Bartlett Graduate School, University College London.

Chapter 15
Fate of the Jewish Doctors – Members of the Jewish Chamber of Physicians in the Warsaw Ghetto (1940–1943)

Maria Ciesielska

Abstract The first organized Chamber of Physicians in Poland was formally constituted only after Polish Independence in 1921. The Chamber of Physicians, based in Warsaw, was an autonomous self-governing institution and included representatives from all the Polish regional chambers of Physicians. During the German occupation in Poland, district medical chambers – health councils operating on the territories of particular districts – were established. They were subjected to the German administration and were commanded to prepare a thorough record of the physicians living and working in the German-occupied territories. This facilitated the expulsion of the Jewish physicians from membership in the Chambers of Physicians, as well as the establishment of a separate Chamber of Physicians for Jewish doctors – the so called National Group of Jewish Physicians. In November 1939 Jews lost their right to health benefits and medical aid, and Jewish patients were removed from city hospitals. In the mid-1940s Jewish doctors were banned from treating non-Jewish patients, and in the fall of 1940 the Warsaw Ghetto was created. The Germans confiscated the equipment and the medicines from the doctors' and dentists' offices and these physicians and dentists were forced to move into the Ghetto. Issues related to health services in the Ghetto were entrusted to the Health Department of the Jewish Council that was chaired by Dr. Izrael Milejkowski, MD. At the same time, welfare organizations were formed to provide medical help for the Ghetto's inhabitants, such as: the Society for Safeguarding the Health of the Jewish Population (Towarzystwo Ochrony Zdrowia Ludności Żydowskiej w Polsce; TOZ) and the Jewish Welfare Association (Żydowska Samopomoc Społeczna). It is estimated that at the time of closure of the Ghetto's gates in November 1940, there were between 750 and 1000 physicians residing within the Ghetto's walls. The conditions of the patients in the Ghetto hospitals deteriorated from week to week due to almost total lack of medications, bandages and medical equipment, as well as because of the ever-present hunger and the constant Typhus epidemics. With time, medical treatment became limited to only the basic wound dressing in the constantly shrinking number of medical institutions. Under these extremely difficult

M. Ciesielska (✉)
Regional Chamber of Physicians, Warsaw, Poland
e-mail: maria.ciesielska@lazarski.pl

M. Moskalewicz et al. (eds.), *Jewish Medicine and Healthcare in Central Eastern Europe*, Religion, Spirituality and Health: A Social Scientific Approach 3,
https://doi.org/10.1007/978-3-319-92480-9_15

conditions, physicians participated in clandestine training of medical students and also performed research on hunger disease. The majority of the physicians who survived the Great Action in the ghetto in the summer of 1942 and the Warsaw Ghetto Uprising in November 1943 were, transported to the concentration camp at Majdanek (KL Lublin). Barely several dozen Jewish physicians reported to the Chamber of Physicians in Warsaw in 1945 in order to register, or, in the subsequent years, to obtain a copy of their medical diplomas. It is estimated that during the Second World War over 5000 Polish physicians lost their lives, mostly in Ghettos and death camps.

Keywords Jewish chamber of physicians · Jewish physicians · Warsaw ghetto · Healthcare in the Warsaw Ghetto

15.1 The First Chambers of Physicians and Its Regional Branches

The first Chambers of Physicians in the Polish territories were established already under the partitions (Austrian and Prussian partitions) (Nasierowski 1992). However, fully organized self-governing Polish institutions were established only after gaining independence. In 1921, based on the parliament law "on the system and scope of operation of chambers of physicians" the Chambers of Physicians were established (The law on the system and scope of operation of Chambers of Physicians, from December 2nd 1921. Law Gazette no 105, position 763). The Chamber of Physicians, based in Warsaw, included representatives of the regional chambers who met for the first time on June 9th, 1923 to elect their leaders (Kordel 2012), and constituted the permanent representation of the self-governing body to regulate all the regional chambers of physicians in the country. Originally, there were seven Chambers of Physicians established on Polish territory: the Chamber of the city of Warsaw, the Chamber Białystok (since 1922 together as the Chamber of Warsaw and Białystok), the Chamber of Łódź, the Chamber of Poznań and Pomerania, the Chamber of Lviv, the Chamber of Cracow and the Chamber of Lublin. In 1925, based on the ordinance of the Minister of the Interior, the Chamber of Physicians of Vilnius and Nowogród was established, and in 1934, the Chamber of Physicians of Silesia was established. In June 1923, the representatives of the existing chambers, elected the chair of the Chamber of Physicians for the first time. Dr. Jan Bączkiewiczbecame the chair of the Chamber of Physicians for the first and second terms of office (1923–1928), and was a long-term chair of the Association of the Polish Physicians and a member of the Warsaw Medical Association. Dr. Witold Chodźkowas the chair of the third and fourth terms of office (1929–1934) and Professor Mieczysław Michałowicz.was elected for the fifth term of office (1935–1939).

The Chamber of Physicians of Warsaw and Białystok based in Warsaw on 37 Koszykowa Street encompassed the area of the city of Warsaw, Warsaw region, and Białystok region (Szkudaj 1998). Its chairs were as follows: Dr. Leon Babiński,

MD – GP, Dr. Jerzy Bujalski, MD – surgeon and gynecologist (killed in KL Auschwitz), and Dr. Władysław Szenajch, MD – pediatrician. In 1938, based on the ordinance of the Minister of Social Care Marian Zyndram-Kościałkowski, dentists were included into the mechanisms of the self-government of physicians, and the area of the Polish state was divided into 4 district chambers of physicians and dentists (The law for the creation of Chambers of Physicians and Dentists. Law Gazette no 34, position 296). One of the tasks of those District Chambers of Physicians was to prepare a comprehensive register of all physicians living and working in the territories subjected to a Chamber and working in the sectors: of the state, of the self-government, of the social insurance, and in the private sector. With the help of the health departments of the provincial offices issuing the right to practice the profession, as well as with the help of the county physicians, the lists of doctors were gradually completed. Every physician who arrived in a Chamber's area was obliged to register as a member of the Chamber. Another task of the regional council was to regulate the prices for both public and private medical services After the First World War, the majority of physicians, due to the low income, tried to obtain jobs with the national or self-governmental health service, and in the following years also in the health insurance services and school hygiene services (The act about the public obligatory insurance of employees. Law Gazette no 44, position 271 from 1920. Based on the ordinance of the Minister of Social Care from December 30th 1933. Law Gazette no 103, position 819). It is also worth emphasizing that in the interwar period physicians working in the national or self-governmental health services were employed as state officials. The Chamber of Physicians also played an important role as a self-governing institution representing physicians and negotiating salaries with the Ministry of Labor. In addition to the representatives of the Chamber of Physicians, members of the Association of Physicians of the Polish State, the Association of Physicians of Health Insurance Office, of the Association of Physicians of Social Insurance Office in Warsaw as well as directors the Offices of Health Insurance participated in the discussions..

After the end of warfare in September 1939, the chamber of physicians began their activity (Wiśniewski 2007). On November 1939 the chief public health adviser Jost Walbaum (*Gesundheitsführer*) and the reporter Werner Kroll (*Referent für Ärztekammern*) sent a circular to the counties' physicians, in which they described the scope of responsibilities of those physicians as representatives of the chambers of physicians. Polish chambers of physicians, according to that circular, had to function as before but now also had to be subject to the German administration. The letter also announced that the Chamber of Chemists had to be included in the Chamber of Physicians. The counties' physicians, according to the announcement, had to prepare, and constantly update, lists of professionally active physicians. The lists had to include the name, surname, address and nationality of the physicians. The letter also described in detail the scope of operation of the hitherto Chambers of Physicians that remained within the borders of the General Government: the Chamber of Cracow, the Chamber of Lublin and the Chamber of Warsaw and Białystok. An annual member fee of 21 zloty was introduced. Jewish Physicians, according to the announcement, had to be "evacuated" from their practices and apartments. The abandoned practices and apartments could then be overtaken by unemployed Polish physicians – Aryans (251/1 JHI).

15.2 The Health Chamber (*Distriktgesundheitskammer*)

Before the German administration was established, the chair of the Chamber of Warsaw and Białystok was Adam Julian Huszcza. On January 26th 1940 Johann Kamiński (*Distriktarzt*) sent a notice to the Chamber of Physicians about its liquidation (Kordel 2012). Based on the ordinance of the governor Hans Frank from February 28th 1940, the Health Chamber whose headquarters were in Cracow (*Gesundheitskammer im Generalgouvernement*) was established on the territory of the General Government (Zdrowie i Życie Gazette. 1940 no 2 p. 16). The chair of Health Chamber was Jost Walbaum, whowas also the chair of the Health Department in the Office of General Government. Werner Kroll, the reporter for the Health Chamber in the Health Department in the Office of General Government, became his deputy. Marian Ciećkiewicz recalled: "On the appointed day, three German officials, accompanied by Dr. Owsiński, the chief city physician in Cracow, came to the Chamber. One of them introduced himself as Dr. Walbaum, the later "minister of health" in G.G. The conversation conducted between these men was only informative and it was not yet known what the German authorities were planning in reference to the Chamber. A few days later, Dr. Kroll, assigned for the position of the chair of the Chamber of Physicians, arrived with his, later well-known, famous von Würzen as a secretary" (Ciećkiewicz 1945).

In October 1940 district health chambers – health councils (*Distriktgesundheitskammern*) operating in the territories of particular districts (Cracow, Lublin, Radom and Warsaw) were created (Zdrowie i Życie Gazette. 1940 no 7 pp. 51–54. Zdrowie i Życie Gazette. 1940 no 8 pp. 57–60). In 1941 the fifth chamber, for the Galicia district, was created in Lviv. Dr. Tadeusz Alkiewicz, as the chair of the District Health Chamber in Warsaw (*Leiter der Gesundheitskammer in Warschau*). In January 1944 he was arrested for conspiracy and sent to the prison at Pawiak. Thanks to the efforts of the General Protective Council he was released on bail. (Bayer 1985).

The newly established health departments included physicians, dentists, chemists, dental technicians, surgeons, midwifes and nurses. There were insurance offices as well as employment agencies within their structures. These institutions were administered by German physicians from the first days of occupation. Werner Kroll ordered an immediate creation of separate accounts for payments and withdrawals of the Mutual Help Insurance Office for Aryan physicians, and a separate account for Jews. One of the first orders imposed in 1939 on all Jews who were at least 10 years old, was the wearing of arm bands bearing the Star of David. In 1940, the Germans imposed a curfew on the Jewish Ghetto inhabitants prohibiting them from leaving their homes after certain hours (it was also forbidden for Jews to use the trains, and, with time, other means of communication) (Adamska et al. 1988).

15.3 Jewish Doctors in the Warsaw Ghetto

In November 1939, Jews were no longer allowed to receive health benefits and medical help, and Jewish patients were removed from city hospitals. In mid-1940, Jewish doctors were prohibited from treating non-Jewish patients. On October 2nd 1940, an ordinance announcing the creation of a Jewish district in Warsaw was issued by the Warsaw governor Ludwig Fisher. During the following weeks approximately 138,000 Jews were forced to move into the Ghetto and 113,000 Poles were moved out of the area. (Adamska et al. 1988). The relocation of large groups of people in the Ghettos allowed for the looting of property which was not permitted to be taken by the people being forced to move. The Germans confiscated the medical and dental equipment from the private practices of the physicians and dentists and from the pharmacies as well.

A lawyer Heinz Auserwald became the commissioner of the Jewish district, and the authority of the district's mayor was given to Adam Czerniaków who became the chairman of the Jewish Council (*Judenrat*). Gradually, the issues pertaining to healthcare in the Ghetto were taken over by the Health Department that was created by the Jewish Council and chaired by Dr., Izrael Milejkowski (Leociak 2001). At the same time public health and welfare organizations were formed to aid in the provision of medical help for the Ghetto inhabitants, such as: The Healthcare Association for Jewish People in Poland (Towarzystwo Ochrony Zdrowia Ludności Żydowskiej w Polsce) and the Jewish Welfare Association (Żydowska Samopomoc Społeczna).

The Ghetto's gates were closed in mid-November of 1940. It coincided with the creation of a separate Department for Jewish Physicians – the so called National Group of Jewish Physicians – based on 11 Tłomackie street, and later on 3 Leszno street. Dr. Izrael Milejkowski became its chairman. Attached to the Health Department was a Health Commission chaired by Dr. Anna Braude-Hellerowa ((Leociak 2001). Dividing physicians into two separate national groups was possible because every physician who registered in the Department was obliged to fill in a "Medical Profession Questionnaire" (*Fragebogen zur erstmaligen Meldung der Heilberufe*), which included details of the physicians" backgrounds According to the racial laws proclaimed by the German Reichstag on September 15th 1935, including the law concerning the citizenship of the Reich, everyone who had three or more grandparents who had been "racially complete Jews", was considered a Jew. A descendant of two Jewish grandparents was considered "a first degree half-breed", and one who had one Jewish grandparent "a second degree half-breed". Religious belief was not taken into consideration. In addition, in occupied Poland on July 24th 1940, the, Germans introduced a decree, according to which "the first degree half-breeds", as well as those who before September 1st belonged to a Jewish community or had a Jewish spouse, were also considered Jews. The construction of the questionnaire reflected this decree. In order to practice and treat Aryan patients in the General Government, one had to prove the existence of a pure Aryan background and fill out the appropriate sections concerning the background and religion of the grandparents. If a physician was considered a Jew, this fact was recorded on the first page of the questionnaire by an marked with the word *Jude*. Jewish physicians could officially practice in the Jewish institutions such as the Jewish hospitals, Jewish organizations [among others, Society for Protection of Health of Jewish People (Towarzystwo Ochrony Zdrowia Ludności Żydowskiej)], or in their own offices, treating only Jewish patients. This significantly limited the medical practice of a large group of assimilated physicians, who had for years been working in the city hospitals or had conducted their own profitable offices.

In 1940, all registered Jewish physicians were required to complete a two-page declaration detailing their income from 1938 until 1940 – "the questionnaire regarding a one-off payment for all medical professionals belonging to the Health Chamber" (*Fragebogen zur Erhebung einer einmaligen Umlage*). This was in accordance with the ordinance decreed by the General Governor for the occupied Polish territories on February 28th, 1940. The calculation of the fee that was to be paid into the Chamber's account was based on the collected data. Dr. Joachim Kamiński was the accountant in charge in Warsaw.

At the same time, every Jewish physician who had been practicing in Warsaw, had to apply for a permit to practice in the area of the "Jewish residential district". According to the Official Register of Physicians, there were 2815 (USL 1939) physicians registered in Warsaw in 1939 who worked in the 23 hospitals, institutes and clinics of the University of Warsaw, 11 health centers, 23 ambulatories and emergency rooms, as well as in their private medical practices. When the gates of the Ghetto were closed in November 1940, there were around 750–1000 physicians residing within its walls (Leociak 2001).

The area of the Ghetto gradually diminished and the number of inhabitants increased due to the constant influxes of Jewish refugees from outside Warsaw. In January 1941, about 400,000 people were cramped in an area of some 307 hectares (Adamska et al. 1988). The conditions under which the ill, both those hospitalized and those visiting the out-patient clinics deteriorated from week to week due to the omnipresent lack of medicines, dressings, medical equipment and in addition due to the all-pervading hunger. As the time passed, treatment of patients was reduced to basic medical care only in the constantly decreasing number of medical institutions. Despite these extremely difficult conditions, physicians also secretly taught medical students and performed research on hunger disease.

15.4 Jewish Hospitals in the Warsaw Ghetto

The outbreak of the war and the siege of Warsaw resulted in enormous destruction of the city. The hospital buildings were also damaged, despite being marked with a red cross. The surgical department of the Jewish Hospital, together with its operating rooms, as well as other wards and the hospital kitchen, were seriously damaged. The situation was also difficult because most of the medical staff were drafted into army. Soon, the Germans ordered the transfer of the hospital that was under the management of the Jewish Community which was permitted to accept Jewish patients only.. Therefore, all the non-Jewish patients were removed from the hospital, and the Jewish staff and patients from other Warsaw hospitals were relocated to the Czyste Hospital (Wysocki 2015). In February 1941, the Germans ordered the relocation of the Jewish Hospital to inside the Ghetto walls. However, they could not find an adequate building to accommodate the whole hospital. Therefore, the hospital was divided into separate departments that were located in different places in the Ghetto. The surgical department was located on 1 Leszno Street. Initially, it contained 300 beds. Dr. Aleksander Wertheim, Dr. Dawid Amsterdamski and Dr. Ignacy Borkowski were in charge there. The medical staff consisted mostly of young physicians and medical students. The departments were constantly overcrowded, lacking in medical supplies and food. Due to the Typhus epidemics in the Ghetto, the departments of infectious diseases were overcrowded as well. Physicians encountered countless difficulties from city officials as well as from the occupation authorities. Many patients were admitted to the emergency room because they had been shot while trying to cross the Ghetto's wall.

The Berson and Bauman Children's Hospital located at 51 Śliska/60 Sienna street was initially located within the Ghetto's borders and functioned until the liquidation of the Small Ghetto on August 1942. Previously, already in October 1941, a branch of the Children's Hospital was created inside the Large Ghetto on Żelazna 86/88 street. Arthur M. Stupay wrote in his book: "Headed by Dr. Anna Braude-Heller, it specialized in children's diseases. The hospital did not have to undergo the heartrending moves that befell Czyste, and had beds, instruments and trained staff. Later, when Czyste and the health clinics were closed, Berson-Bauman functioned to treat infectious diseases. After the final *Aktion,* it became an aid station for anyone seeking medical help. The Germans destroyed it in the Warsaw Uprising of April 1943" (Stupay 2014, 192).

15.5 The Great Liquidation Actions in July and September, "The Kettle" 1942

On July 22nd 1942, the Great Liquidation Action began in the Warsaw Ghetto. In mid August 1942, aSS 1 staff led by Hermann Höfle was formed and it was decided to liquidate the Jewish hospitals. One of the students working in the surgical department on Leszno street, Marek Balin, remembered the day of the liquidation of the surgical department this way: "It was on August 15th 1942, in the surgical department of the "Czyste" Hospital in the Warsaw Ghetto, on Leszno street. Our hospital, that normally could accommodate about 200 patients, currently had nearly 600 people, and, in addition, it was going to admit more patients from different departments. (...) Since July 1942, Germans were deporting thousands of Jews to the death camps, and together with the deportations of the unlucky, they were decreasing the Ghetto's area, they were cutting off streets and even districts, including hospitals. So, there was now a big inflow of patients and staff who managed to save themselves from constant blockades and bestial roundups. Unfortunately for us, Germans found out that Jews were gathering at our place. They knew that most of these Jews were seriously ill, unable to work, unable to walk, because our department, in general, was admitting the victims of shootings. Like in other institutions, the management of the hospital was said to reduce patients to the minimum – down to 150 people and to send the bigger part of people to the East "to work". Of course, healthy people, who learned about this reduction, immediately dispersed or hid. But what to do with the ill? How could the management deliberately send them to their deaths? How to conduct the selections and consciously sentence them to death, or to protect others for the time being? According to which criteria to choose? An overwhelming inability to do anything prevailed in the hospital. All of us were conscious of the tragedy. Jews themselves had to send their brothers to their deaths, also disabled people, invalids, or those who terribly suffered. Not only Jews had to select Jews, but physicians – patients, people who they had been taking care of and whose sufferings they were trying to mitigate, whose pain they were trying to ease, whose limbs they were trying to save and whose senses they were trying to restore. The patients for whom they were not sleeping at nights, in order to perform complicated, life-giving surgeries in the light of smoky carbide lamps. (…) In the evening of that unforgettable day, the death sentence for most of the patients was going to be signed by the physicians. Usually in the evening there was a round of a physician on duty and of some volunteers. This time it was a round of the directors: Dr. Rothaub, Dr. Borkowski and Dr. Szenicer from our department, as a substitute of Dr. Amsterdamski who had been taken hostage by the Germans. The committee of these excellent doctors had to judge. They had a list with names. Next to a name, the "+" mark meany deportation, death, and the "–" mark meant remainingin the hospital. The doctors stopped next to each bed for a long time. They whispered or talked quietly. They held the discussions in in Latin, so that the patient would not guess anything or comprehend the horror of this unusual consultation. The directors entered the orthopedic ward where there were frail bodies with stiff limbs, suspended on blocks and splints. Dr. Szenicer used his handkerchief repeatedly.… He avoided not only the gaze of the patients, but he was also embarrassed in front of his colleagues. He walked around the wards with his head down, and part of his face was always covered with the handkerchief. (…) Other doctors too reached for their handkerchiefs. At a certain moment the head of our department left the ward. I never saw him again. The committee did not complete its work. The next day the Germans carried out the selection themselves and they filled five wagons with patients which they then deported. Those, who were lucky to stay, did not remain in the hospital on Leszno street for long, because 2 days later the hospital was emptied and all of us were rushed to the Umschlagplatz" (301/7017 JHI).

According to German sources, over 250,000 Jews were deported from the Warsaw Ghetto during the 46 days of the Great Liquidation Action (Młynarczyk 2004). Apart from a few exceptions, all of them were killed in the gas chambers of the death camp in Treblinka. The medical staff that remained in the Ghetto was formed into the so called Combined Jewish Hospital, which occupied the buildings on Sawki street on the so- called Umschlagplatz. Dr.r Józef Stein and Dr. Anna Braude-Hellerowa were in charge of this institution.

During and after the liquidation action in July, some Jewish physicians managed to escape to the Aryan side, hiding in the Warsaw area with the help of Polish physicians, who were usually their prewar friends. Some Jews hid in hideouts – most often various secretly made recesses that had been prepared in advance. People who were discovered were usually shot to death immediately. It would often happen that urban sewers or bunkers that had been dug in advance became hideouts. During the liquidation action in July, many people with "good" Aryan looks", and those who had the financial means, managed to get through to the Aryan side, hiding in the Warsaw area and beyond. It is estimated that 8000 people escaped from the Ghetto in that way. The next great wave of escapes took place in the winter of 1943, after the end of the January liquidation action, and before the outbreak of the Warsaw Ghetto Uprising in April 1943. In general, physicians leaving the Ghetto were assimilated Jews with Aryan acquaintances and friends. Thanks to those prewar friendships it was much easier for them to find help and to hide on the other side of the wall. For example, Professor Edward Loth helped his assistant Dr. Ludwik Stabholz to escape from the Ghetto; Dr. Andrzej Trojanowski saved, among others, Dr. Stanisław Szenicer; and Dr. Feliks Kanabus helped Dr. Mieczysław Tursz. However, life on the Aryan side was wrought with danger, one of which was the presence of so called "szmalcownicy" – people who would follow their victims on leaving the Ghetto and then blackmail them demanding money for their "silence". When these victims could not meet their demands a "szmalcownik" would denounce them and the Poles who had hid them to the Gestapo. Occupied Poland was the only country where one could face a death sentence for hiding Jews (Paulsson 2007).

The final reduction in the number of the hospital staff took place on September 6th 1942, when a few of the doctors received so-called "life- tickets" enabling them to remain in the Ghetto. Less than a week later 950 patients of the hospital and 50 of the staff were deported from the *Umschlagplatz* to Treblinka. Those doctors who did not get the "life tickets" remained in the Ghetto in hiding.

15.6 Jewish Physicians During and After the Ghetto Uprising of 1943

In April 1943, the Warsaw Ghetto Uprising broke out. By that time the hospitals in the Ghetto were hardly functioning.. Physicians in the remaining hospital on 6/8 Gęsia Street hid in the shelter that had been prepared in advance, but which was discovered by the Germans on May 13th. These who survived the Uprising were deported to the concentration camp at Majdanek (KL Lublin) and most of them were murdered during the November liquidation action called "Dożynki" (Aktion Erntfest). Only a few lucky survivors who were transported to other camps before November 3rd 1943, survived (Kranz 2004).

After suppressing the Ghetto Uprising on May 16th 1943, the Germans started searching for the Jews who were hiding on the Aryan side. When found these Jews were caught and imprisoned in Pawiak Prison. They were not registered. They were held in separate empty cells in the basement, which previously had served as a potato storage room. The Jews would be held there for hours to days until they were executed.. These executions were a daily occurrence in the Ghetto ruins surrounding the prison. One of the Jewish physicians who was executed was Dr. Aleksander Jerzy Wertheim, the head of the surgical department of the Jewish Hospital at "Czyste. He was arrested because of his activity in the Home Army and was executed in the Ghetto ruins together with his two sons Bronisław and Stanisław. His wife Janina also died executed at Pawiak (Hasselbusch and Ciesielska 2015).

A gynecologist, Dr. Alina Brewda, who managed to survive in the bunkers until the outbreak of the Ghetto Uprising, was, according to her testimony, transported to the concentration camp at Majdanek in April 1943 (Minney 1966). In May, other physicians who had remained in the Ghetto, and who were hiding in the shelter under the so -called residual hospital on Gęsia street, were also transported to Majdanek.

In the postwar account of Dr. Henryk Fenigstein, we can find a testimony about the fate of Jews deported from the Ghetto: "The first transport of deportees was sent on Tuesday April 20th 1943 to Trawniki where a number of Warsaw Jews from Szulc Factory, and others who had decided to go, were already staying. The second transport, with the workers of "Werterfassung", left on Friday April 23rd for Flugplatz Lublin, where a temporary camp, Dulag, had been formed. The third transport left on Saturday April 24th to Flugplatz. Two wagons from this transport, with about 300 men, who were locksmiths and shoemakers as well as families of the Jewish council of Lichtenbaum and of Stolzman and others, were directed to Poniatów, where, similarly to Trawniki, Toebbens' Jews were already residing. On Friday April 23rd members of the Jewish Council namely the engineers Marek Lichtenbaum, A. Stilzman and Szereszewski and the advocate Wielikowskiwere shot and killed at the courtyard of Umschlagplatz" (301/7017 JHI).

There were very few Jewish doctors who remained in the Ghetto. Among them was Dr. Anna Braude-Hellerowa. Although her friends prepared fake documents to help her and prepared a safe hiding place on the Aryan side, she refused their help. Dr. Adina Blady-Szwajger recalled that in her last letter sent from the Ghetto in March 1943 Dr. Braude-Hellerowa wrote: "Do not worry about me, I have different plans" (Blady-Szwajger 2010, 116). She most likely died under the ruins of the hospital on Gęsia Street together with her patients. A few physicians, for example Dr. Henryk Beck, managed to survive in specially prepared bunkers who together with his non-Jewish wife 33 other people, hid in a bunker between 7 Śliska street and 22 Sienna street. They remained there for 110 days until the Red Army arrived (Jaworska 1982).

15.7 Conclusions

It is estimated that over 5000 Polish physicians died during the Second World War, mostly in the Ghettos and death camps (Kordel 2012). In 1945, only a few dozen Jewish physicians reported to the Chamber of Physicians in Warsaw in order to register, or, in the following years, to obtain a copy of their medical diplomas. Reconstruction of the fate of the Jewish physicians during the war based on the archival documentation is virtually impossible because Jews were considered by the Germans as *numbers* only who were destined for extermination within the genocidal system. Even if any documentation was prepared, it was completely destroyed in the last days of war. Therefore, reconstruction of a picture of the past has to be based on memoirs and postwar accounts of survivors. Using these materials we are able to reconstruct the history of the professional group of physicians as an example of the fate of the Jews in Poland during Second World War.

Translated by Anna Topolska

Bibliography

251/1 Chamber of Health in the General Government (Gesundheitskammer im Generalgouvernement), Ordinances and circulars, The Archive of Jewish Historical Institute (JHI), Warsaw.

301/509 The Fate of the Jews After Their Displacement from Warsaw, April 1943, reported by Chryzman Ilie, Fenigstein Henryk, Feinstein Marek, Grunberg Aleksander, 1945, Collection of Holocaust Survivor Testimonies, Jewish Historical Institute (JHI), Warsaw., DOI: https://doi.org/10.17226/9566

301/7017, Testimony of Dr. Marek Balin. (n.d.). *Collection of Holocaust Survivor Testimonies*. Jewish Historical Institute (JHI), Warsaw.

Adamska, J., Kaźmierska, J., & Sakowska, R. (1988). *Tak było... Sprawozdania z warszawskiego getta 1939–1943*. Warszawa: Zetpress.

Bayer, S. (1985). *Służba zdrowia Warszawy w walce z okupantem*. Warszawa: Wydawnictwo Ministerstwa Obrony Narodowej. https://doi.org/10.1128/AAC.28.6.781.

Blady-Szwajger, A. (2010). *I więcej nic nie pamiętam*. Warszawa: Świat Książki.

Ciećkiewicz, M. (1945). *Izba lekarska w dobie okupacji niemieckiej i na przełomie*. Accessed 23 Aug 2016.

Hasselbusch, R., & Ciesielska, M. (2015). *Dance among swords. Polish medical personnel in Pawiak during German occupation 1939–1944*. Warszawa: Muzeum Niepodległości w Warszawie.

Jaworska, J. (1982). *Henryka Becka "Bunkier 1944 roku"*. Wrocław: Zakład Narodowy im. Ossolińskich.

Kordel, P. (2012). *Geneza, struktura i funkcjonowanie samorządu lekarskiego w Polsce w latach 1989–2009*. Doctoral dissertation written under direction of Professor Michał Musielak in the Chair of Social Sciences of the Department of Health Studies at the Medical University of Karol Marcinkowski in Poznań. https://doi.org/10.1039/c2nr30305a.

Kranz, T. (2004). Obóz koncentracyjny na Majdanku a „akcja Reinhardt". In D. Libionka (Ed.), *Akcja Reinhardt. Zagłada Żydów w Generalnym Gubernatorstwie* (pp. 233–247). Warszawa: Instytut Pamięci Narodowej. https://doi.org/10.3200/JRLP.138.4.371-384.

Leociak, J. (2001). Służba zdrowia. In B. Engelking & J. Leociak (Eds.), *Getto warszawskie. Przewodnik po nieistniejącym mieście*. Warszawa: IFiS PAN.

Minney, R. J. (1966). *I shall fear no evil – The story of Dr Alina Brewda*. London: William Kimber.

Młynarczyk, J. A. (2004). Treblinka – obóz śmierci "akcji Reinhardt". In D. Libionka (Ed.), *Akcja Reinhardt. Zagłada Żydów w Generalnym Gubernatorstwie* (pp. 217–232). Warszawa: Instytut Pamięci Narodowej.

Nasierowski, T. (1992). *Świat lekarski w Polsce. II połowa XIX w. – I połowa XX w. Idee, postawy, konflikty* (pp. 23–32). Warszawa: Okręgowa Izba Lekarska.

On March 19th 1920 the act „about the public obligatory insurance of employees" was resolved (Law Gazette no 44, position 271 from 1920). Based on the ordinance of the Minister of Social Care from December 30th 1933 (Law Gazette no 103, position 819) beginning with January 1st 1934 the health insurance offices were transformed into social insurance offices.

Paulsson, G. S. (2007). *Utajone miasto. Żydzi po aryjskiej stronie Warszawy (1940–1945)* (E. Olender-Dmowska, Trans.). Kraków: Centrum Badań nad Zagładą Żydów and Wydawnictwo Znak. https://doi.org/10.1111/j.1365-2141.2007.06644.x.

Stupay, A. M. (2014). *Hope expired life persists*. Łódź: Łódź University Press.

Szkudaj, T. (1998). Izba lekarska Warszawsko-Białostocka w zbiorach specjalnych Głównej Biblioteki Lekarskiej. *Medycyna Nowożytna, 5/2*, 117–125.

The act about creating Chambers of Physicians and Dentists. Law Gazette no 34, position 296.

The act about the system and scope of operation of chambers of physicians, from December 2nd 1921. Law Gazette no 105, position 763.

The ordinance of the Minister of Public Health from March 15th 1922 about creating Chambers of Physicians. The ordinance of the Minister of Public Health from September 22nd 1922 about changing the ordinance about creating Chambers of Physicians.

Urzędowy Spis lekarzy, lekarzy-dentystów, farmaceutów, felczerów, pielęgniarek, położnych uprawnionych i samodzielnych techników dentystycznych (USL). (1939). Warszawa.

Wiśniewski, Z. (2007). *Lekarze i Izby Lekarskie w drugiej Rzeczypospolitej*. Warszawa: Naczelna Izba Lekarska.

Wysocki, A. (2015). Lekarze polscy w czasie II wojny światowej. In W. Noszczyk & J. Supady (Eds.), *Dzieje medycyny w Polsce* (Vol. 2, pp. 587–641). Warszawa: PZWL.

Chapter 16
Coping with the Impossible: The Developmental Roots of the Jewish Medical System in the Ghettos

Miriam Offer

Abstract This chapter examines the coping patterns used by the Jewish medical staff to deal with the spread of epidemics through the ghettos as a result of the German decrees. In most of the ghettos, the Jews, as persecuted victims, independently established medical systems drawing on modern, professional perceptions, in subhuman conditions, seen only in cases of genocide and mass atrocities. In the ghettos, the Jewish physicians and leaders fought to maintain the health and save the lives of the inmates by rapidly establishing extensive medical systems, notwithstanding the limitations. During the Holocaust itself, the ghetto physicians and leaders recognized the phenomenon as unique and memorialized it as such in their writings. The chapter addresses the characteristics of the medical system and its developmental roots. The central claim is that the medical system was not created out of a void, but was possible because of processes that had occurred in the Jewish society prior to the Holocaust, including modernization and secularization, on the one hand, and anti-Semitism, on the other. Alongside the Jewish cultural tradition of the sanctification and the saving of life and longstanding nurturing of the medical profession, the Jewish society adopted the principles of modern medicine, including a public health perception and the cultivation of social medicine and preventive medicine. The escalation in anti-Semitism during the interwar period intensified the need for wide-scale national ethnic Jewish medical systems, which offered advanced, professional medical services for the millions of Jews, particularly in Poland (which had the largest Jewish population) and in other European countries generally. A study of this background is a prerequisite to understanding the Jewish reaction pattern during the Holocaust in general, and the Jewish medical reaction pattern in the ghettos, in particular.

Keywords Holocaust · Public health · Modernization · Secularization · Nazism · Anti-Semitism · Collective ethnic identity

M. Offer (✉)
Western Galilee College, Akko, Israel
e-mail: miriamo@wgalil.ac.il

M. Moskalewicz et al. (eds.), *Jewish Medicine and Healthcare in Central Eastern Europe*, Religion, Spirituality and Health: A Social Scientific Approach 3,
https://doi.org/10.1007/978-3-319-92480-9_16

16.1 The Jewish Medical Activity in the Ghettos – As Reflected in Writings of Ghetto Physicians During the Holocaust Itself

Many of the Jewish medical staff in the ghettos documented the medical activities actually as they unfolded. Moreover, despite their short-lived perspective from the epicenter of the Holocaust, their documentation of Jewish medicine reveals that even then, they appreciated the unique nature of these extraordinary activities in the context of human history, and that the study of them would be necessary after the war.

Thus, for example, Dr. Aaron Pik from the Shavli Ghetto in Lithuania (Siauliai), in several writings throughout the ghetto period, until his death in July 1944, documented many aspects of coping with medical issues. Among the descriptions in his diary, "Notes from the Valley of Slaughter," written in Hebrew, is how, in the Shavli Ghetto, the Jews set up a Jewish hospital in the cemetery, which was within the ghetto boundaries [my emphasis, M.O.]:

> "… The evening before last, we had a celebration in the ghetto – **the first anniversary of the establishment of the hospital in the ghetto.** A year ago, the Christian physicians were forbidden to treat Jewish patients and the *Netzigut* [The Jewish Representation—Judenrat, M.O.] received an order to remove all Jewish patients immediately from all urban hospital departments, despite the fact that they included patients at risk with 40-degree temperatures. This cruelty and evil after all the troubles they had suffered did not astonish us … Undeterred, the *Netzigut* found a solution to the calamities, even if not so easily: They immediately began to establish … a place … to house the patients …
>
> The only place was the morgue in the cemetery. The room with the cement floor was very cold during this last winter, which was particularly chilly. There was no equipment, furniture, or bricks … **but the necessity, the initiative, the generosity of spirit, the adjustment, the courage of the *Netzigut*, overcame all the obstacles,** and in a short space of time, the morgue and the rooms that had served the Chevra Kadisha [the Jewish burial society] were turned into a hospital; very small, but under the present conditions, quite splendid and proper …
>
> … **Even in the ghetto, despite our enemies' endeavors to depress the surviving remnant and to make them like the dust in a threshing, our strength has not yet waned, and our initiative and the ability to adjust are still as good as new. Those who are tormented and oppressed, those who are broken and weakened, will shake off the dust and lift up their heads once again to live and to act**, when they will surely be saved and redeemed from the ghetto and their loathsome adversary. **This was the first celebration since the Germans burst into our city, and left a pleasurable impression, like a chink of light in the darkness of night** (Pik 1942, p. 177; Offer 1993).

Dr. Aaron Pik calls attention to an important insight; that the very establishment of a hospital of professional standards in the ghetto cemetery, an unforeseen location intended to bury the dead, symbolizes the phenomenon's uniqueness: of initiative, forbearance, and overcoming the obstacles with which the Nazis confronted the Jews. From his descriptions and evaluation, it is clear that the Jewish inmates of the ghetto also saw the medical activity as a sign of moral strength. So great was the ghetto inmates' appreciation of the hospital in the ghetto cemetery, that they saw fit to mark its first anniversary with a celebration; the first celebration since the German occupation of Shavli. In the words of Pik, this "left a pleasurable impression, like a chink of light in the darkness of night." The medical activity was a source of Jewish pride, instilling hope, like a candle in the dark.

Another example of the Jewish physicians' perception of the medical activity in the ghettos as a unique phenomenon can be found in the writings of Dr. Israel Milejkowski, who headed the medical system in the Warsaw Ghetto, in the winter of 1941–1942, two and a half years after the Germans occupied Poland.

The initiators of the clandestine "Oneg Shabbat" archive in Warsaw, the largest of the ghettos, headed by Dr. Emanuel Ringelblum, asked public figures in the ghetto to fill in a questionnaire about ghetto life from the perspective of their public roles.

In Dr. Milejkowski's response, he describes the medical activity in the ghetto [my emphasis, M.O.]:

> "[...] My ramified and highly intensive work in the area entrusted to me; the health board and hospitalization for the Jewish community in the Ghetto, takes up all my waking hours and leaves me no free time ... my impressions are therefore gleaned from my daily work and practice ... we have organized Jewish life where the true Jewish qualities of mercy and charity find realization ... quiet, unassuming work also is going on ... public kitchens, **medical courses, as well as other studies in order to equip the younger generation to grow up with skills. Intensive medical research being carried out on death through starvation and typhus. The entire clinical—and unfortunately plentiful—material is being studied and conclusive medical results of the research are being prepared. ...We simply want later—after the cataclysm—to be able to show the world that even these terrible trials could not break us. At this very moment a small scientific conference is in progress at the hospital, devoted to Ghetto problems, which includes scientific papers, addresses, and discussion. The result of this work will later be published and will be, I hope, of great interest the world over** ... (Milejkowski 1942, in Kermish 1986, pp. 741–744).

In his response, Dr. Milejkowski noted initiatives of professional medical courses to continue to educate the next generation of physicians, in addition to research on typhus and hunger. As these were prevalent phenomena in genocide conditions and had scarcely been observed in normal times, the physicians lacked sufficient professional knowledge to cope with them; hence the research. Milejkowski assessed the activity of the Jewish physicians in the ghetto as exceptional under its gradual conditions of extinction, and thus predicted that the establishment of the Jewish medical system in the ghetto would attract tremendous interest around the world with publication of its details after the cataclysm. Prior to his deportation to Treblinka during the *Aktion* of January 1943, he collated the findings of the hunger study conducted on his own initiative by physicians and scientists in the Warsaw Ghetto, and in the Introduction, expressed the uniqueness of the phenomenon:

> "... the subject of our study is the usual, day-to-day hunger; hunger in the physical sense. From this point of view, we, the Jews, are outside the fence of competition, especially now. We are honored to encase this subject in a scientific problem to satisfy our hunger for science and to gain knowledge at the expense of our physical hunger ... if the discussion is from a historical viewpoint, then this is ... truly "The Unfinished Symphony of Jewish Physicians from 1942" (Milejkowski 1942, in Schechter 1953a, b).

In other words, the physicians took the initiative to turn the insurmountable state of physical hunger enforced on the Jews into a situation of coping from an unexpected place: a scientific study of the hunger rampant in the ghetto by physicians who, themselves, were starving and ill.

Milejkowski perceived this medical activity as "the only answer to the murderers," a kind of "medical rebellion." Several months before he was taken by the Germans, he even managed to convey the understanding that the medical activity initiated by the ghetto physicians was unique on any scale.

> … A last few words to honor you, the Jewish doctors. What can I tell you, my beloved colleagues and companions in misery. You are a part of all of us. Slavery, hunger, deportation, those death figures in our ghetto were also your fate. And you by your work **could give the henchman the answer "non omnis moriar," "I shall not wholly die"** (Milejkowski 1942, in Winick 1979).

Thus, both Dr. Pik from the Shavli Ghetto and Dr. Israel Milejkowski in Warsaw, who undertook important roles in the medical system in the ghetto, expressed their appreciation of this activity while it was actually happening and memorialized it in writing. They both emphasized their admiration for the Jewish medical organization in the ghetto and acknowledged it as a reaction pattern initiated by the Jewish victims themselves, independently, with no external assistance from the Germans or anyone else. They noted the quintessentially professional components on which the system was based, established by the ghetto physicians and leaders against all the odds in the ghetto's subhuman conditions.

This admiration was penned by other physicians in the ghettos during the Holocaust period as well as in the writings of physicians who survived the horrors. One example is Dr. Mark (Meir) Dworzecki, who was a physician in the Vilna Ghetto and later became a historian and one of the pioneers in the research of medicine during the Nazi period. As early as 1946, he pointed to the uniqueness of the Jewish medical activity in the ghettos and wrote a kind of memorial "lament" to the physicians:

> Among the masses of martyrs who fell victim to the Nazi slaughter, the Jewish people will remember the doctors who risked their lives in a dual struggle—against the murderers' sword and against epidemics and disease.
>
> Jewish physicians in exile have performed an exalted public mission since time immemorial. Many served in the courts of kings and ministers, many—towering personalities in Torah and science—served as exilarchs, leaders, and teachers of Jewish law for the dispersed. Their deeds transcended the bounds of professional medicine and extended to all areas of life, particularly social assistance and public hygiene.
>
> Amid the assault on the body and soul of the European Diaspora, doctors took up a special place, knowing how to preserve the human image amid the agonies of the ghetto and to instill hope and comfort in hearts until the last moment. [...] (Mark Meir Dworzecki, 1946)

Dworzecki described the Jewish physicians as warriors, who "risked their lives in a dual struggle" against the Nazis and against morbidity and epidemics. He wrote that the Jewish physicians earned themselves a unique position in the history of the Holocaust, and that their efforts to comfort and encourage the public were a shining example of how to preserve the human image and values. He also wrote that the activity of the Jewish physicians during the Holocaust must be examined in light of the place of Jewish medicine throughout history.

Having seen that the ghetto physicians appreciated the uniqueness of their medical activity under ghetto conditions in the historical context and the history of medicine in particular, I will now examine how it is reflected in post-Holocaust research, and will observe this medical activity from different angles.

16.2 The Jewish Medical Systems in the Ghettos: The Phenomenon and Its Scope As Reflected in Research

Studies of the Jewish medical activity in the ghettos to date reveal a broad, comprehensive, historical picture of the medical system in the "enforced Jewish community" ("Zwangsgemeinschaft" in German) under the Nazi regime (Offer 2014a, b, 2015, 2015–2016). The Jews, living under impossible conditions imposed by the Nazis, nevertheless organized intensive medical activities in many ghettos. In those unimaginable circumstances, physicians and ghetto leaders created hospitalization possibilities for patients, where they received ongoing medical treatment. They provided academic and in-service training for physicians and nurses. Even medical research continued in the ghettos, attempting to provide responses to diseases that developed under genocide conditions. These activities have been relatively extensively described, especially regarding the largest ghetto, the Warsaw Ghetto. This medical system served the ghetto population of over 400,000 people. The ghetto was divided into six areas, and a medical center was set up in each. Two major hospitals served the entire ghetto population. One was the Czyste Jewish Hospital for adults with over 1200 beds, which had existed during the interwar period, but was not situated within the ghetto boundary. However, the Jews reestablished it in various dilapidated buildings scattered throughout the ghetto. The other one was the Berson and Bauman Children's Hospital, which was located within the confines of the ghetto, but had to move around, from time to time, during the ghetto period. Alongside the comprehensive network of clinics, pharmacies for distributing medicines, and a first-aid station, a chemical and bacterial laboratory was established by scientists in the ghetto. The staff of the Judenrat and the Society for Safeguarding the Health of the Jewish Population (Towarzystwo Ochrony Zdrowia Ludności Żydowskiej, TOZ, which operated in the interwar period) shouldered most of the burden of running this medical system. Among the prominent figures involved were the aforementioned Dr. Israel Milejkowski, who was head of the Judenrat Health Department; Adam Czerniakow, Head of the Judenrat, perceived as an assimilated Jew, who cooperated with all the legal and clandestine activities to advance the medical system in the ghetto, and Prof. Ludwik Hirszfeld, a well-known scientist who had converted to Christianity. About 800 physicians in the ghetto collaborated and made a large contribution, despite their diversity in outlook and beliefs (Offer 2015).

In the medical education field, for example, the school of nursing that existed during the interwar period continued to operate in the Warsaw Ghetto, when approximately 80 students enrolled over four separate intakes (Blum-Bielicka 1961; Blum 2014). In addition, an underground faculty of medicine provided pre-clinical and clinical studies for junior and senior medical students. During the ghetto period, approximately 500 novice medical students were taught by senior physicians and scientists at high academic levels including the preparation of assignments and sitting examinations (Roland 1992; Offer 2015, 2016a).

Noteworthy among the research projects was the aforementioned underground study of hunger disease, carried out by approximately 30 physicians. Hunger was a syndrome afflicting all ghetto inmates, and caused thousands of deaths. As explained in their writings, these physicians undertook to turn the physical hunger that was rife in the ghetto into "hunger for the sake of science." Hence, they perpetuated the Nazis' heinous acts through a scientific study of the effects of hunger on the human body. This research was perceived by some as a medical spiritual rebellion. Most of it survived and, many decades later, is receiving wide acclaim among medical researchers (Winick 1979). Another comprehensive research project was conducted on typhus, a disease that claimed many victims in the ghetto (Penson 1946; Muszkowska-Penson 2002; Rutkowski 2004).

This phenomenon of Jewish medical systems was observed in many other ghettos. A few brief examples are mentioned here. Out of approximately 35,000 inmates in the Theresienstadt Ghetto at any particular time (with a total of about 140,000 throughout the entire ghetto period), about 10% worked in the medical system, including about 700 physicians. There were four hospitals, many smaller rooms for patients close to the crowded barracks, and about 36 clinics for patients who were taken there from the barracks for treatment. Prof. Hermann Strauss, an inmate of the ghetto and former Director of the Jewish Hospital in Berlin, developed an in-service training system for the ghetto's medical staff. Research was carried out on anemia, polio, and meningitis, among other diseases (Nadav 2009; Jenss and Reinicke 2014; Offer 2016b). The Vilna Ghetto had a well-organized and professional medical system, including the Jewish Hekdesh hospital, which operated in the interwar period and was situated in the ghetto. Since its equipment was confiscated by the Nazis, the hospital continued medical and other activities largely by improvisation and initiative. Underground activities were also carried out at the hospital. Alongside all the professional medical departments, Dr. Kalman Shapira established a preventive medical education system. For example, the "Folks-Gezunt," a popular pre-war medical-scientific journal for the broad Jewish public was continued in the form of a "live journal:" lectures on hygiene were given every 2 weeks for the ghetto residents in the theater (Prais 1998; Longacre et al. 2015).

The Kovno Ghetto was unique in that the "Council of Elders" (Aeltestenrat) was headed by a physician. The leaders and community begged the esteemed physician Dr. Elhanan Elkes to take on the role, and he agreed. He worked to preserve and strengthen the health services in the ghetto, which included a hospital established at the end of 1941, after its predecessor was burned down. About 130 physicians worked in the Kovno Ghetto. Clinics and a pharmacy, which also served as an underground station, were established. Classes and lectures on medical topics were offered, as well as courses for nurses and a medical library (Nadav 2009).

Seven hospitals operated in the Lodz Ghetto, the largest of which had approximately 400 beds. They included a maternity hospital with 300 beds, and a hospital for infectious diseases, where sophisticated laboratory tests were carried out. Studies were conducted on nutrition and growth disorders resulting from malnutrition.

In addition, a scientific council, headed by the woman physician Dr. Klausenberg, operated in the ghetto as an in-service training system for physicians. After the *Aktion* in which one of the Lodz hospitals was burned down in September 1942 and in which 1700 hospitalized patients were deported to Chelmno, the Jews in the ghetto changed their medical service system. Physicians were allocated to all the factories, and the ghetto was divided into 40 sub-areas with one physician in charge of each (Sandhaus 1993; Urbach and Urbach 1976, 1978).

It can thus be said that the Jews under the Nazi regime established highly professional and impressive medical and health systems. Moreover, the medical and health services set themselves public health challenges according to modern conceptions, including concern for sanitation, immunization, disease quarantine, and quality control of food and water (Holland et al. 1984, p. 20; Borowy and Hardly 2008).

It is both interesting and astonishing that the medical system dealt not only with emergency medicine or treating patients, but also, during this difficult time, continued with typical "routine" activities: studies, in-service training, and research (Offer 2016a). Conspicuously also, this system was set up **by the Jews, who were the persecuted society**, not by the Germans, and certainly not by external authorities. Moreover, they used the most modern approaches, despite many limitations (Offer 2015).

16.3 The Medical Systems in the Ghettos: Prominent Characteristics

The details described above raise the question of what generated such a widespread, parallel response pattern observed in so many different locations. During the Holocaust, without external assistance, the Jewish victims of persecution established an independent Jewish medical system according to modern professional standards.

Even before the extermination began, in 1941, and certainly subsequently, the Jews in the ghettos were a persecuted, segregated society, subjected to difficult life conditions, resulting in terrible sanitation and disease, similar to the conditions of other persecuted populations in situations of genocide or ethnic cleansing. In just about every location where there was a high concentration of Jews under Nazi occupation, the Jews themselves established a health system, evidenced by the existence of a health department in many Judenräte (Trunk 1977; Elkin 1992; Ben-Sefer 2001; Weisskopf 2008; Goldschmidt 2007; Nadav 2009). The medical staff, despite being persecuted victims themselves, independently established medical services. One of the prominent characteristics of the Jewish medical system in the ghetto was the medical infrastructure created to provide a service not only to a chosen few, but to the entire ghetto population, according to public health criteria.

The medical systems included the three main public channels of medical activity:

(1) preventing the spread of disease and epidemics (2) creating alternative medical services to those destroyed by the Germans, which were not included in the ghetto boundaries, including medical treatment in the form of hospitalization and day-clinics, and (3) continuing to advance medical research and education to find cures for diseases that spread under the genocide conditions in the ghetto, hitherto not experienced by the Jewish community.

These components were the outcome of modern developments in Europe since the end of the eighteenth century. Two central processes in the creation of the modern states are important for this study. The first is the formation of the centralized state with a single centralized government and laws applying to the entire country, which directs and shapes organizational systems in the various areas of life. The second is the emergence of the concept of the modern "nation," on which comprehensive literature exists (e.g., Hobsbawm 1991). In light of this, since the end of the eighteenth century, in the modern states, the perception of the government as responsible for public health (Volksgesundheit, santé publique) was developed. This was applied by the distribution of public authorities, which dictated various modes of behavior to the public, broadened the study and training systems for medicine and nursing, and conducted experimental research to find cures for diseases, and especially solutions for epidemics (Thomson 1970). As the promotion of public health became a state-supported national goal, the government also encouraged medical research. Hence, the development of the public health systems in a particular society seems to be an indicator of its general advancement.

The medical services established in the ghettos in different regions and their modus operandi were a direct reflection of these modern perceptions pertaining to public health. From the description thus far, it is clear that, despite all the limitations and difficulties, the medical and health services in the ghettos unmistakably reflected a modern society. The phenomenon begs historical explanation.

16.4 The Medical System in the Ghettos: As a Continuation of the Jewish Society in the Pre-Holocaust Era

To comprehend the reasons for the creation of this historical phenomenon of establishing the professional Jewish medical systems in the terrible conditions in the ghettos, I will focus on the characteristics of the Jewish society during the interwar period and the place of the medical profession within it. These characteristics influenced the reaction patterns of physicians and public leaders in the ghettos.

The Jewish medical activity in the ghettos was the outcome of several processes that unfolded among the Jewish society in Europe between the two world wars, particularly in Poland, which was the largest Jewish center.

16.5 An Independent Jewish Medical System for the Jewish Population in Poland

The Jewish medical services in the ghettos were organized rapidly. Far from being created out of a void, their scope, diversity, and relatively high standards drew on Jewish medicine developed previously, mainly during the interwar period. The uniqueness of the Polish Jewish collective in the fields of health and medicine, when Poland was independent between the two world wars, lies in the development of a separate, independent system of medical institutions and Jewish professional medical frameworks, designed primarily to serve the country's large Jewish population. Thus, this national-ethnic system required Jewish practitioners in all the branches of medicine. Even though the disproportionately high number of Jewish physicians in the population was a longstanding, two-thousand-year-old tradition ("Medicine," Encyclopedia Judaica 1971), the additional feature in Poland, in this period, was a partial, state-like infrastructure.

This picture of the medical system raises the following question: What were the reasons behind the emergence of such a well-developed, independent Jewish medical system in Poland during the interwar period? This can be attributed to three central causes.

(1) **Modernization and secularization processes**

This was a period of accelerated secularization and modernization of the Jews of Poland, starting in the second half of the nineteenth century, both in the Austro-Hungarian Empire (in the Galicia region) and in Czarist Russia (in the Congress Poland region). It was manifested in many ways, but primarily through the rise of political parties. However, of prime importance for the topic at hand was the development of a Jewish secular education system, reflecting modern standards, as Jews flocked also to non-Jewish schools, and later, applied to universities (Mendelsohn 1983). As a result of the aims to advance professionally and climb the socioeconomic ladder, there was an increase in those pursuing a medical career, which was traditional among Jews. In 1931, there were 4488 independent Jewish physicians in Poland, which constituted 56% of all independent physicians, and 2256 pharmacists and laboratory workers, which amounted to 24.1% of the total number in the general population (Mendelsohn 1983; Mahler 1968). Approximately 800 of the physicians were in Warsaw at the time of the German occupation, which is a large number, even considering that the Warsaw Ghetto held over 400,000 inmates at the peak.

(2) **Collective awareness and ethnic Jewish medical institutions**

Whereas in Central European countries and further West, the secularization and modernization were the cause of integration and assimilation, spearheaded, to a large extent, by the Jewish white collar workers, this was not the case in Poland. According to Ezra Mendelsohn, in independent Poland, which was apparently entangled in severe internal and external problems, only nationalism and the church served as unifying forces. Both the Polish nationalism and Catholicism were, by nature, exclusive, anti-pluralistic, and anti-Semitic. The lack of liberalism in independent Poland meant extreme intolerance of minority groups, and the Jews suffered especially. As a result, despite all the internal conflict among the Jews, a separate, independent Jewish culture and collective awareness developed (Mendelsohn 1983). The Polish state even granted the Jews minority status, as it did to other communities, mainly by order of the President of the Republic, on October 14th 1927. This order granted the communities the right to engage in charitable activities and to establish benevolent institutions, thus rendering the communities as central to the development of hospitals and health organizations (Bacon 2004). Some of these institutions received support and involvement from American Jews of Polish origin, as well as from the Joint Distribution Committee (JDC), through which the Jews of Poland received innovative professional ideas from the new world.

(3) **Anti-Semitism and the development of the Jewish ethnic medical services**

The process was accelerated by the escalating anti-Semitism during this period. Specifically, due to anti-Semitism in higher education institutions, a considerable number of Jews went to study medicine outside Poland, mainly in Germany and Austria. They brought back the most up-to-date approaches in medicine and nursing, and began implementing them in the Jewish medical institutions, thus promoting modern medicine.

I have thus far described the central place of Jewish physicians and the highly developed state of Jewish medicine during the interwar period as a background to understanding the Jewish society's capability to establish professional medical services in the ghettos. However, other issues still need to be addressed.

16.6 The Jewish Medical Activity in the Ghettos in Light of Nazi Policy

Some of the Jewish medical activity in the ghettos was performed in secret and some was performed legally; the latter relating mainly to preventive measures against the spread of epidemics. The field of medical education and the promotion of research into the unique morbidity of the ghettos were usually clandestine and concealed from the Germans. However, the legal activities of the Jewish medical services present a huge paradox regarding the perpetrators' point of view and the rationale leading to the Holocaust. The Nazis' anti-Jewish policy was based on racist anti-Semitic ideology, which did not recognize the Jews as human beings with any value, but as subhuman (Bachrach 1956; Burleigh and Wipperman 1991), or "parasites," as termed by Hitler in ***Mein Kampf***. According to this assertion, concentrating a multitude of Jews in an enclosed space would be expected to lead to rapid collapse and uncontrollable disease.

Nonetheless, the crowded ghettos were set up after the invasion of Poland and the Warsaw Ghetto was the largest in Europe. As Browning showed, the ghettoization was not a result of an orderly, organized process. There is no doubt that the ghettos were perceived as a short-term transitional phenomenon, but they outlived those expectations.

When the ghettos became a long-term phenomenon, the question of how to deal with them arose. From the ideological and pseudoscientific point of view, the German physicians in the Generalgouvernement followed the racial-anti-Semitic line: the Jews were the carriers and disseminators of epidemics, and therefore should be isolated. According to most of them, the Jews should be sentenced to death either by starvation in the ghetto, or by shooting. In 1940, the physicians in Warsaw, for example, recommended enclosing the Jews in the ghetto for health reasons (Browning 2004), supposedly to prevent them from spreading disease, even though the epidemics among the Jewish population had ceased. As the German physicians knew, the establishment of the ghetto would not provide a solution to the epidemics, and the living conditions there would lead to further dissemination. The Director of the Department of Health in the Generalgouvernement established health services for the Poles to prevent the infection of the Germans. However, he did not establish such a system for the Jewish public. The Jews were required to set up their own separate governing body and to deal with health issues under impossible conditions and without the provision of resources. Nevertheless, since the Jewish medical and health services in the ghettos were allowed to develop as they did, it seems that the German authorities, who were terrified of epidemics, implicitly assumed that the Jews would be capable of coping with their medical problems. Thus, alongside the humiliating and suspicious attitude toward the Jews in the ghettos because of the risk they posed to "Aryan" public health, it was assumed that they would be able to cope reasonably with medical, health, and sanitation problems, since there were so many physicians and nurses among them. This evaluation was based on their knowledge of the Jewish reality, recognizing not only that they were human beings, but also talented.

The temporary nature of the state of affairs described above, and why it was such, should be noted. The German authorities allowed the organization of medical services to take place and even relied on them, to a large extent, until 1942, when the die was cast for the Final Solution. As is well known, current scholars have abandoned the intentionalist approach, which was dominant in the early days of Holocaust research. According to that view, anti-Jewish Nazi policy was a deliberate, linear development, and the ghettos were viewed as a preparatory phase for the Final Solution (Michman 2003). The ghettos were a temporary framework to preserve the Jews until a solution could be found for them "elsewhere." Hence, medical services were necessary, at first, also from the Germans' point of view, even if they did not provide concrete assistance for their establishment and maintenance. When the ghetto phenomenon became prolonged, the Germans overseeing them argued about what to do with the inmates; whether to use them for productive needs or to starve them to death. However, decisions were made locally, and not from the top, by the central government in Berlin (Browning 2004). Once the Final Solution crystallized and implementation began, there was no hope for the medical services, which rapidly dwindled.

Having addressed the Nazi "logic" that allowed the Jews to create certain medical services, other questions arise from the perspective of the study of Jewish life in the ghettos. How can the tremendous effort invested by the Jewish medical staff be explained, when one considers the ongoing, gradual annihilation in the ghettos? Was this effort not almost futile?

16.7 The Jewish Medical Activity in the Ghettos from the Perspective of the Jewish Medical Staff

From the Jewish point of view, the living conditions created by the Germans in the ghettos, which led to death by starvation and disease, were perceived as deterioration in their treatment. Even though it is now known that ghettoization was not part of the "Final Solution to the Jewish Problem," the Jewish leaders, physicians, underground journalists, and others perceived the establishment of the ghettos as motivated by the idea that the Jews could be exterminated by creating conditions that would "naturally" diminish their numbers (by reducing fertility and promoting mortality).

Various forces in the Jewish society attempted to fight against this. The teacher, Chaim Kaplan, gives a good description of this in his diary, as early as March 10th 1940:

> It is our good fortune that the conquerors failed to understand the nature and strength of Polish Jewry. Logically, we are obliged to die. According to the laws of nature, our end is destruction and total annihilation …
>
> … The Jews of Poland – oppressed and broken, shamed and debased, still love life, and do not wish to leave this world before their time. Say what you like, the will to live amidst terrible suffering is the manifestation of some hidden power whose nature we do not yet know. It is a marvelous, life-preserving power that only the most firmly established and strongest of the communities of our people have received as a blessing […]
>
> We have remained naked. But as long as that secret power is concealed within us, we shall not yield to despair. The strength of this power lies in the very nature of the Polish Jew, which is rooted in our eternal tradition that commands us to live… (Kaplan 1940, Jewish Virtual Library website)

The Jewish medical enterprise in the ghettos may be examined from the perspective expressed by Kaplan, namely the values and mental strength that motivated the Jewish society. What reason did the medical staff in the ghettos have for investing so much of the meager strength and resources at their disposal in treating patients who, in any event, were destined to die? The question is reinforced in light of the vibrant and varied medical activity in the ghettos, as described earlier. What was the point of all the institutions, the activities, and the research, if all the inhabitants of the ghetto were condemned to death? The obvious answer is that, until the deportations started, the ghetto inhabitants did not foresee the active, systematic extermination process, but only gradual destruction. According to the prevailing view, despite the high mortality rate, disease and starvation, many would survive if they only received the appropriate treatment. Apparently, this was the rationale for maintaining an active medical system.

However, this answer does not explain everything, because even before the mass deportations began, there were those who doubted the rationality of helping the sick. Dr. Chaim Einhorn expresses such doubts in his memoir:

> As can be understood from what we have told, we frequently asked ourselves whether it was worth continuing to provide medical care under such conditions. But in spite of everything, up to the last moment in the ghetto, the doctors continued their tasks. And I am certain that many continued with their medical duties until the very last moment of their lives, whether in the bunkers, the concentration camps, or in the ghetto uprising (Einhorn, in Dworzecki 1958, p. 13).

The answer is apparently related to the basic survival instinct in most human beings, who wish to continue functioning for as long as they breathe the breath of life. Moreover, the opportunity to relieve people's suffering does not lose its value just because the patients will die or suffer in the long-term. If this were the case, there would be no reason to care for the very old who are approaching the end of their lives. In addition, the Jewish tradition played a role by emphasizing the supreme value of life and the saving of life. This served as the impetus to provide care and reduce suffering.

A physician who worked in the Berson and Bauman Children's Hospital in the ghetto, Adina Blady-Szwajger, describes the sense of obligation to perform medical tasks in the ghetto:

> We weren't there to look at the horror, only to treat the sick or help with a quiet death. But, above all, to save lives because, even though times were bad, as bad as could be, we still wouldn't accept that it was of no use and thought that if we pulled through, we'd save those children and they would survive to the end. So we tried to save them with those scraps of food, medicines and injections and some of them got better. And when they began to get better, when, from those terrible, swollen lumps, skeletons began to emerge, we'd sometimes even see something resembling a smile … (Blady-Szwajger 1990, p. 42).

The nurse Sabina Garfunkel-Glocer, who worked in the operating theater, describes the shortage of medical equipment and supplies alongside the sense of meaning and of mission:

> We had many patients, Jews who needed our help … Jews who had been injured by the Germans because they were Jews. It was a question of our honor to help patients, and with joint forces, to save those who could still be saved (Garfunkel-Glocer 1959, in Gruenbaum, p. 589).

16.8 Conclusion

In this chapter, I have shown that, during the Holocaust itself, the Jewish physicians were already documenting the Jewish medical activity in the ghettos as a phenomenon with unique characteristics; considering the professional challenges with which they were forced to cope under the subhuman conditions imposed by the Germans, which led to the outbreak of epidemics and gradual annihilation. The Jewish physicians in the ghettos and Holocaust-survivor physicians described the medical systems established by the Jews in the ghettos as unprecedented in history and worthy of special attention. Clearly, their activities were unable to change the appalling conditions and thousands of Jews in the ghettos died of diseases even before the systematic implementation of the Final Solution. Nonetheless, from the comparative research to date in the field of the study of Jewish medicine in the ghettos, I have pointed to several prominent characteristics of the medical activities set up by the Jews in the ghettos under Nazi occupation, which have not, thus far, received adequate attention. In addition, I have raised some points for consideration in explaining how this reaction pattern developed simultaneously in such a large number of ghettos. In other words, I have offered an historical explanation for this unique phenomenon. I drew attention to the fact that from the study of the Jewish medical activity during the Holocaust, the Jewish communities in the ghettos may be seen as a persecuted victim with modern characteristics. Therefore, despite the genocide conditions to which this society was subject, it succeeded in independently establishing a professional medical system using modern perceptions based on public health principles. This response pattern was an expression of the continued developments within the Jewish society in the modern era, which accelerated especially during the interwar period. Whereas, on the one hand, secularization and academization increased the numbers of Jewish students in general, and Jewish medical students in particular, on the other hand, anti-Semitism was on the rise. The limitations imposed on the inclusion of Jewish physicians in academic and public health institutions and medical associations expedited the development of the independent Jewish medical services for the entire Jewish community in Poland, which was the largest Jewish center in Europe between World Wars I and II. During that period, Jewish physicians and medical institutions were at the forefront of modern medicine. This was in keeping with the longstanding important role of medicine and physicians in Jewish society and tradition, which places great emphasis on the preservation of life as a commandment and fundamental principle. This apparently explains how the medical activity in the ghettos developed in such inconceivable conditions, in a unique manner unparalleled in history.

It is appropriate to end this chapter by quoting the historian Emanuel Ringelblum's evaluation of the Jewish medical activity in the Warsaw Ghetto: [My emphasis, M.O].

> Earlier we mentioned the quiet, passive heroic stand of the educators and primarily of Dr. Korczak. We have told how they went willingly to their deaths, accompanying the children whom they had educated for years. **The conduct of the doctors and nurses at the Jewish**

hospital was similar. It bordered onto [the Umschlagplatz. M.O.] and the screams and cries of the deportees could be heard. Everyone knew that the deportation would not bypass the hospital. There were dozens of such examples from the provincial towns, from Lublin, and other places. Therefore, some of the physicians and nurses left. **But a few dozen doctors and nurses stood guard and did not abandon the patients until the very last moment. When the tragic moment arrived and more than 1000 patients were loaded onto the train cars, a small number of doctors and nurses went with them. Such was the behavior of the people who were viewed as subhuman by the Nazis.** (Ringelblum 1942, in Kermish and Gutman 1994)

References

Bachrach, Z. (1956). *Man's perception of Nazi ideology.*, Especially ch. 10. Tel Aviv: Ministry of Defense, Broadcast University.

Bacon, G. (2004). Partial success, great achievements: On Jewish autonomy in interwar Poland. In I. Bartal (Ed.), *Kehal Yisrael: Jewish self-rule through the ages* (pp. 351–371, especially 359 and 361). Jerusalem: Zalman Shazar Center for Jewish History. In Hebrew.

Ben-Sefer, E. (2001). Gemilat chesed and moral behavior at Westerbork: Lessons from the past to remember for the future. In E. Maxwell & J. Roth (Eds.), *Remembering for the future: The Holocaust in an age of genocide* (pp. 35–53). Basingstoke: Palgrave.

Blady-Szwajger, A. (1990). *I remember nothing more: The Warsaw children's hospital and the Jewish resistance* (p. 42, T. Darowska & D. Stok, Trans.). New York: Pantheon Books.

Blum, A. (2014). The nursing school in the Warsaw Ghetto. In M. A. Grodin (Ed.), *Jewish medical resistance in the Holocaust* (pp. 173–177). New York/Oxford: Berghahn.

Blum-Bielicka, L. (1961). Szkoła Pielęgniarstwa przy Szpitalu Starozakonnych w Warszawie (1923–1943). *Biuletyn Z.I.H, 40*, 66–77 (In Polish).

Borowy, I., & Hardly, A. (Eds.). (2008). *Of medicine and men, biographies and ideas in European social medicine between the world wars.* Frankfurt/Main: Peter Lang.

Browning, C. R. (2004). *The origins of the final solution: The evolution of Nazi Jewish policy, September 1939–March 1942* (pp. 125, 151). Lincoln/Jerusalem: University of Nebraska Press/Yad Vashem.

Burleigh, M., & Wipperman, W. (1991). *The racial state: Germany 1933–1945.* Cambridge: Cambridge University Press.

Dworzecki, M. (Ed.). (1946). The struggle for existence and health in the ghettos during the Nazi occupation. *Dappim Refui'im, 6*, 2 (In Hebrew).

Einhorn, C. (1958). The physician in the Warsaw Ghetto. In M. Dworzecki (Ed.), *Hama'avak lekiyum velebriyut biymei Hashoah* [The struggle for survival and health during the Holocaust] (p. 13). Jerusalem: World Jewish Congress and Association of Physician Holocaust Survivors. In Hebrew.

Elkin, R. (1992). The Jewish Hospital in Berlin. *Yalkut Moreshet, 53*, 53–91 (In Hebrew).

Encyclopedia Judaica. (1971). vol. 11, col. –1211. Jerusalem.

Garfunkel-Glocer, S. (1959). The fate of the Czyste Hospital. In Y. Gruenbaum (Ed.), *Diaspora encyclopedia* (p. 589). Jerusalem: Enzyklopedia shel Galuyot (In Hebrew)

Goldschmidt, A. (2007). The health services, morbidity and mortality in the Theresienstadt Ghetto in comparison to other ghettos. In *Seventh nahariya conference on morbidity and medicine in the Holocaust, Book of abstracts*, 2.

Hobsbawm, A. (1991). *Nations and nationalism since 1870.* Cambridge: Cambridge University Press.

Holland, W., et al. (Eds.). (1984). *Oxford textbook of public health* (Vol. 1, p. 20). Oxford: Oxford University Press.

Jenss, H., & Reinicke, P. (2014). Der Arzt Hermann Strauss, 1868 – 1944. In *Autobiographische Notizen und Aufzeichnungen aus dem Ghetto Theresienstadt*. Berlin: Hentrich & Hentrich.

Jewish Virtual Library Website. *Kristallnacht: Discussions by the Nazi authorities following Kristallnacht.* November12, 1938. Retrieved from: http://www.jewishvirtuallibrary.org/jsource/Holocaust/krisdis.html

Kaplan, C. A. (1940). The Warsaw Ghetto diary of Chaim A. Kaplan, Translation retrieved from the Jewish Virtual Library Website: http://www.jewishvirtuallibrary.org/warsaw-ghetto-diary-of-chaim-a-kaplan

Kermish, J. (Ed.). (1986). *To live with honor and die with honor! Selected documents from the Warsaw Ghetto underground archives "O.S." [Oneg Shabbath]*. Jerusalem: Yad Vashem.

Longacre, M., Beinfeld, S., Hildebrandt, S., Glantz, L., & Grodin, M. (2015). Public health in the Vilna Ghetto as a form of Jewish resistance. *American Journal of Public Health, 105*(2), 293–301. https://doi.org/10.2105/AJPH.2014.302312.

Mahler, R. (1968). *Poland's Jews between the two world wars* (p. 159). Tel Aviv: Dvir (In Hebrew).

Mendelsohn, E. (1983). *The Jews of East Central Europe between the world wars* (p. 82). Bloomington: Indiana University Press.

Michman, D. (2003). *Holocaust historiography: A Jewish perspective. Conceptualizations, terminology, approaches and fundamental Iissues*. London: Vallentine Mitchell.

Milejkowski, I. (1942). *Responses of Jewish intellectuals and scholars to a survey concerning appraisal of the situation of the Jewish population in the Warsaw Ghetto*, ŻIH, ARI/86.

Muszkowska-Penson, J. (2002). Jakub Penson (1899–1971) part 1: Pre-war and the Second World War occupation time till academic habilitation. Annales Academiae Medicae Gedanensis, 32, 321–330. In Polish.

Nadav, D. S. (2009). *Medicine and Nazism* (pp. 53–64). Jerusalem: Hebrew University Magnes Press 74–84.

Offer, M. (1993). *Medicine and doctors in the Shavli Ghetto following the discovery of the diary of Dr. Aaron Pik: Notes from the valley of slaughter – Written notes from the Shavli Ghetto in Lithuania from the years 1942, 1943, 1944*. Master's thesis, University of Haifa, Haifa, Israel (In Hebrew).

Offer, M. (2014a). Medicine in the Shavli Ghetto: In light of the diary of Dr. Aaron Pik. In M. A. Grodin (Ed.), *Jewish medical resistance in the Holocaust* (pp. 164–172). New York/Oxford: Berghahn.

Offer, M. (2014b). The study of Jewish medicine during the Holocaust: Its creation and preliminary steps. *Korot, The Israel Journal of the History of Medicine and Science, 22*, 129–154.

Offer, M. (2015). *White coats inside the ghetto: Jewish medicine in Poland during the Holocaust.* (pp. 257–271; 478–479). Jerusalem: Yad Vashem 478–479 (In Hebrew). [English edition (2018)].

Offer, M. (2015–2016). The study of Jewish medicine during the Holocaust: Trends and characteristics of its development. *Korot, The Israel Journal of the History of Medicine and Science, 23*, 111–149.

Offer, M. (2016a). The study of medicine and medical research in the ghettos: How and why the underground medical school was established in the Warsaw Ghetto. *Bishvil Hazikaron Journal, 26*, 30–38 (In Hebrew).

Offer, M. (2016b). Book review Der Arzt Hermann Strauss, 1868–1944. *Social History of Medicine, 29*(4), 853–855.

Penson, J. (1946). Clinical features of typhus epidemic in 1940 and 1941–1942 in Warsaw: Studies on the nitrogen metabolism with special regard to kidneys and method of early detection. *Polski tygodnik lekarski, 52*, 1399–1404, 1478–1487, 1553–1569. In Polish.

Pik, A. (1942, 1943, 1944/1998). *Notes from the Valley of Slaughter – Written notes from the Shavli Ghetto in Lithuania from the years 1942, 1943, 1944* (p. 177). Tel Aviv: Association of Lithuanian Jews in Israel. (In Hebrew).

Prais, L. (1998). *Between productivity and the Final Solution: Medical and health care problems in the Lithuanian ghettos-Wilna, Kovna and Shavli, 1941–1944.* Master's thesis. The Hebrew University of Jerusalem, Jerusalem, Israel. (In Hebrew).

Ringelblum, E. (1942). Gutman, I., & Kermish, J. (1994). *Last writings: Polish–Jewish relations January 1943–April 1944* (p. 37). Jerusalem: Yad Vashem and Ghetto Fighters' House (In Hebrew).

Roland, C. G. (1992). *Courage under siege: Starvation, disease and death in the Warsaw Ghetto* (pp. 186–198). New York: Oxford University Press.

Rutkowski, B. (2004). Jakub Penson and his studies in acute renal failure during typhus epidemics in the Warsaw Ghetto. *Journal of Nephrology, 17*, 175–179.

Sandhaus, Y. (1993). Medicine in the Lodz Ghetto: Organization and essence. *Dapim Leheker Tekufat Hashoah, 10*, 157–169 (In Hebrew).

Schechter, M. (1953a). Health and morbidity at times of hunger and shortage in the world [Dedicated to the memory of the Warsaw Ghetto physicians on the 10th anniversary of the liquidation of the ghetto]. *Harofe Haivri, 1–2*, 76–89 (In Hebrew).

Schechter, M. (1953b). The unfinished symphony of Jewish physicians from 1942. *Harefuah, 44*, 202–205 (In Hebrew).

Thomson, D. (1970). *Europe since Napoleon* (pp. 113, 355). Hamonsworth/Middlesex: Penguin.

Trunk, I. (1977). *Judenrat: The Jewish Councils in Eastern Europe under Nazi occupation.* New York: Stein & Day Ch. 4 and 7.

Urbach, V., & Urbach, M. (1976). Morbidity and health services in the Lodz Ghetto. *Korot, 7*, 172–246 (In Hebrew).

Urbach, V., & Urbach, M. (1978). Health services in the Lodz Ghetto. *Korot, 7*, 461–488 (In Hebrew).

Weisskopf, V. (2008). The Jewish hospital in Budapest under the Nazi occupation. *Harefuah, 147*(A), 81–84 (In Hebrew).

Winick, M. (Ed.). (1979). *Hunger disease: Studies by the Jewish physicians in the Warsaw Ghetto* (M. Osnos, Trans.). New York: Wiley.

The manufacturer's authorised representative in the EU is Springer Nature Customer Service Centre GmbH, Europaplatz 3, 69115 Heidelberg, Germany. If you have any concerns regarding our products, please contact ProductSafety@springernature.com

Printed and bound by CPI Group (UK) Ltd, Croydon, CR0 4YY
05/10/2026
02228135-0009